Sustainable Agriculture in the Era
of Climate Change

Sustainable Agriculture in the Era of Climate Change

Edited by Edison Harrison

www.statesacademicpress.com

States Academic Press,
109 South 5th Street,
Brooklyn, NY 11249, USA

Visit us on the World Wide Web at:
www.statesacademicpress.com

ISBN: 978-1-63989-749-0

Cataloging-in-Publication Data

Sustainable agriculture in the era of climate change / edited by Edison Harrison.
   p. cm.
Includes bibliographical references and index.
ISBN 978-1-63989-749-0
1. Sustainable agriculture. 2. Climatic changes. 3. Crops and climate. 4. Agricultural ecology.
I. Harrison, Edison.
S494.5.S86 S87 2023
338.1--dc23

Table of Contents

Preface

This book was inspired by the evolution of our times; to answer the curiosity of inquisitive minds. Many developments have occurred across the globe in the recent past which has transformed the progress in the field.

Sustainable agriculture is a farming practice that seeks to fulfill the present food needs of the society without compromising the needs of the future generations. Agricultural productivity is constrained by exposure to extreme vulnerability of climate in the form of floods, droughts, heat, cold waves, unseasonal rains and cyclones. Efficient management of natural resources and developing climate resilient crop varieties are some major strategies that can be adopted for overcoming the challenges of climate change. One of the major approaches which focuses on sustainable agriculture to combat climate change is climate-smart agriculture. It seeks to achieve three prominent outcomes, namely, increasing productivity, enhancing resilience and reducing emissions. This book includes some of the vital pieces of works being conducted across the world, on various topics related to sustainable agriculture and climate change. It consists of contributions made by international experts. This book aims to equip students and experts with the advanced topics and upcoming concepts in this area of study.

This book was developed from a mere concept to drafts to chapters and finally compiled together as a complete text to benefit the readers across all nations. To ensure the quality of the content we instilled two significant steps in our procedure. The first was to appoint an editorial team that would verify the data and statistics provided in the book and also select the most appropriate and valuable contributions from the plentiful contributions we received from authors worldwide. The next step was to appoint an expert of the topic as the Editor-in-Chief, who would head the project and finally make the necessary amendments and modifications to make the text reader-friendly. I was then commissioned to examine all the material to present the topics in the most comprehensible and productive format.

I would like to take this opportunity to thank all the contributing authors who were supportive enough to contribute their time and knowledge to this project. I also wish to convey my regards to my family who have been extremely supportive during the entire project.

Editor

Opportunities for Green Energy through Emerging Crops: Biogas Valorization of *Cannabis sativa* L. Residues

Carla Asquer [1], Emanuela Melis [1,*], Efisio Antonio Scano [1] and Gianluca Carboni [2]

[1] Biofuels and Biomass Laboratory, Renewable Energies Facility, Sardegna Ricerche – VI strada ovest Z.I. Macchiareddu, 09010 Uta, Italy; asquer@sardegnaricerche.it (C.A.); efisioas@tin.it (E.A.S.)

[2] Agris Sardegna, Viale Trieste 111, 09123 Cagliari, Italy; gcarboni@agrisricerca.it

* Correspondence: emanuela.melis@sardegnaricerche.it or emymelis@hotmail.com

Abstract: The present work shows the experimental evidence carried out on a pilot scale and demonstrating the potential of *Cannabis sativa* L. by-products for biogas production through anaerobic digestion. While the current state-of-the-art tests on anaerobic digestion feasibility are carried out at the laboratory scale, the here described tests were carried out at a pilot-to-large scale. An experimental campaign was carried out on hemp straw residues to assess the effective performance of this feedstock in biogas production by reproducing the real operating conditions of an industrial plant. An organic loading rate was applied according to two different amounts of hemp straw residues (3% wt/wt and 5% wt/wt). Also, specific bioenhancers were used to maximize biogas production. When an enzymatic treatment was not applied, a higher amount of hemp straw residues determined an increase of the median values of the gas production rate of biogas of 92.1%. This reached 116.6% when bioenhancers were applied. The increase of the specific gas production of biogas due to an increment of the organic loading rate (5% wt/wt) was +77.9% without enzymatic treatment and it was +129.8% when enzymes were used. The best management of the biodigester was found in the combination of higher values of hemp straw residues coupled with the enzymatic treatment, reaching $0.248~\mathrm{Nm^3 \cdot kg_{volatile~solids}}^{-1}$ of specific biogas production. Comparisons were made between the biogas performance obtained within the present study and those found in the literature review coming from studies on a laboratory scale, as well as those related to the most common energy crops. The hemp straw performance was similar to those provided by previous studies on a laboratory scale. Values reported in the literature for other lignocellulosic crops are close to those of this work. Based on the findings, biogas production can be improved by using bioenhancers. Results suggest an integration of industrial hemp straw residues as complementary biomass for cleaner production and to contribute to the fight against climate change.

Keywords: renewable energy technologies; sustainability; clean energy; bioenergy; biogas; industrial hemp; anaerobic digestion

1. Introduction

Industrial hemp (*Cannabis sativa* L.) is a valuable crop, and all parts of the plant can be used in many ways. Recent surveys carried out in the past few years (e.g., [1,2]) suggested that industrial hemp is a niche crop of increasing interest for its properties and versatility. New uses and innovative products appear on the market (more than 25,000 products have been discerned [3]), thus *Cannabis sativa* L. is becoming a very attractive crop on a global scale.

In Europe, hemp cultivation is mainly a multi-purpose crop. The market interest for hemp seeds and the need for attaining maximum economic viability of the related supply chains are stimulating a progressive shift of interest from traditional stem fiber use (textile, pulp or paper) towards multi-purpose cultivations. Indeed, in recent years, an increasing interest for new products obtainable as food or feed from seeds and for phyto-based cosmetics from inflorescence is emerging [4].

Hemp is a crop with fast growth, high biomass production at low inputs (fertilisers/pesticides), good CO_2 capture per hectare (about 2.5 t/ha), and soil protection due to the length of its roots, suitable for many industrial processes [5,6]. Appropriate soils for hemp are deep, show pH between 6.0 and 7.5, and have a good availability of nutrients and water holding capacity [7]. Moreover, hemp requires proper preparation of the seedbed, especially on clay soils, for a homogenous emergence due to its particular sensitivity to waterlogging. Sandy soils are less suitable for this cultivation, because of its poor water holding capacity determining greater water requirements [8]. It depends on climatic conditions. Indeed, in the South Mediterranean environment, higher irrigation volumes are required, with respect to the North Mediterranean one [9,10], but hemp water requirements are lower [11] compared to other specialized and common crops, such as maize, which are also cultivated for biogas production in Europe.

Industrial hemp cultivation is growing over time. A great increase was recorded from 2013 to 2017 in Europe [2], because of the introduction of policies and local incentives to the hemp industry [12].

As a result of the Italian Regulation [13], industrial hemp cultivation and processing assumed an increased national relevance. The regulation supports (also by including economic incentives) and promotes the development of integrated supply chains valuing research findings and pursuing local integration, as well as effective environmental and economic sustainability.

During the first decades of the 20th century, Italy was one of the most important producers on a global scale. In 1940, cultivated areas exceeded 100,000 ha, corresponding to more than 80,000 tons of hemp fibers [14].

The extension of the cultivated areas in Italy from 1961 to 2017 are reported in Figure 1 (sources of data: [15–17]).

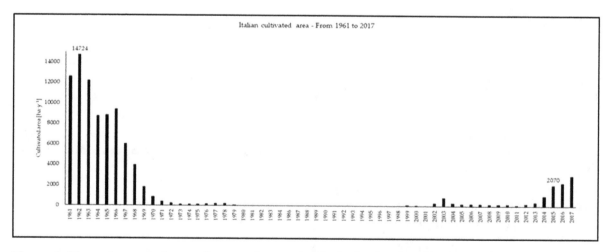

Figure 1. Hemp harvested areas (hectares by year) in Italy, from 1961 to 2017 (source of data: [15–17]).

Cultivated areas were significant during the 1960s and 1970s, and cultivations stopped in the 1980s and 1990s, mainly due to strict policies and regulations against the use of narcotic and psychotropic drugs. From 1999 to 2018, a new interest in hemp cultivation was developed, supported by national and European funding (the Italian Ministry of Agricultural Food and Forestry Policies financed a project to promote industrial hemp supply chains; during the same period, the European Union funded 3-year projects to reintroduce this feedstock for multiple purposes and to differentiate crops).

Due to the global resurgence of hemp cultivation needed to meet the requirements of the hemp sectors widespread today (building construction, food/animal feed, pharmaceutical, paper, textile, etc.), the recovery of hemp fiber and hurd residues should be addressed.

In this regard, some research and development projects funded by the European Union, such as MultiHemp [18] and GRACE [19], were already developed to demonstrate the sustainability of hemp-derived products according to the biorefinery concept.

However, another possible recycling path, aimed at increasing the economic and environmental benefits in a circular economy perspective, is the conversion of the agro-industry by-products into energy carriers.

Despite its thousands of uses, hemp by-products' potential as an energy feedstock is yet to be examined in depth. To date, few works have identified industrial hemp as an energy crop (for instance, a potential energy crop to produce bioenergy in [20]; ethanol production in [21,22]; methyl ester production in [23]; pyrolysis feedstock in [24]; biomass for thermochemical processes in [5]; combustion in [25,26]; co-firing in coal and peat power stations in [27]; and gasification or co-firing in [28]).

A few scientific works related to industrial hemp as a potential energy crop for biogas production can be found in the literature, going from 1990 to 2014. Rehman et al. [29] give an overall perspective of using hemp as a bioenergy crop in Pakistan, including biogas production.

Kreuger et al. [30], Heiermann et al. [31], Adamovics et al. [32], Mallik et al. [33] and Kaiser [34] provided data from anaerobic digestion trials carried out on a laboratory scale (a few liters-capacity reactors) (Table 1). As a pre-treatment, hemp was ground to a few mm or powder size. Since the grinding size influences the digestion kinetics [35] and, more generally, particle size affects the hydrolytic phase of the biodegradation of lignocellulosic feedstock [36], this factor should be considered when analyzing and comparing studies based on a laboratory scale. However, fine grinding is not reasonably achievable and economically affordable in industrial applications processing huge amounts of biomass.

Table 1. Main experimental conditions and information related to the scientific works found in the literature on anaerobic digestion of hemp, based on a laboratory scale.

Experiment	Hemp Cultivar	Country	Thermal Conditions	HRT	Specific Biogas Yield	Specific Methane Yield	Methane Content
[30]	Futura75	Sweden	50 °C	30 days	-	234 ± 35 m^3·t^{-1} VS (mean $\pm$ std.dev. [1])	-
[31]	Fedora19	Germany	35 °C	35 days	$453 \div 567$ L$_N$·kg^{-1} VS	$259 \div 301$ L$_N$·kg^{-1} VS [2]	$53 \div 57$ (%vol)
[32]	Futura75 (among other cultivars)	Latvia	38 ± 1 °C	53 days	$0.357 \div 0.370$ L·g^{-1} VS (coarse particles); $0.470 \div 0.530$ L·g^{-1} VS (fine particles)	$0.172 \div 0.185$ L·g^{-1} VS (coarse particles); $0.240 \div 0.270$ L·g^{-1} VS (fine particles)	-

[1] std.dev. means "standard deviation"; [2] N means "normal", VS means "volatile solids".

For the present work, the study of Mallik et al. [33] was excluded from the comparison reported in Table 1, because of a lack of information about biogas production/methane production performance. In [34], experiments conducted in a batch digester were presented, where industrial hemp was co-digested with other vegetable wastes and poultry litter. These three types of biomass, fed to 10-L reactors, had the same size (2 cm), which did not allow the authors to consider the relationships between chemical composition and size, and how they influenced anaerobic digestion. It was difficult to assign biogas/methane production performances to each biomass making up the admixtures (with attention to hemp) to make comparisons with the results obtained in the present study.

The present work goes beyond the past research approaches to anaerobic digestion of hemp. It focuses on assessing actual opportunities of a large-scale use of industrial hemp straw residues. Indeed, in the south of Italy, hemp is primarily cultivated for seed production while hemp straw residues are ordinarily left in the field, due to their scarce economic value as well as to the limited industrial interest and knowledge about this by-product. This study aimed to provide a more comprehensive knowledge of this residual lignocellulosic biomass.

The screening carried out on hemp straw residues for biogas generation completes the major gaps identified in the related state-of-the-art, by offering in-depth knowledge of the effective performance of Cannabis sativa L. residues in biogas production.

This work considers an alternative use of hemp straw residues with respect to the already developed market sectors (see, for instance, [29]) and, consequently, it suggests new market opportunities for hemp-derived products. The outcomes show the effective potential of developing a new supply chain, based on an emerging lignocellulosic crop for biofuel production. These aspects will be economically relevant both for farmers and contractors in biogas/biomethane sectors.

As pointed out by [37], lignocellulosic crops are not a common source of biomass for biogas production. The authors emphasize that the most significant constraints to hemicellulose/cellulose digestion are related to the lignin content, crystallinity of cellulose, and particle size. These limits may be reduced through optimization of the methodologies and technologies supporting biogas and subsequent biomethane production, for more sustainable use of crop residues for energy purposes. Among other techniques, the use of specific enzyme systems should be considered to reduce the lignocellulose's recalcitrance to anaerobic digestion. An in-depth presentation of the chemical and biological mechanisms between recalcitrant biomass and enzymes was provided by [38].

As reported by [39], commercial bioenhancers are not thoroughly characterized, but the positive results provided by the recent literature (+30% increase circa, as reported by [40]) related to biogas production from biomass with a complex lignocellulosic structure stimulate further applications and studies.

This work includes treatment with a commercial preparation of bioenhancers developed to improve biogas production through anaerobic digestion of cellulose and hemicellulose in lignocellulosic crops, like *Cannabis sativa* L. residues, to contribute to an advance in the field.

Additionally, by assessing the current state of industrial hemp usage and deployment, it emerged that a synergistic approach along the entire supply chain should be adopted, by integrating high-value components of hemp and other parts of the plant into a well-designed biorefinery, in order to support the local economy in a more sustainable way.

2. Materials and Methods

In 2017, an experimental hemp crop on the *Cannabis sativa* L. cultivar "Futura75" was carried out at San Giovanni Suergiu (pilot site located in the south-western side of the Sardinian Island, Italy). This trial is part of the CANOPAES project (the Italian acronym for "CANapa: OPportunità Ambientali ed Economiche in Sardegna", focused on the environmental and economic opportunities of hemp in the Sardinian Island).

Futura75 was chosen for its diffusion in Europe and its ability to produce both seeds and biomass.
According to the available long-term data, the San Giovanni Suergiu's climate is typically the Mediterranean. During the crop cycle, the thermopluviometric trend was characterized by maximum temperatures above the average, and rainfall was equal to about one-third of the seasonal average. Hemp was sown at a density of 120 plants·m^{-2} and at a depth of 0.02 m.

In addition, 60 kg·ha^{-1} as urea were top-dressed at about one month after emergence. Irrigation was performed by sprinklers with 75% ETm (maximum evapotranspiration) restitution and no weed control was required. Hempseeds were harvested by an ordinary combine. After that, the by-product straw naturally dried on the field (moisture <15% on a wet basis). Then, straw was raked and baled for transportation to the pilot plant. The green biomass yield was about 20 t·ha^{-1} while naturally dried straw was about 3.7 t·ha^{-1}.

Based on the assumption that different uses of hemp straw can coexist, though the specific features of the local market drive the types of use (as stated by [6] and [7], the dual-purpose oil-fiber of *Cannabis sativa* L. is dominant in the European territory), this work assumed a hypothetical scenario made of a dual-purpose supply chain: Hempseeds were harvested by a combine, to be used for oil extraction, while residues were processed for energy carrier generation (specifically, biogas).

Then, single-step digestion was performed in the pilot plant described below. The duration of the experiment was of 423 days (from March 2018 to June 2019).

2.1. Feedstock Characterization and Pre-Treatment, Admixture Preparation, and Pilot Plant

Since the chemical composition and physical characteristics (e.g., moisture content M) were used to define the admixtures proportion, to manage the process stability and to optimize anaerobic digestion, proximate analysis and ultimate analysis of hemp straw residues were performed.

Samples were prepared by drying hemp straw at 105 ± 2 °C in a thermostatic oven (Memmert GmbH, Schwabach, Germany), and by shredding and mixing the material through a cutter.

The proximate analysis was conducted using a thermogravimetric analyzer (TGA701, LECO Corporation, St. Joseph, MI, USA) following [41], to determine the moisture content (M), volatile solids on a dry basis ($VS_{d.b.}$), ash content, and fixed carbon (FC) (reported as percentage by mass [%wt]).

Total carbon, hydrogen, total nitrogen, and sulphur were determined by conducting the ultimate analysis through a CHNS analyzer (Truspec, LECO Corporation, St. Joseph, MI, USA) in accordance with [42].

Fiber composition (ADF: Acid detergent fiber, NDF: Neutral detergent fiber, ADL: Acid detergent lignin) of the lignocellulosic feedstock was used to determine the daily intake of enzymes (see Section 2.2). Values were obtained by using a fiber analyzer ANKOM 2000 (ANKOM, Macedon NY, USA), following the Van Sœst methodology [43–46]. Concerning the hemicellulose and cellulose contents, those values were estimated by subtracting ADF from NDF and ADL from ADF [47,48].

Chemical and physical characteristics (with their standard deviations) of hemp residues are listed in Tables 2–4.

Table 2. Proximate analysis of hemp residues. The cultivar "Futura 75".

Proximate Analysis	
[%wt]	
$M^1_{d.b.}$	7.71 ± 0.01
$VS^2_{d.b.}$	81.37 ± 0.08
$Ash_{d.b.}$	2.50 ± 0.25
$FC^3_{d.b.}$	16.13 ± 0.35

[1] M means "moisture"; [2] VS means "Volatile Solids"; [3] FC means "Fixed Carbon".

Table 3. Ultimate analysis of hemp residues. The cultivar "Futura 75".

Ultimate Analysis	
[%wt]	
$Carbon_{d.b.}$	47.41 ± 0.04
$Hydrogen_{d.b.}$	6.52 ± 0.10
$Nitrogen_{d.b.}$	1.64 ± 0.02
$Sulphur_{d.b.}$	0.18 ± 0.00

Table 4. Fiber composition (mean values, [% dry matter]) of hemp residues. The cultivar "Futura 75".

Chopped Hemp, Reproductive Stage		
ADL [1] [%wt]	NDF [2] [%wt]	ADF [3] [%wt]
7.87	59.16	44.40

[1] ADL means "Acid detergent lignin"; [2] NDF means "neutral detergent fiber; [3] ADF means "acid detergent fiber"

The feedstock characterization did not include parameters, such as starch and sugar contents, because of the composition of hemp straw residues mainly characterized by the lignocellulosic structure.

The anaerobic digester used in the present work was a tubular, horizontal reactor of 1.13 m³ total volume. It is 2.25 m long and its external diameter is 779 mm. It was radially mixed using a mechanical stirrer. The reactor was fed via a pneumatic pump, conveying the substrate previously introduced into a 250-kg-capacity feeding hopper. The reactor was tested by filling it with 960 L of digestate (corresponding to about 85% of the total volume).

The digestate produced during the process was discharged into a 200-kg-capacity tank, by using a pneumatic pump. The reactor was heated by three electrical resistances located in its center, loading, and discharging sides.

Sampling operations for the reactor sludge were performed using two valves located in the loading and discharging sides of the reactor.

Operations and parameters settings were managed and controlled by a programmable logic controller (PLC).

The feedstock pre-treatment consisted in mechanical milling, by shredding hemp straw residues using a 20-L-capacity cutter (dry cut). Then, coarse particles (maximum size: 1 cm) were mixed with the recirculated digestate in a 40-L-capacity cutter. When necessary, different amounts of water were added.

The operative settings were changed during the experimental period to investigate different process conditions (see Section 2.2).

2.2. Feeding Phases

The reactor was filled with digestate coming from an anaerobic digestion industrial plant treating corn silage and triticale. The digestate was used as received from the industrial plant. Then, a start-up phase was performed. During this phase, the temperature was increased by 2 °C daily until a constant value of 39 °C (mesophilic conditions) was reached.

Subsequently, the daily feeding rate of admixtures (Q, $[m^3_{substrate} \cdot d^{-1}]$) was increased during the first phase to reach an adaptation of the bacterial consortium to the specific substrate. After the start-up phase, the hemp to digestate proportion (hereinafter: percentage of new hemp straw in the admixture C, [% wt/wt]), Q, and digestate recirculation ratio (R, adimensional) were changed during the experimental period.

C and R values were chosen keeping in account the fluid-dynamics behavior of the admixtures. An increase of the hemp straw amount, indeed, could make the pumping of the admixtures itself difficult.

Two reference values of C (C1: 3% wt/wt; C2: 5% wt/wt) were set to evaluate the process. These conditions were tested by considering the presence/absence of specific bioenhancers. Treatments were randomly applied during the entire experiment.

Concerning the ranges of the organic loading rate (OLR, $[kg_{VS} \cdot (m^3_{reactor} \cdot d)^{-1}]$), hydraulic retention time (HRT, [d], dependent on Q), R, and C, different regimes were defined.

The abovementioned variables were analyzed along with the specific gas production (SGP, $[m^3_{biogas\ or\ methane} \cdot kg_{VS}^{-1}]$) and the gas production rate (GPR, $[m^3_{biogas\ or\ methane} \cdot (m^3_{reactor} \cdot d)^{-1}]$).

A commercial enzymatic preparation (Micropan Biogas ®from Eurovix, IT) was applied to reduce the current supply of hemp straw residues and to maximize biogas production at the same time. It is made of microbial enzymes containing cellulase, lipase, xylanase, active principles of *Fucus Laminariae*, algae *Lithothamnium calcareum*, natural nutrients/grow factors, selected yeast, mineral biological catalysts rich in oligo elements, and selected microorganisms (facultative anaerobic bacteria, such as: *Bacillus subtilis*, *Bacillus maceraus*; strictly anaerobic bacteria genus *Methanobacterium*). The specific gravity was 0.8 t/m³.

The daily intake was introduced into the reactor by dissolving the powder into water (1:4 wt/v).

The dosage was divided into two parts, depending on the fiber composition of lignocellulosic feedstock (Table 4), C, and Q. The daily intake was defined by multiplying the percentage of cellulose and hemicellulose of hemp residues (see Table 4) with the hemp mass in Q and a coefficient of 0.05, as suggested by the producers. Thus, 20 g per day were obtained.

The first enzymatic inoculum of the reactor sludge was calculated by dividing the working volume of the reactor by Q, and by multiplying this value with the daily intake (500 g of bioenhancers were introduced into the reactor).

2.3. Management of the Reactor and Process Stability

Both the process stability and reactor features were controlled and managed by using a set of parameters.

Management of the reactor: As regards the reactor, the following were considered:

- TS (total solids), $VS_{d.b.}$, determined via proximate analysis on a weekly/sub-weekly basis for the new admixture introduced into the feeding hopper, the material in the hopper/in the digestate tank, and the sludge inside the reactor;

- HRT [d], calculated as:

$$HRT = \frac{V}{Q},\qquad(1)$$

where V is the total digester volume [$m^3_{reactor}$] and Q is the daily feeding rate [$m^3_{substrate}\cdot d^{-1}$];

- OLR ([$kg_{VS}\cdot(m^3_{reactor}\cdot d)^{-1}$]), calculated as:

$$OLR = \frac{V\cdot S}{Q},\qquad(2)$$

where S is the VS concentration on a wet basis in the feeding admixtures [$\%wt_{w.b.}$].

Management of process stability: The process stability was monitored by considering the below listed parameters:

- pH and FOS/TAC ratio (volatile fatty acids content/buffer capacity) of the reactor sludge (daily measures by means of, respectively, a multi-parametric analyzer Orion Versa Star (ThermoScientific Inc., Waltham, MA, USA) and an automatic titrator T70 (Mettler Toledo International Inc., Columbus, OH, USA));

- Biogas production [$m^3_{biogas}\cdot d^{-1}$] (daily values provided by a biogas flow meter);

- Biogas composition daily (CH_4, CO_2, O_2 [%wt]; NH_3, and H_2S [ppm]), determined using a portable gas analyzer GA2000 (Geotechnical Instruments UK Ltd., Coventry, UK). Biogas composition is strictly related to SGP and GPR;

- Temperature of the reactor sludge, measured through three temperature probes located in the center, the loading, and discharging sides of the reactor, and monitored through the PLC.

Performance parameters: The two main performance parameters considered in the anaerobic digestion trials carried out on hemp straw residues are:

- SGP [$Nm^3_{biogas\ or\ methane}\cdot kg_{VS}^{-1}$], calculated as:

$$SGP = \frac{G}{Q\cdot S},\qquad(3)$$

where Q and S were already described, G is the daily production of biogas/methane [$m^3_{biogas\ or\ methane}\cdot d^{-1}$];

- GPR ([$Nm^3_{biogas\ or\ methane}\cdot(m^3_{reactor}\cdot d)^{-1}$]), calculated as the daily production of biogas/methane per m^3 of sludge accumulated in the reactor.

The abovementioned parameters are related to:

- C (percentage of new hemp in the admixture, [% wt/wt]), calculated as:

$$C = \frac{Mass_{hemp}}{Mass_{hemp} + Mass_{digestate} + Mass_{water}}\cdot 100,\qquad(4)$$

where: $Mass_{hemp}$, $Mass_{digestate}$, and $Mass_{water}$ are the mass of the hemp, digestate, and water composing the admixtures;

- R (digestate recirculation ratio, adimentional), following Equation (5):

$$R = \frac{\sum_i Mass_{digestate}}{\sum_i Mass_{hemp}},$$
(5)

where $\sum_i Mass_{hemp}$ and $\sum_i Mass_{digestate}$ [g·10^3] are the cumulative amounts of new hemp or digestate composing hemp-digestate admixtures in a specific time window.

2.4. Statistical Analyses

Statistical analyses were performed on biogas composition, SGP, and GPR using Statgraphic Centurion XVI [49]. The mean, standard deviation, maximum and minimum values, and median were calculated for each feeding phase composing the experimental period. Also, the skewness and kurtosis were calculated to determine which kind of statistical analysis should be performed.

Depending on this first data analysis and with a specific focus on the skewness and kurtosis, any evenness emerged among variances. Consequently, non-parametric tests were applied.

Two non-parametric tests were considered: Mann–Whitney [50,51] and Kruskal–Wallis [52]. The former was used to determine if two ordinal and independent random samples (feeding phases) were part of the same population. The latter was useful to compare median values of the different groups to identify whether they belonged to a population characterized by the same median.

Statistical analyses were developed by considering $p < 0.05$. Graphical representations were provided by using box-and-whisker diagrams.

3. Results and Discussion

3.1. Feeding Phases

Based on the outcomes of the experiment, it was divided into phases. The main characteristics of the feeding phases, except for the start-up phase, are reported in Table 5. Indeed, the start-up phase was characterized by high instability of the main process parameters as well as by a rapid variation of admixture feeding rates over time. Thus, the remaining feeding phases were labeled from 1 to 7.

Table 5. Main parameters (with their standard deviations) of anaerobic digestion trials on hemp residues.

Phase [-]	Description [-]	Duration [d]	OLR [1] [kg$_{VS}$·m^{-3}·d^{-1}]	HRT [2] [d]	C [3] [% wt/wt]	R [4] [-]
1	No enzymatic treatment	45 (day 98–day 143)	2.8 ± 0.6	29 ± 2	2.3 ± 0.0	16.3 ± 1.1
2	No enzymatic treatment	27 (day 144–day 171)	1.3 ± 0.2	34 ± 4	3.0 ± 0.9	18.9 ± 0.4
3	No enzymatic treatment	55 (day 172–day 227)	2.9 ± 0.8	34 ± 7	2.9 ± 1.8	20.7 ± 0.8
4	Enzymatic treatment	34 (day 228–day 262)	3.8 ± 0.8	31 ± 7	2.5 ± 1.7	22.0 ± 0.3
5	Enzymatic treatment	35 (day 263–day 298)	3.2 ± 0.8	30 ± 6	5.1 ± 0.2	22.0 ± 0.4
6	No enzymatic treatment	36 (day 299–day 335)	3.1 ± 0.9	33 ± 9	5.2 ± 1.0	20.9 ± 0.3
7	Enzymatic treatment	87 (day 336–day 423)	3.1 ± 1.0	29 ± 3	4.4 ± 2.0	20.3 ± 0.2

[1] OLR means "organic loading rate"; [2] HRT means "hydraulic retention time"; [3] C means "percentage of new hemp in the admixture"; [4] R means "digestate recirculation ratio".

Feeding phases were classified according to enzymatic treatment, OLR, HRT, C, and R (Table 5).

The two C reference values applied in this study are close to the organic loadings used in the industrial plants of anaerobic digestion.

3.2. Management of the Reactor and Process Stability

Management of the reactor -TS and $VS_{d.b.}$ trends of the material in the feeding hopper, in the reactor sludge, and the digestate tank are reported in Figures 2 and 3.

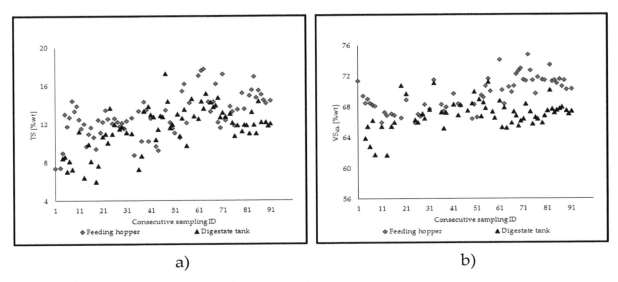

a)   b)

Figure 2. TS (total solids) (**a**) and $VS_{d.b.}$ (volatile solids on a dry basis); (**b**) trends of the material in the feeding hopper and the digestate tank, concerning consecutive sampling.

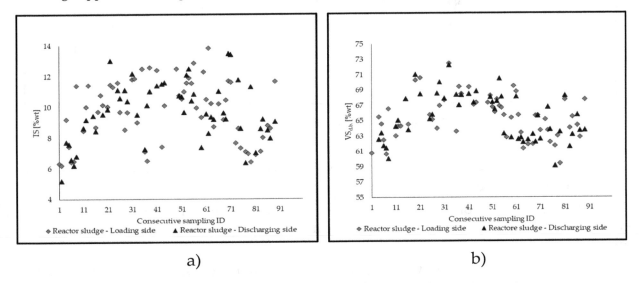

a)   b)

Figure 3. TS (total solids) (**a**) and $VS_{d.b.}$ (volatile solids on a dry basis); (**b**) trends of the reactor sludge, concerning consecutive sampling.

By comparing the TS and $VS_{d.b.}$ trends in the feeding hopper and digestate, it can be seen that from the 70th sampling, these parameters were notably lower in the discharged sludge than in the corresponding feeding mixtures. The 70th sampling corresponds to the beginning of the 5th feeding phase and this occurrence emerged from the observations related to the 6th and 7th phases as well. In the previous feeding phases, a distinction between the TS and $VS_{d.b.}$ evolution in the fed slurry and the corresponding digestate cannot be seen. This result is related to the higher reference value of C (C2), which was better than the other one (C1).

With regard to Figure 3, the reactor sludge did not show relevant differences in terms of TS and $VS_{d.b.}$ by comparing the loading side to the discharging side, mainly due to a certain mixing of the sludge along the longitudinal section of the reactor. TS and $VS_{d.b.}$ seemed to increase from phase "1" to phase "4" and to decrease in the subsequent phases (characterized by the reference value, C2).

Management of process stability: Concerning the main design and operation process parameters of the reactor, the trends of HRT and OLR are shown in Figures 4 and 5.

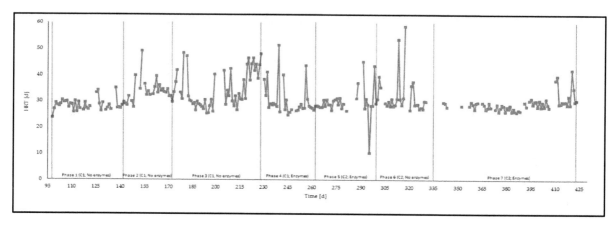

Figure 4. Hydraulic retention time (HRT) trend.

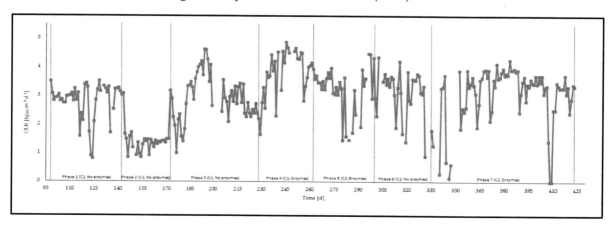

Figure 5. Organic loading rate (OLR) trend.

HRT did not show significant variations during the entire experiment (Figure 4) and it was excluded from the statistical analysis.

OLR (Figure 5) was monitored through the daily determination of TS and $VS_{d.b.}$ in the feeding admixture and adjusted by setting the hemp share in the feeding admixture and the loading flow rate (Figure 2).

Regarding the process stability parameters, the trends of FOS/TAC and pH, biogas production, and biogas composition (CH_4, CO_2, NH_3, H_2S) are reported in Figures 6–9.

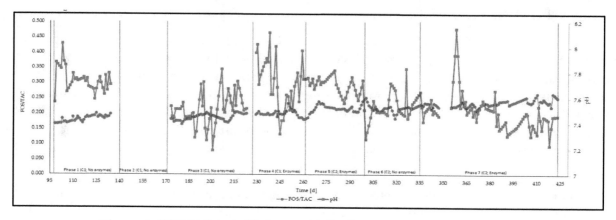

Figure 6. FOS/TAC (volatile fatty acids/buffer capacity) and pH trends.

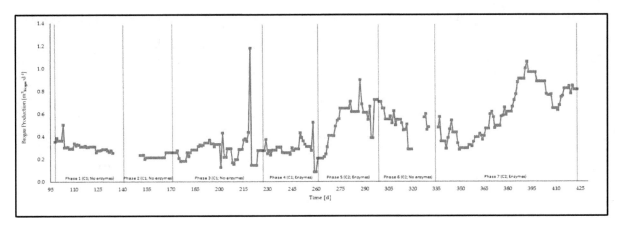

Figure 7. Biogas production trends (GPR) (via biogas metering).

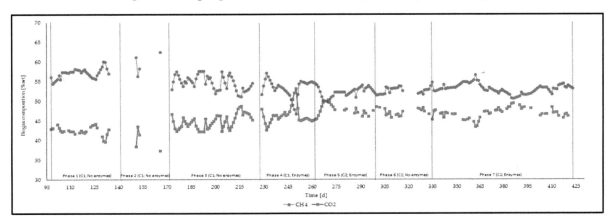

Figure 8. Biogas composition (CH_4 and CO_2, [%wt]) detected by the portable gas analyzer during the experimental period.

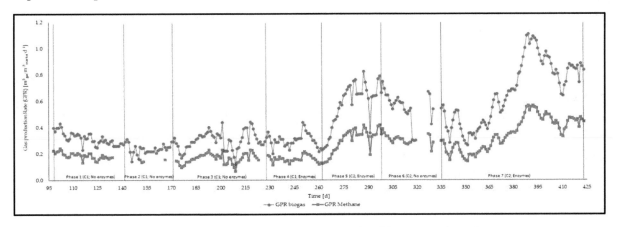

Figure 9. Gas production rate (GPR) trends of biogas and methane.

By considering the entire experimental campaign, pH varied between 7.2 and 8.0, accompanied by higher values during the first phase.

The FOS/TAC ratio increased over time, from about 0.170 to values close to 0.270. These values are very similar to the typical ones of the industrial plants treating lignocellulosic biomass, such as corn silage.

Overall, the FOS/TAC trend is consistent with the increase in the percentage of new hemp in the feeding admixtures (C2 treatment) (see Table 5) from the fifth feeding phase to the seventh one. It can be supposed that the introduction of higher amounts of hemp provoked a shift of the anaerobic digestion microbial dynamics towards the predominance of acidogenic reactions.

Daily biogas production (Figure 7) was characterized by high variability during the experimental campaign. It showed a significant increase of the biogas produced during the experimental phases from "5" to "7" (C2), accompanied by a rapid increase during the fifth phase and a decrease over the sixth one (without enzymes). The last feeding phase showed rising values for most of the days (to reach about 1.1 m^3 biogas per day), corresponding to the coupling of higher C and enzymatic treatment. It can be considered as the most suitable among all the treatments applied. On average, the CH_4 content [%wt] in the biogas produced during the entire experimental campaign was 53.8 ± 2.2. CO_2 content [%wt] was 45.6 ± 2.4. CH_4 concentration showed a slow reduction from the phase "1" to the phase "4". From the phase "5", a gradual increase of its concentration was detected. CH_4 and CO_2 trends (Figure 8) did not show any peak attributed to organic overload.

Performance parameters: The performance parameters, SGP and GPR, of the anaerobic digestion trials are shown in Figures 9 and 10.

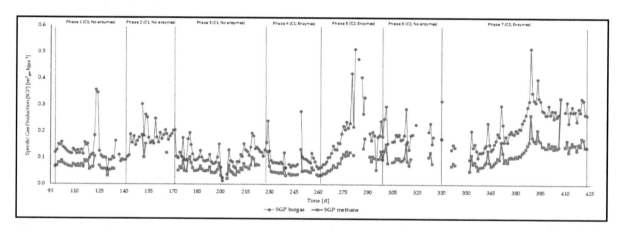

Figure 10. Specific gas production (SGP) trends of biogas and methane.

SGP of biogas/methane is the energy yield of an anaerobic digestion system regardless of OLR, thus it plays an important role when assessing the energy yield of the considered process.

Both trends showed an increase starting in the fifth feeding phase. It was more evident during the seventh period.

By comparing these outcomes with those obtained on a laboratory scale [30,32], it can be pointed out that the present work provided higher SGP values both for biogas and methane than the previous literature on anaerobic digestion trials of hemp straw. Overall, the management of the reactor and the process stability was enhanced with respect to biogas production, especially in the last part of the experiment, characterized by the combination of higher values of C and the application of enzymes.

SGPs of the present work are lower compared to those obtained in the same pilot plant, using vegetable feedstock characterized by high degradability (lower lignin contents and higher starch content). Due to the lignocellulosic nature of this crop, the specific biogas/methane production of hemp straw was lower than those related to raw potatoes (0.68 Nm^3 of biogas $\cdot kgVS^{-1}$ and 0.37 Nm^3 of methane $\cdot kgVS^{-1}$) [53], potato chips (0.81 Nm^3 of biogas $\cdot kgVS^{-1}$ and 0.47 Nm^3 of methane $\cdot kgVS^{-1}$) [54], and fruit and vegetable wastes (0.78 Nm^3 of biogas $\cdot kgVS^{-1}$ and 0.43 Nm^3 of methane $\cdot kgVS^{-1}$) [55].

Experiments carried out on the same pilot digester using admixtures made of different kinds of feedstock (admixture composition: 30%wt of shredded corn, the remaining part consisting in whey, vegetable water, pomace pitted, and manure to maintain the OLR between 2.5 and 3.5 $kgVS\cdot m^3{}_{reactor}\cdot d^{-1}$) reported SGP_{biogas} from 0.623 ± 0.212 $Nm^3\cdot kgVS^{-1}$ to 0.768 ± 0.227 $Nm^3\cdot kgVS^{-1}$ and $SGP_{methane}$ from 0.281 ± 0.160 $Nm^3\cdot kgVS^{-1}$ to 0.438 ± 0.096 $Nm^3\cdot kgVS^{-1}$ [56]. Also, hemp straw residues' performance in anaerobic digestion was found to be similar to other lignocellulosic crops [57], the specific methane production of which is between 0.17 and 0.39 $Nm^3\cdot kgVS^{-1}$. Results are comparable to those reported by [58] for other agricultural crops, such as oats, flax, and sorghum, but lower than the hemp energy yields considered there.

Thus, hemp straw residues used in the same pilot plant showed lower SGP of biogas and methane probably because of the lignocellulosic composition. The results obtained by the residues considered in this work could be affected by the harvesting time, which is the reproductive stage when the lignin content is higher than in the vegetative stage. By considering that the reproductive stage is the core of the supply chain scenario assumed in this study, related to the extraction of oil from seeds, the results already discussed about the potential of biogas production from straw residues suggest the recovery of this low-value by-product to energy conversion. By considering the ultimate analysis, the carbon:nitrogen ratio is useful to define the biomass suitability for biochemical (ratio <30) or thermochemical processes (ratio >30). The ratio of hemp straw residues is 28.9 (Table 3), thus it can be considered for both, but with slightly higher suitability for biochemical conversion.

3.3. Statistical Analysis

Results of the data analysis (mean, standard deviation, maximum and minimum, skewness, kurtosis, and median values) carried out on the most significant process parameters and outcomes are reported in Tables 6 and 7.

Table 6. Main statistical variables of the biogas composition and biogas and methane gas production rate (GPR) and specific gas production (SGP).

Process Parameter	Variable	Unit	Feeding Phase	No. of Values	Mean ± Std.dev.	Minimum	Maximum
Biogas composition	CH_4	[%wt]	1	37	57.1 ± 1.2	54.4	60.0
			2	4	59.5 ± 2.8	56.3	62.4
			3	51	54.7 ± 1.9	51.0	57.6
			4	34	53.3 ± 2.3	46.6	57.1
			5	36	52.1 ± 1.2	49.7	54.1
			6	28	52.6 ± 0.9	51.4	54.7
			7	88	53.0 ± 1.2	50.5	56.6
	CO_2	[%wt]	1	27	42.4 ± 1.1	39.6	44.2
			2	3	41.1 ± 2.6	38.4	43.5
			3	51	44.7 ± 1.8	42.2	48.7
			4	34	46.2 ± 2.2	42.6	53.1
			5	20	47.7 ± 1.3	45.4	50.3
			6	19	47.0 ± 1.0	45.2	48.5
			7	52	46.8 ± 1.3	43.4	49.4
GPR	GPR Biogas	[$Nm^3 \cdot d^{-1}$]	1	46	0.317 ± 0.047	0.228	0.433
			2	28	0.232 ± 0.035	0.140	0.291
			3	56	0.299 ± 0.070	0.119	0.441
			4	35	0.301 ± 0.052	0.199	0.438
			5	36	0.581 ± 0.162	0.238	0.238
			6	26	0.552 ± 0.115	0.297	0.297
			7	88	0.668 ± 0.036	0.262	1.109
	GPR CH_4	[$Nm^3 \cdot d^{-1}$]	1	37	0.186 ± 0.025	0.132	0.241
			2	4	0.147 ± 0.010	0.135	0.155
			3	51	0.164 ± 0.038	0.068	0.231
			4	34	0.160 ± 0.025	0.114	0.215
			5	36	0.303 ± 0.086	0.129	0.418
			6	23	0.307 ± 0.036	0.218	0.388
			7	88	0.353 ± 0.123	0.141	0.424
SGP	SGP Biogas	[$Nm^3 \cdot kgVS^{-1}$]	1	45	0.129 ± 0.054	0.054	0.358
			2	27	0.191 ± 0.037	0.148	0.304
			3	54	0.110 ± 0.035	0.025	0.194
			4	33	0.097 ± 0.047	0.059	0.277
			5	32	0.207 ± 0.113	0.055	0.514
			6	23	0.198 ± 0.048	0.132	0.316
			7	75	0.250 ± 0.119	0.095	0.825
	SGP CH_4	[$Nm^3 \cdot kgVS^{-1}$]	1	35	0.069 ± 0.013	0.032	0.105
			2	4	0.140 ± 0.037	0.103	0.186
			3	49	0.060 ± 0.019	0.013	0.104
			4	31	0.050 ± 0.023	0.033	0.133
			5	27	0.092 ± 0.033	0.041	0.174
			6	20	0.104 ± 0.027	0.071	0.173
			7	75	0.132 ± 0.062	0.052	0.439

Table 7. Skewness, kurtosis, and median values of the biogas composition, and biogas and methane gas production rate (GPR) and specific gas production (SGP).

Process Parameter	Variable	Unit	Feeding Phase	No. of Values	Skewness	Kurtosis	Median
Biogas composition	CH_4	[%wt]	1	37	0.041	0.571	57.3
			2	4	−0.167	−1.191	−
			3	51	−0.297	−1.401	44.7
			4	34	−3.061	2.19	53.8
			5	36	−1.043	−0.722	52.2
			6	28	0.773	−0.717	52.5
			7	88	0.543	0.326	53.1
	CO_2	[%wt]	1	27	−1.652	1.25	42.3
			2	3	−0.383	−	−
			3	51	0.934	−1.069	43.6
			4	34	3.808	3.712	45.7
			5	20	0.478	−0.446	48.2
			6	19	−0.257	−0.962	46.6
			7	52	−1.431	0.653	46.8
GPR	GPR Biogas	[$Nm^3{\cdot}d^{-1}$]	1	46	1.231	−0.350	0.311
			2	28	−1.027	1.151	0.232
			3	56	−0.473	−0.079	0.295
			4	35	1.028	0.835	0.303
			5	36	−2.012	−0.395	0.639
			6	26	−2.098	0.995	0.582
			7	88	0.161	−2.276	0.674
	GPR CH_4	[$Nm^3{\cdot}d^{-1}$]	1	37	−0.197	0.073	0.188
			2	4	−0.444	−1.202	−
			3	51	−0.761	−0.559	0.168
			4	34	0.588	−0.33	0.157
			5	36	−1.944	−0.643	0.339
			6	23	−0.172	1.138	0.310
			7	88	0.000	−2.399	0.351
SGP	SGP Biogas	[$Nm^3{\cdot}kgVS^{-1}$]	1	45	8.717	16.069	0.118
			2	27	3.400	2.619	0.182
			3	54	1.225	0.748	0.104
			4	33	6.655	9.930	0.084
			5	32	2.940	1.566	0.193
			6	23	2.462	0.869	0.185
			7	75	8.592	16.019	0.248
	SGP CH_4	[$Nm^3{\cdot}kgVS^{-1}$]	1	35	0.240	1.828	0.069
			2	4	0.375	−0.860	−
			3	49	1.195	0.743	0.054
			4	31	6.431	9.402	0.046
			5	27	0.570	0.032	0.080
			6	20	2.450	0.980	0.095
			7	75	9.078	17.583	0.130

As already mentioned in Section 2.4, skewness and kurtosis indices show that, except for SGP of biogas, the considered parameters can be attributed to a normal distribution, but variances of the two groups differ. Thus, comparisons were made on medians instead of mean values.

The outcomes of the Kruskal–Wallis test performed on the feeding phases "3", "4", "5", and "6", which are the most representative phases with respect to the treatments applied in this work (C1 and C2; presence/absence of enzymes), are shown in Figure 11.

The main outcomes of the two comparisons, "3" and "6", and "4" and "5", are described below.

Methane content [%wt] showed a tendency to lower median values when C is higher (C2) and higher values for the lower C regime (C1). This is confirmed by the Kruskal–Wallis test. Conversely, the behavior of the median value of the CO_2 content in biogas is opposite to the methane content, and it was higher for the higher C regime (C2). This is due to the addition of new raw material that modifies the reactions towards acidogenic conditions and promotes an increase of the CO_2 content in biogas with respect to the effect of the digestate recirculated, containing an amount of undigested substrate that is less reactive.

An increase of C led to an increase of GPR (median values: Phase "3": 0.295 $Nm^3{\cdot}d^{-1}$; phase "4": 0.303 $Nm^3{\cdot}d^{-1}$; phase "5": 0.639 $Nm^3{\cdot}d^{-1}$; phase "6": 0.582 $Nm^3{\cdot}d^{-1}$). When the enzymatic treatment was not applied (phase "3" and phase "6"), the higher C regime (C2) corresponded to an increase of the GPR_{biogas} median values of 92.1%. This was about 116.6% when bioenhancers were applied.

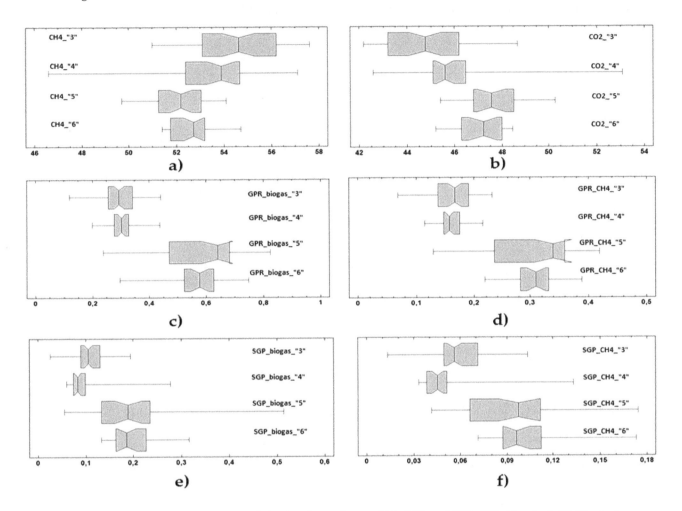

Figure 11. Box-and-whisker plots of the feeding-phases "3"–"6" and "4"–"5". Kruskal–Wallis test on the statistic variables: CH_4 content in biogas (**a**), CO_2 content in biogas (**b**), gas production rate (GPR) biogas (**c**), GPR methane (**d**), specific gas production (SGP) biogas (**e**) and SGP methane (**f**).

$GPR_{methane}$ showed similar trends observed for GPR_{biogas}: Higher values in the phases characterized by the C2 regime (median values: Phase "5": 0.339 $Nm^3 \cdot d^{-1}$; phase "6": 0.310 $Nm^3 \cdot d^{-1}$) than in those related to the C1 treatment (median values: Phase "3": 0.168 $Nm^3 \cdot d^{-1}$; phase "4": 0.157 $Nm^3 \cdot d^{-1}$). Hence, the increase due to the higher C regime was, respectively, +115.9% and +84.5% with and without enzymes.

With regard to the SGP of biogas/methane, an increase of C values led to higher energy yields.

The increase of SGP_{biogas} due to the increment of C was +77.9% without enzymatic treatment and +129.8% with enzymatic treatment. Thus, the coupling of a higher C regime with the addition of enzymes allowed the best management of the pilot plant to be obtained.

Essentially, similar behavior was observed for $SGP_{methane}$: The increasing of C promoted $SGP_{methane}$ (+165.9%) when enzymes were not applied. The increase associated with the enzymatic treatment was +73.9.

Statistically significant differences were found between phases "5" and "6" and phases "3" and "4", for all the parameters considered in the statistical analysis. Thus, it is reasonable to assert that variations of C influence all the parameters contributing to energy yields (SGP, GPR). The enzymatic treatment, instead, showed statistically significant differences in SGP in phases characterized by lower C values ("3" and "4").

The results of the Mann–Whitney test on the feeding phases "6" and "7", performed in order to assess the effect of the enzymatic treatment when the biodigester is managed by applying higher values of C, are reported in the box-and-whisker plot of Figure 12.

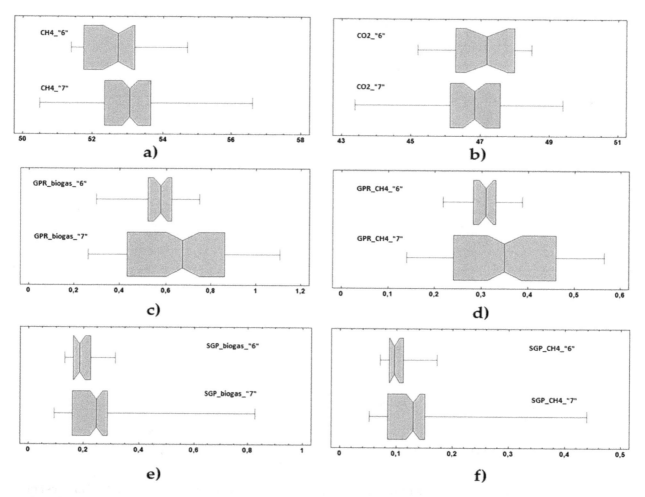

Figure 12. Box-and-whisker plots of the feeding-phases "6" and "7". Mann–Whitney test on the statistic variables: CH_4 content in biogas (**a**), CO_2 content in biogas (**b**), Gas Production Rate (GPR) biogas (**c**), GPR methane (**d**), Specific Gas Production (SGP) biogas (**e**) and SGP methane (**f**).

Skewness and kurtosis (Table 7) showed normal distributions of CH_4 and CO_2 contents in biogas, but SGP and GPR deviated from normality.

The Mann–Whitney test did not show any statistically significant difference in CH_4 content. Similar results were obtained for the CO_2 content in biogas (median of phase "6": 47.2% wt, median of phase "7": 46.8% wt).

SGP and GPR of biogas and methane were higher in the last phase of the experiment (with enzymatic treatment) than the second-last phase (without enzymes). More specifically, GPR and SGP of biogas and methane in the seventh phase reached the maximum values of the entire experimental campaign.

Concerning $GPR_{methane}$, any statistically significant difference was found between the two feeding phases (median value of phase "6": 0.310 $Nm^3 \cdot d^{-1}$, median value of phase "7": 0.351 $Nm^3 \cdot d^{-1}$).

More generally, in the last feeding phase, characterized by the enzymes, SGP and GPR of biogas and methane were higher than in the previous periods. Thus, the best management of the biodigester was characterized by this combination: A higher percentage of new hemp straw in the admixtures (C2) coupled with enzymatic addition.

4. Conclusions

The experimental campaign carried out on the cultivar "Futura 75" grown on the pilot site "San Giovanni Suergiu" (Sardinia, Italy) allowed an assessment of the effectiveness of using *Cannabis sativa* L. straw residues as a substrate for anaerobic digestion at an industrial scale and to enhance the management of the biodigester fed with hemp straw residues.

In this work, the feasibility of using this substrate in anaerobic digestion (which is currently often underutilized) was evaluated.

Results in terms of GPR and SGP of biogas/methane are promising, especially if compared to other vegetable feedstocks commonly used in anaerobic digestion and by considering that industrial hemp is characterized by higher values of lignin, which leads to high recalcitrance [59,60]. However, the SGP of biogas/methane is lower compared to corn silage, commonly used in industrial plants of anaerobic digestion (common values of about 0.7 to 0.85 $Nm^3 \cdot kg_{VS}^{-1}$), but the low cost of hemp straw residues and their behavior in the anaerobic digestion contribute to the definition of this by-product as a good process moderator when using other types of biomass leading easily to process instability.

The comparison between the findings of this work and the literature related to previous experiments carried out on a laboratory scale led to the assertion that biogas and methane yields provided by the trials carried out on hemp straw residues in the Sardinian pilot digester are similar or higher than those provided by the previous studies based on few liters-capacity reactors. It should be considered that differences in the energy yields reported may depend on environmental conditions (climate, soil type, crop management, etc.), the different cultivars used, hemp stage (vegetative versus reproductive stage), and hence, chemical and physical characteristics (such as TS, $VS_{d.b.}$, carbon:nitrogen ratio, etc.).

Results of SGP are close to those of other lignocellulosic crops but lower than those produced by highly degradable vegetable feedstocks studied through the same pilot plant. It suggests conducting additional experimental studies on hemp straws residues as a co-substrate in anaerobic digestion involving one or more easily digestible types of biomass.

Energy yields of anaerobic digestion carried out on hemp straw residues are influenced by different operating conditions: Increased feeding admixture composition (depending on C and R) produced a statistically significant increase in terms of methane content in biogas and of the parameters influencing GPR and SGP. Enzymatic treatments tended to enhance the SGP of biogas/methane.

The fluid dynamics of hemp-digestate admixtures play an important role in digestion kinetics, affected by solid–liquid separation and solid particles' tendency to sedimentation. Thus, further research should pay attention to this topic, to define relationships between the reactor and specific characteristics of admixtures.

These prodromal studies based on pilot-scale experiments on *Cannabis sativa* L. residues should be continued by analyzing more extensive conditions for factors inhibiting anaerobic digestion of hemp (e.g., heavy metals absorbed by roots, straw, leaves, and seeds during plant growth), and to different daily intakes of enzymatic preparations. Further investigations should pay attention to the enzymatic or other chemical additives' effects on energy yields (GPR, SGP, etc.).

The sustainability of hemp straw residues' biogas conversion should be evaluated as well, to define achievable costs and economic benefits. This hypothetical chain based on this emerging crop must be compared to the most commonly used energy feedstock. The main advantage in the energy conversion of hemp straws residues is to use a by-product of a cultivation carried out to obtain seeds as the main product: The consumption of water and fertilizers, however limited, is necessary to obtain seeds and no other input is spent, except for harvest and transport operations from the field to the plant.

In addition, with respect to the hemp-related supply chain, it has to be considered that relevant constraints to industrial hemp market development are fewer innovations in harvesting technologies and processes or processing facilities, as well as transportation/distribution issues (mainly due to the high low bulk density of this type of biomass) [61]. New research should overcome these current limits of industrial hemp exploitation and valorization to ensure more effective development of sustainable supply chains.

The results of this study produce a baseline to stimulate new perspectives of using hemp straw residues in the biogas sector and to inspire its consideration in the biorefinery thinking.

Author Contributions: Conceptualization, C.A., E.M. and E.A.S.; Data curation, C.A., E.M. and E.A.S.; Formal analysis: C.A.; Methodology, C.A., E.M. and E.A.S.; Project administration, G.C.; Supervision, E.A.S.; Validation, C.A., E.M. and E.A.S.; Writing—original draft preparation, E.M.; Writing—review and editing, C.A., E.M., G.C., E.A.S.

Acknowledgments: Gratefully acknowledges Agris Sardegna (Regional Agency for Research in Agriculture) of the Sardinia Autonomous Region for the providing of hemp fibre composition data (Table 4), as part of the project agreement "CANOPAES", under the Regional Law No. 15/2015 of the Sardinia Autonomous Region.

References

1.   Carus, M.; Sarmento, L. *The European Hemp Industry: Cultivation, Processing and Applications for Fibres, Shivs, Seeds and Flowers; Report 2016–05*; European Industrial Hemp Association: Brussels, Belgium, 2016; p. 9. Available online: eiha.org/media/2016/05/16-05-17-European-Hemp-Industry-2013.pdf (accessed on 1 April 2019).

2.   Carus, M. *The European Hemp Industry: Cultivation, Processing and Applications for Fibres, Shivs, Seeds and Flowers*; Report 2017–03-26; European Industrial Hemp Association: Brussels, Belgium, 2017; p. 9. Available online: eiha.org/media/2017/12/17-03_European_Hemp_Industry.pdf (accessed on 1 April 2019).

3.   Johnson, R. *Hemp as an Agricultural Commodity*; CRS Report; Congressional Research Service: Washington, DC, USA, 2014; p. 34.

4.   Carus, M.; Karst, S.; Kauffmann, A.; Hobson, J.; Bertucelli, S. *The European Hemp Industry: Cultivation, Processing and Applications for Fibres, Shivs and Seeds*; European Hemp Industry Association: Brussels, Belgium, 2013; Available online: www.votehemp.com/wp-content/uploads/2018/09/13-03_European_Hemp_Industry.pdf (accessed on 1 April 2019).

5.   Żuk-Gołaszewska, K.; Gołaszewski, J. *Cannabis sativa* L.—Cultivation and quality of raw material. *J. Elem.* **2018**, *23*, 971–984. [CrossRef]

6.   Tang, K.; Struik, P.C.; Yin, X.; Thouminot, C.; Bjelková, M.; Stramkale, V.; Amaducci, S. Comparing hemp (*Cannabis sativa* L.) cultivars for dual-purpose production under contrasting environments. *Ind. Crops Prod.* **2016**, *87*, 33–44. [CrossRef]

7.   Amaducci, S.; Scordia, D.; Liuc, F.H.; Zhang, Q.; Guo, H.; Testa, G.; Cosentino, S.L. Key cultivation techniques for hemp in Europe and China. *Ind. Crops Prod.* **2015**, *68*, 2–16. [CrossRef]

8.   Struik, P.C.; Amaducci, S.; Bullard, M.J.; Stutterheim, N.C.; Venturi, G.; Cromack, H. Agronomy of fibre hemp (*Cannabis sativa* L.) in Europe. *Ind. Crops Prod.* **2000**, *11*, 107–118. [CrossRef]

9.   Amaducci, S.; Amaducci, M.T.; Benati, R.; Venturi, G. Crop yield and quality parameters of four annual fibre crops (hemp, kenaf, maize and sorghum) in the North of Italy. *Ind. Crops Prod.* **2000**, *11*, 179–186. [CrossRef]

10.   Cosentino, S.L.; Riggi, E.; Testa, G.; Scordia, D.; Copani, V. Evaluation of European developed fibre hemp genotypes (*Cannabis sativa* L.) in semi-arid Mediterranean environment. *Ind. Crops Prod.* **2013**, *50*, 312–324. [CrossRef]

11.   Di Bari, V.; Campi, P.; Colucci, R.; Mastrorilli, M. Potential productivity of fibre hemp in southern Europe. *Euphytica* **2004**, *140*, 25–32. [CrossRef]

12.   Vantreese, V.L. Hemp Support. *J. Ind. Hemp* **2002**, *7*, 17–31. [CrossRef]

13.   Italian Republic. *Law n. 242, 2 December 2016. Disposizioni Per la promozione della Coltivazione e della Filiera Agroindustriale della Canapa*; General Series n. 304; Gazzetta Ufficiale della Repubblica Italiana: Rome, Italy, 30 December 2016.

14.   Di Candilo, M.; Ranalli, P.; Liberalato, D. Gli interventi necessari per la reintroduzione della canapa in Italia. *Agroindustria* **2003**, *2*, 27–36.

15.   FAOSTAT. Available online: www.fao.org/faostat (accessed on 23 April 2019).

16.   ISTAT. Available online: www.agri.istat.it (accessed on 23 April 2019).

17.   European Industrial Hemp Association (EIHA). Available online: www.eiha.org (accessed on 29 November 2019).

18. MultiHemp Project. Available online: www.multihemp.eu (accessed on 29 November 2019).

19. GRACE Project. Available online: www.grace-bbi.eu (accessed on 29 November 2019).

20. Tedeschi, A.; Tedeschi, P. The potential of hemp to produce bioenergy. In Proceedings of the 2nd World Conference on Biomass for Energy, Industry and Climate Protection, Rome, Italy, 10–14 May 2004; pp. 148–152.

21. González-García, S.; Luo, L.; Moreira, M.T.; Feijoo, G.; Huppes, G. Life cycle assessment of hemp hurds use in second generation ethanol production. *Biomass Bioenergy* **2012**, *36*, 268–279. [CrossRef]

22. Kuglarz, M.; Gunnarsson, I.B.; Svensson, S.-E.; Prade, T.; Johansson, E.; Angelidaki, I. Ethanol production from industrial hemp: Effect of combined dilute acid/steam pretreatment and economic aspects. *Bioresour. Technol.* **2014**, *163*, 236–243. [CrossRef]

23. Ragit, S.S.; Mohapatra, S.K.; Gill, P.; Kundu, K. Brown hemp methyl ester: Transesterification process and evaluation of fuel properties. *Biomass Bioenergy* **2012**, *41*, 14–20. [CrossRef]

24. Branca, C.; Di Blasi, C.; Galgano, A. Experimental analysis about the exploitation of industrial hemp (*Cannabis sativa*) in pyrolysis. *Fuel Process. Technol.* **2017**, *162*, 20–29. [CrossRef]

25. Rice, B. Hemp as a feedstock for biomass-to-energy conversion. *J. Ind. Hemp* **2008**, *13*, 145–156. [CrossRef]

26. Burczyk, H.; Grabowska, L.; Kołodziej, J.; Strybe, M. Industrial Hemp as a Raw Material for Energy Production. *J. Ind. Hemp* **2008**, *13*, 37–48. [CrossRef]

27. Finnan, J.; Styles, D. Hemp: A more sustainable annual energy crop for climate and energy policy. *Energy Policy* **2013**, *58*, 152–162. [CrossRef]

28. Hanegraaf, M.C.; Biewinga, E.E.; van der bijl, G. Assessing the ecological and economic sustainability of energy crops. *Biomass Bioenergy* **1998**, *15*, 345–355. [CrossRef]

29. Rehman, M.S.U.; Rashid, N.; Saif, A.; Mahmood, T.; Han, J.-I. Potential of bioenergy production from industrial hemp (*Cannabis sativa*): Pakistan perspective. *Renew. Sustain. Energy Rev.* **2013**, *18*, 154–164. [CrossRef]

30. Kreuger, E.; Prade, T.; Escobar, F.; Svensson, S.-E.; Englund, J.-E.; Björnsson, L. Anaerobic digestion of industrial hemp–Effect of harvest time on methane energy yield per hectare. *Biomass Bioenergy* **2011**, *35*, 893–900. [CrossRef]

31. Heiermann, M.; Ploechl, M.; Linke, B.; Schelle, H.; Herrmann, C. Biogas Crops-Part I: Specifications and Suitability of Field Crops for Anaerobic Digestion. *Agric. Eng. Int. CIGR J.* **2009**, *11*, 1–17.

32. Adamovics, A.; Dubrovskis, V.; Platace, R. Productivity of industrial hemp and its utilisation for anaerobic digestion. In Energy Production and Management in the 21st Century, Vol. 2. *WIT Trans. Ecol. Environ.* **2014**, *190*, 1045–1055.

33. Mallik, M.K.; Singh, U.K.; Ahmad, N. Batch digester studies on biogas production from Cannabis sativa, water hyacinth and crop wastes mixed with dung and poultry litter. *Biol. Wastes* **1990**, *31*, 315–319. [CrossRef]

34. Kaiser, F.; Diepolder, M.; Eder, J.; Hartmann, S.; Prestele, H.; Gerlach, R.; Ziehfreund, G.; Gronauer, A. Biogas yields from various renewable raw materials. In Proceedings of the 7th FAO/SREEN Workshop, Uppsala, Sweden, 30 November–2 December 2005.

35. Dumas, C.; Silva Ghizzi Damasceno, G.; Barakat, A.; Carrere, H.; Steyer, J.-P.; Rouau, X. Effects of grinding processes on anaerobic digestion of wheat straw. *Ind. Crops Prod.* **2015**, *74*, 450–456. [CrossRef]

36. Lynd, L.R.; Weimer, P.J.; Van Zyl, W.H.; Pretorius, I.S. Microbial cellulose utilization: Fundamentals and biotechnology. *Microbiol. Mol. Biol. Rev.* **2002**, *66*, 506–577. [CrossRef] [PubMed]

37. Merlin Christy, P.; Gopinath, L.R.; Divya, D. A review on anaerobic decomposition and enhancement of biogas production through enzymes and microorganisms. *Renew. Sustain. Energy Rev.* **2014**, *34*, 167–173. [CrossRef]

38. Xu, N.; Liu, S.; Xin, F.; Zhou, J.; Jia, H.; Xu, J.; Jiang, M.; Dong, W. Biomethane production from lignocellulose: Biomass recalcitrance and its impacts on anaerobic digestion. *Front. Bioeng. Biotechnol.* **2019**, 1–12. [CrossRef]

39. Čater, M.; Zorec, M.; Marinšek Logar, R. Methods for improving anaerobic lignocellulosic substrates degradation for enhanced biogasp. *Springer Sci. Rev.* **2014**, *2*, 51–61. [CrossRef]

40. Herrero Garcia, N.; Benedetti, M.; Bolzonella, D. Effects of enzymes addition on biogas production from anaerobic digestion of agricultural biomasses. *Waste Biomass Valor.* **2019**, *10*, 3711–3722. [CrossRef]

41. ASTM D7582-15. Standard Test Methods for Proximate Analysis of Coal and Coke by Macro Thermogravimetric Analysis. Available online: https://www.astm.org/Standards/D7582.htm (accessed on 30 June 2017).

42. ASTM D5373-16. Standard Test Methods for Determination of Carbon, Hydrogen and Nitrogen in Analysis Samples of Coal and Carbon in Analysis Samples of Coal and Coke. Available online: https://www.astm.org/Standards/D5373.htm (accessed on 2 January 2018).

43. Van Sœst, P.J.; Robertson, J.B.; Lewis, B.A. methods for dietary fiber, Neutral Detergent Fiber, and nonstarch polysaccharides in relation to animal nutrition. *J. Dairy Sci.* **1991**, *74*, 3583–3597. [CrossRef]

44. ANKOM Technologies. *Acid Detergent Fiber in Feeds—Filter Bag Technique (for A200 and A200I)*; ANKOM Technologies: Macedon, NY, USA, 2011.

45. ANKOM Technologies. *Method 8—Determining Acid Detergent Lignin in Beakers*; ANKOM Technologies: Macedon, NY, USA, 2005.

46. ANKOM Technologies. *Neutral Detergent Fiber in Feeds—Filter Bag Technique (for A200 and A200I)*; ANKOM Technologies: Macedon, NY, USA; Available online: www.ankom.com/sites/default/files/document-files/Method_6_NDF_A200.pdf (accessed on 29 November 2019).

47. Jung, H.-J.G. Analysis of forage fiber and cell walls in ruminant nutrition. *J. Nutr.* **1997**, *127*, 810S–813S. [CrossRef]

48. Theander, O.; Aman, P.; Westerlund, E.; Andersson, R.; Pettersson, D. Total dietary fiber determined as neutral sugar residues, uronic acid residues, and Klason lignin (the Uppsala method). *J. Assoc. Anal. Chem. Int.* **1995**, *78*, 1030–1044.

49. Statgraphics. Available online: www.statgraphics.com (accessed on 15 September 2019).

50. Kruskal, W.H. Historical Notes on the Wilcoxon Unpaired Two-Sample Test. *J. Am. Stat. Assoc.* **1957**, *52*, 356–360. [CrossRef]

51. Neuhäuser, M. Wilcoxon–Mann–Whitney Test. In *International Encyclopedia of Statistical Science*; Springer: Berlin/Heidelberg, Germany, 2011.

52. Kruskal, W.H.; Wallis, W.A. use of ranks in one-criterion variance analysis. *J. Am. Stat. Assoc.* **1952**, *47*, 583–621. [CrossRef]

53. Pistis, A.; Asquer, C.; Scano, E.A. Anaerobic digestion of potato industry by-products on a pilot-scale plant under thermophilic conditions. *Environ. Eng. Manag. J.* **2013**, *12*, 93–96.

54. Asquer, C.; Pistis, A.; Scano, E.A.; Cocco, D. Energy-oriented optimization of an anaerobic digestion plant for the combined treatment of solid and liquid wastes in a potato chips industrial plant. In Proceedings of the 22nd EUBCE, Hamburg, Germany, 23–26 June 2014.

55. Scano, E.A.; Asquer, C.; Pistis, A.; Ortu, L.; Demontis, V.; Cocco, D. Biogas from anaerobic digestion of fruit and vegetable wastes: Experimental results on pilot-scale and design of a full-scale power plant. *Energy Convers. Manag.* **2014**, *77*, 22–30. [CrossRef]

56. Scano, E.A. Trattamento di Biomasse Vegetali e Algali Finalizzato All'Ottenimento di Energia. Potenziali Sviluppi in Sardegna. Ph.D. Thesis, University of Cagliari, Cagliari, Italy, 2016. Available online: http://hdl.handle.net/11584/266883 (accessed on 29 November 2019).

57. Frigon, J.-C.; Guiot, S. Biomethane production from starch and lignocellulosic crops: A comparative review. *Biofuels Bioprod. Bioref.* **2010**, *4*, 447–458. [CrossRef]

58. International Energy Agency. Biogas from Crop Digestion. Bioenergy Task 32. 2011. Available online: http://www.ieabioenergy.com/publications/biogas-from-energy-crop-digestion/ (accessed on 27 September 2019).

59. Ghosh, S.; Henry, M.P.; Christopher, R.W. Hemicellulose conversion by anaerobic digestion. *Biomass* **1985**, *6*, 257–269. [CrossRef]

60. Brodeur, G.; Yau, E.; Badal, K.; Collier, J.; Ramachandran, K.B.; Ramakrishnan, S. Chemical and physicochemical pretreatment of lignocellulosic biomass: A review. *Enzyme Res.* **2011**, *2011*, 17. [CrossRef]

61. Fortenbery, T.R.; Bennett, M. Opportunities for Commercial Hemp Production. *Rev. Agric. Econ.* **2004**, *26*, 97–117. [CrossRef]

2

Social Vulnerability Assessment by Mapping Population Density and Pressure on Cropland in Shandong Province in China during the 17th–20th Century

Yu Ye [1,2,*], Xueqiong Wei [1], Xiuqi Fang [1] and Yikai Li [1]

[1] School of Geography, Faculty of Geographical Science, Beijing Normal University, Beijing 100875, China; weixueqiong1988@126.com (X.W.); xfang@bnu.edu.cn (X.F.); liyikai2016@foxmail.com (Y.L.)

[2] Key Laboratory of Environment Change and Natural Disaster, Ministry of Education, Beijing Normal University, Beijing 100875, China

* Correspondence: yeyuleaffish@bnu.edu.cn

Abstract: Cropland area per capita and pressure index on cropland are important parameters for measuring the social vulnerability and sustainability from the perspective of food security in a certain region in China during the historical periods. This study reconstructed the change in spatial distribution of cropland area per labor/household and pressure index on cropland during the 17th–20th century by using historical documents, regression analysis, pressure index model, and GIS (geographic information system). Following this, we analyzed the impacting process of climate change and sustainability of cropland use during the different periods. The conclusions of this study are as follows: (i) there was an obvious spatial difference of labor/household density, as there was higher density in three agricultural areas, which had the same pattern as cropland distribution during the same periods; (ii) Cropland area per capita was relatively higher during the 17th–18th century, which were above 0.4 ha/person in the majority of counties and were distributed homogenously. Until the 19th century and the beginning of 20th century, cropland area per capita in a considerable proportion of regions decreased below 0.2 ha/person, which embodies the increase in social vulnerability and unsustainability at that time; (iii) The pressure index on cropland also showed a spatial pattern similar to cropland area per capita, which presented as having a lower threshold than nowadays. During the 17th–18th century, there was no pressure on cropland. In comparison, in the 19th century and at the beginning of 20th century, two high-value centers of pressure index on cropland appeared in the Middle Shandong and the Jiaodong region. As a result, pressure on cropland use increased and a food crisis was likely to have been created; (iv) A higher extent of sustainable cropland use corresponded to the cold period, while a lower extent of sustainable cropland use corresponded to the warm period in Shandong over the past 300 years. The turning point of the 1680s from dry to wet was not distinctively attributed to the decrease in the extent of sustainable cropland use in Shandong. Since the beginning of the 20th century, the increasing pressure on the sustainability of cropland use finally intensified the social conflict and increased the probability of social revolts.

Keywords: pressure on cropland; labor/household density; Shandong Province in China; 17th–20th century

1. Introduction

Human interference has occurred with the climate system, with climate change posing a threat to natural systems and human sustainable development. The core concept of the fifth assessment report of Work Group II of the Intergovernmental Panel on Climate Change (IPCC WGII AR5) is the theme of impact, adaptation, and vulnerability related to climate change. It illustrates that climate-related risk results from the interaction of natural hazards (including hazardous events and trends) with the vulnerability and exposure of human and natural systems [1,2]. In the traditional agricultural society of historical China, climate change first impacts the level of food production, which hinders the improvement in living standards and social development by a transmission of forcing–responding chain. The forcing–responding chain means that the impact of climate change passes on from climate change to agriculture harvest to food supply, and finally to famine and social stability [3]. Overall, the extent of the impact on society by climate change depends on the social vulnerability and human adaptation actions. Socially sustainable development depends on whether the contradiction between human and land is resolved.

In the majority of current studies on the impact of historical climate change and human adaptation, climate change and societal stability have been well studied. For example, some typical researchers analyzed the impact of climate change on violent conflict in Europe over the last millennium [4]; the relationship between climate and the collapse of Maya civilization [5]; North Atlantic seasonality and implications for Norse colonies [6]; the relationship between sun, climate, hunger, and mass migration [7]; linkage of climate with Chinese dynastic change [8], and so on. However, some intermediate factors in this influencing process (e.g., population, agricultural production, and policy adjustment) were not fully considered. In addition, social vulnerability has not been stressed in these similar international studies.

In China, some researchers discussed the Chinese population and cropland area mainly from the perspective of historical geography or agricultural history. For example, Ge [9] studied the history of Chinese demographic composition, population change, and distribution; Li [10] analyzed the impact of climate change on several instances of Chinese historical population fluctuation; He [11] firstly evaluated the ancient land data in China; and other researchers [12–16] evaluated Chinese historical land data. They have produced methods for the data estimation of Chinese population and cropland area. Recently, scientists working in the field of global change have reconstructed the spatial distribution of historical cropland cover in China [17,18] or regions in China [19–22]. However, to understand the dynamics of climate-related risks, it would be better to combine agricultural production with the impacts of historical climate change. Pressure index on cropland was first put forward by Cai et al. [23], and has been extensively used to evaluate food security in certain regions [24,25]. In addition, the indexes (e.g., population density, cropland area per capita, pressure index on cropland, and so on) are important parameters that represent societal vulnerability from the perspective of food security. They are available to be used for research on the impact of historical climate change and social sustainability development.

North China is located in the northern temperate monsoon belt. The variability of temperature and precipitation is significant here. It has both higher sensitivity and certain adaptation ability to the impact of climate change. In addition, it was the administrative center of the traditional agricultural area in China during the Qing Dynasty. The impacting and responding processes of climate change in this region directly relate to the social stability, which is often preferentially considered by the central government. There have been many studies on the impact of climatic disasters and its response in this region. They include the analysis of relationship between revolt and drought–flood in Shandong Province during middle and late Qing Dynasty [26]; a case study on the impact of extreme climate events on migration and land reclamation in the early Qing Dynasty [27]; various types of responses in Northeast China to climatic disasters in North China over the past 300 years [28]; revolts frequency in the North China Plain during 1644–1911 and its relationship with climate [29]; social responses in Eastern Inner Mongolia to flood/drought-induced refugees from the North China Plain during 1644–1911 [30], and so on. In these similar national studies, the impacting and responding processes of historical climate change were mostly based on the method of time series comparison, with less attention paid to social vulnerability from the perspective of food security.

This article explores the spatial difference of factors such as the labor/household density, cropland area per capita, and pressure index on cropland in Shandong Province during the 17th–20th century. It would be used to estimate change of social vulnerability and sustainability from the perspective of food security in this region during the historical periods. It also provides fundamental data for research on historical climate change impact and adaptation.

2. Research Area

This paper takes the modern Shandong Province in China as the research area. It is located in the mid-latitude area of the northern hemisphere, within the range of 34°22′52″ N 114°19′53″ E–38°15′02″ N 122°42′18″ E, including 110 cities or counties. In the Qing dynasty, Shandong Province had 10 districts (named as Fu), 3 states directly under the central government (named as Zhili states), 8 scattered states under Fu (named as San states), and 96 counties [31]. The administrative boundaries of some counties in Shandong Province have changed, although this was mainly attributed to the split and combination of counties. Therefore, for ease of comparison with modern results, we converted historical data into the following indexes based on modern county boundaries (the base map comes from the 1:400 base data of China including administrative boundaries, rivers, roads, cities, etc.).

Shandong Province is located on eastern coast area of China, which is the lower reach of the Yellow River (the middle and northern part of Beijing). The Hangzhou Great Channel runs through Shandong Province. Shandong mainly consists of plain and hilly area, which occupies 55% and 28.7% of the total land area, respectively. This is located in the Middle and South Shandong. The northwest area includes the Northwest Plain of Shandong, an alluvial plain formed by the Yellow River. The eastern peninsula is mostly gently fluctuating hilly areas (Figure 1). Shandong has a semi-humid monsoon climate in the warm temperate zone. The climate is mild, and four seasons are discernible. The mean annual temperature is 11–14 °C. The mean annual precipitation is 550–950 mm. Shandong Province is an important agricultural production area in China. Cropland is distributed extensively, being found mainly on the Northwest Plain, Southwest Plain, and Jiaolai Plain. Forest and grassland are mainly distributed on the mountain, and hilly areas in Middle and South Shandong, Jiaodong Peninsula, and the Yellow River Delta. Wetland occurs mainly along the coast (Figure 2).

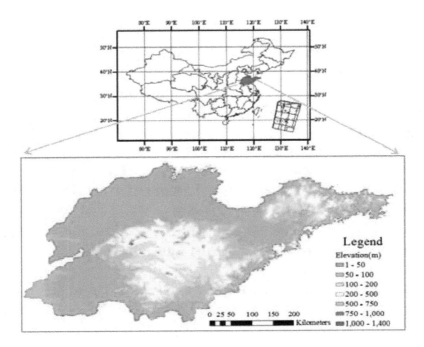

Figure 1. Location of Shandong Province.

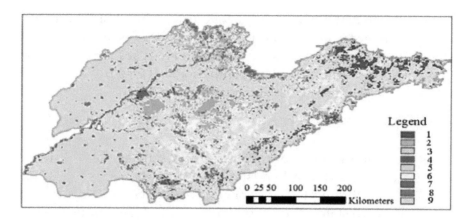

Figure 2. Land use/cover in Shandong Province. 1 = Evergreen coniferous forest; 2 = Deciduous broadleaved forest; 3 = Shrub; 4 = Coastal wetland; 5 = Grassland; 6 = Meadow; 7 = Urban land; 8 = River and lake; and 9 = Cropland.

3. Data Sources and Methods

3.1. Sources of Historical Climatic Data

The temperature data were sourced from the decadal mean temperature change series of North China from 1380s to 1980s [32] and that of Eastern China over the past 1000 years [33]. Precipitation data were sourced from precipitation (drought/flood) change series of North China over the past 2000 years [33]. These climatic data were all based on historical documents.

3.2. Sources and Processing of Population Data

Population data were sourced from gazetteers of counties in Shandong Province during the Qing Dynasty (1644–1911) and the period of Republic of China (1912–1949). There was a total of 244 volumes, which cover the 110 cities or counties in the research area. For one county, there are sometimes 2–4 versions of gazetteers in different periods. The records on the amount of labor, households, and population in different versions of gazetteers were validated with each other. The units of labor and household amount are "Ding" and "Hu", respectively. Ding and Hu are the population tax units in China during the historical periods.

First, we estimated the calculation ratio of labor, household, and population. "Ding" is mostly defined as an adult male aged from 16–60, who must pay the labor tax. One household has 2–3 laborers and includes 5–6 persons in Shandong Province generally during the 17th–20th century. The ratios of labor, household, and population were calculated from county data recording these numbers in 244 volumes of gazetteers, which is relatively reasonable at that time.

Secondly, we reconstructed the numbers of laborer of the 17th century and 18th century as well as the number of households of the 19th century and 20th century by interpolation. The number of records for labor, households, and population during the past four centuries used in this paper are listed in Table 1. The correlation analysis results used for interpolation are as follows:

(1) $Y = 1.444x$, $R^2 = 0.843$; (x = laborer in the 17th century; Y = laborer in the 18th century)
(2) $Y = 0.461x$, $R^2 = 0.869$; (x = laborer in the 18th century; Y = households in the 19th century)
(3) $Y = 1.14x$, $R^2 = 0.802$; (x = household in the 19th century, Y = household in the 20th century).

Table 1. The number of records for labor, household, and population during the past 400 years.

	17th Century	18th Century	19th Century	20th Century
Laborer (Ding)	53	52		
Laborer/Household (Ding/Hu)			40	46
Population (Person)			26	44

3.3. Spatial Distribution Change of Labor/Household Density and Cropland Areas per Capita

According to the above data for labor, household, and cropland area [34] in each county in Shandong during the 17th–20th century, by the equations of 1 Hu = 2–3 Ding, 1 Hu = 5–6 person the labor/household density and cropland area per capita in the four time-sections were calculated and spatially analyzed using the inputs of 1 Hu = 2–3 Ding and 1 Hu = 5–6 people (Figures 3 and 4). To represent the social vulnerability from the perspective of food security, we used 0.05 ha as the basic unit of division referring to the warning line of cropland area per capita (0.053 ha) put forward by Food and Agriculture Organization of United Nations (FAO) [35]. The legend of cropland area per capita is expressed as the segmentations separated by 1, 2, 4, 8, 16, 32 times 0.05 ha.

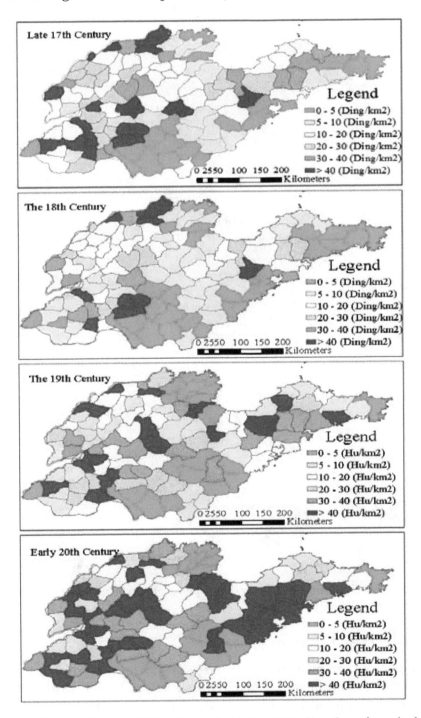

Figure 3. Spatial distribution change of labor/household density in Shandong from the late 17th century to the beginning of 20th century.

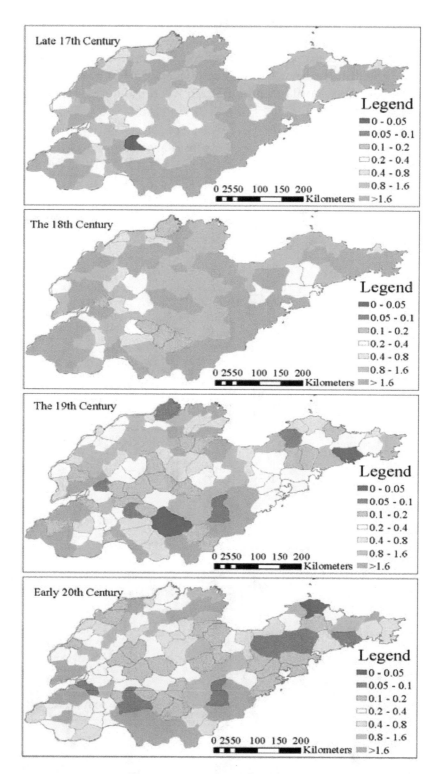

Figure 4. Spatial distribution change of cropland area per capita in Shandong from the late 17th century to the beginning of 20th century (Unit: ha).

3.4. Spatial Distribution of Pressure Index on Cropland

Pressure index on cropland measures the degree of shortage of cropland resources in a certain region. It also reflects the pressure on cropland and the social vulnerability from the perspective of food security during the historical periods. The minimum cropland area per capita represents the necessary cropland area to satisfy a person's basic food consumption at the normal living level in a certain region. Since the Qing dynasty, the "warning line" of cropland area per capita in Shandong has changed, as crop yield had been improved from about 1500 kg/ha to 6000 kg/ha [36,37]. Dietary structure based on grain has most likely remained the same. To make historical research easier, the model put forward by Cai et al. (2002) [23] is simplified by assuming the minimum cropland area per capita in four time-sections to be four times greater than those in modern times, while minimum cropland area per capita in modern times applies the warning line of cropland area per capita (0.053 ha) put forward by FAO (Wang, 2001) [35]. Pressure index on cropland (K) is the ratio of minimum cropland area per capita (S_{min}) and actual cropland area per capita (S_a). Its formulation is:

$$K = S_{min}/S_a \qquad (1)$$

Assume cropland area per labor and cropland area per household are Sd and Sh, respectively. According to the ratios of the numbers of laborer, household, and population discussed above (1 Hu = 3 Dings, 1 Hu = 6 people), it can be obtained that:

$$K = (0.053 \times 4)/S_a = (0.053 \times 4)/(S_d \times 6/3) = 0.424/S_d \text{ or}$$
$$K = (0.053 \times 4)/S_a = (0.053 \times 4)/6S_h = 1.272/S_h$$

The pressure index on cropland in each county in four time-sections was calculated. Following this, we produced the spatial distribution map of K-values by the ArcGIS software (ArcGIS is a word-leading application platform which can be used for collecting, organizing, managing, analyzing, communicating, and releasing geographic information) and analyzed its change (Figure 5). Finally, K value is interpolated by the inverse distance weighted method to identify the vulnerable center in the 19th century and at the beginning of 20th century, which is shown as the brown area in Figure 6. We divided K into 1–6 grades, which are 0–0.8, 0.8–1.6, 1.6–3.2, 3.2–4, and >4 by considering the actual discrete distribution of K-value and lower historical agricultural production level. The higher index grade represents a heavier pressure on sustainable cropland use and a larger possibility of an impending food crisis.

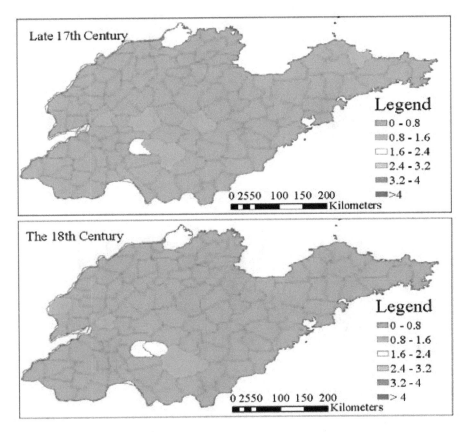

Figure 5. *Cont.*

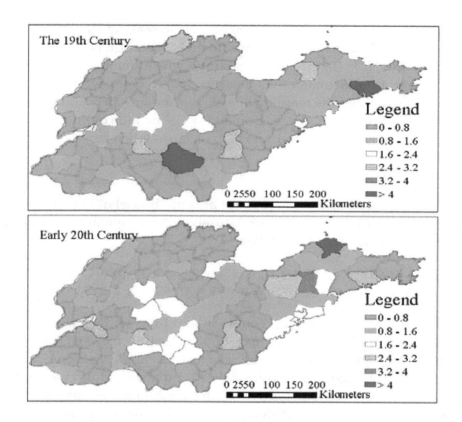

Figure 5. Spatial distribution change of pressure index on cropland in Shandong from the late 17th century to the beginning of the 20th century.

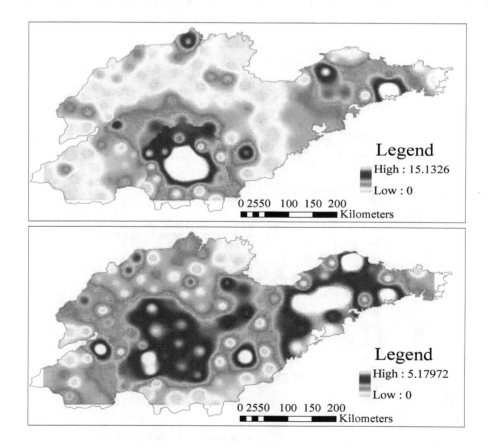

Figure 6. Spatial pattern of pressure index on cropland in Shandong during the 19th century (**up**) and at the beginning of the 20th century (**down**).

3.5. Analysis of Climate Change and Sustainable Cropland Use

The sustainability of cropland use can be measured by the above indexes, including population density, cropland area per capita, and pressure index on cropland. We compared climate change phases with the extent of sustainable cropland use in Shandong during different periods over the past 300 years. By the linkage of some intermediate elements, such as population, agricultural production, policy adjustment, and so on, society vulnerability and food sustainability during different periods were analyzed and discussed.

4. Results of Analysis

4.1. Climate Change in North China over the Past 300 Years

It shows that five regions in Eastern China (including North China) all had two distinctive cold periods (1620s–1710s and 1800s–1860s). The warmest period occurred in the 20th century during the past 500 years. The annual average temperature in the coldest hundred years (1800s–1900s) was lower than that of 20th century by 1.0 °C, with the coldest 30 years having happened in 1650s–1680s (Ge et al., 2012 [33]. The series of average temperatures in North China since the 1380s show that two cold periods occured (1550s–1690s and 1800s–1860s) (Wang et al., 1991 [32]).

The climate in Eastern China during the Qing Dynasty (1644s–1911s) was generally humid, although decadal variation was very distinctive. There were continuous droughts in 1720, 1785, 1810, and 1877. Climate in the 20th century tended to be dry but fluctuating, with the middle of 1940s being wetter than the middle of the 1960s. After this, the weather tended to be dry since the 1980s. It also shows that three sub-regions in Eastern China had a high consistency in the dry/wet change since the 1680s (all humid relatively), although the change in the dry/wet trend of North China (dry relatively) during 1520s–1680s was opposite to that of Jianghuai and Jiangnan regions (wet relatively) (Ge et al., 2012 [33]).

4.2. Spatial Distribution Change of Labor/Household Density in Shandong over the Past 300 Years

From the spatial distribution map of labor/household density (Figure 3), it was found that there existed an obvious spatial difference in labor/household density, which showed a similar pattern to cropland area in the corresponding periods. In essence, there was a relatively greater proportion of the population in the agricultural area appropriate for cultivation, while only a minority of the population settled in the regions not appropriate for cultivation. This embodies the impact of land suitability for cultivation on population distribution. In agricultural areas such as Northwest and Southwest Shandong as well as the Jiaolai Plain, the density of labor/household in the majority of cities/counties was above 10 Ding/km^2 during the 17th–18th century. At the beginning of the 20th century, most of cities/counties reached above 10 Hu/km^2. In comparison, there was a smaller distribution of the population in the hilly areas of middle and south Shandong, the Jiaodong Peninsula, and coastal swamp area. The density of labor/household in many cities/counties in these regions was below 5 Ding/km^2 or 5 Hu/km^2 from the 17th century to the beginning of the 20th century.

The population density of Shandong Province over the past 300 years has been increasing, especially in the three agricultural areas. Furthermore, the spatial difference of population density decreased from the 19th century to the beginning of the 20th century. In the Northwest and Southwest agricultural areas, the labor densities were above 10 Ding/km^2 during the 17th–18th century. Until the 19th century and the beginning of 20th century, the population densities in the whole research area were increasing, and reached 20 Hu/km^2 in many regions of the three agricultural areas.

4.3. Spatial Distribution Change of Cropland Area per Capita in Shandong over the Past 300 Years

From the spatial distribution map of cropland area per capita in Shandong from the late 17th century to the beginning of the 20th century (Figure 5), it was found that the cropland area per capita was distributed relatively uniformly and the spatial difference was less obvious than the population density.

The values of cropland area per capita during the 17th–18th century were higher, with those of the majority of counties being above 0.4 ha. This is eight times higher than the modern warning line of cropland area per capita put forward by the FAO. However, cropland area per capita in a few counties in Binzhou, Linyi, and Jining was below four times higher than the modern warning line, which means that food security in these regions might be at risk with stronger social vulnerability and would most likely be affected by climatic disasters.

The value of cropland area per capita from the 19th century to the beginning of 20th century generally decreased. The strength of social vulnerability and the possibility of social turbulence resulting from threatened food security increased. During the 19th century, the cropland area per capita in the majority of counties in the middle Shandong and Jiaodong Peninsula was below 0.1 ha, which was two times higher than the modern warning line of cropland area per capita. Food security in Binzhou, Linyi, Jining, Laiwu, Tai'an, Yantai, and Weihai were under threat. At the beginning of the 20th century, the cropland area per capita in the whole research area decreased universally, especially in counties in Jiaolai Plain and middle Shandong, as the area in these places decreased to below 0.2 ha. In Northwest and Southwest of Shandong, it appeared that the cropland area per capita of many counties were lower than 0.4 ha. The numbers of vulnerable regions increased. New areas under threat (e.g., Jinan, Dezhou, and Qingzhou) appeared at the beginning of 20th century with previous areas under threat to food security in the 19th century still having this risk (Figure 5).

4.4. Spatial Distribution Change of Pressure on Cropland in Shandong over the Past 300 Years

From the spatial distribution map of the pressure index on cropland in Shandong from the late 17th century to the beginning of the 20th century (Figure 6), it was found that the pressure index on cropland showed a similar spatial distribution to cropland area per capita, and its threshold was lower than that in modern times. The pressure index on cropland distributed homogeneously. It was relatively lower in the 17th–18th century and increased during the 19th–20th century.

Pressure index on cropland in the majority of counties during the 17th–18th century was below 0.8, which means that there was no pressure on cropland. The exceptions included minority regions in Jinan, Linyi, and Binzhou, which probably had threat to food security with the pressure index being above 0.8 (Figure 6). Until the 19th century and the beginning of the 20th century, two high-value centers of pressure on cropland appeared in middle Shandong and the Jiaodong Peninsula (Figure 7), including some cities or counties in Linyi, Tai'an, Jinan, Laiwu, Yantai, Qingzhou, and Weihai, with their pressure index on cropland being above 1.6. There is the possibility of a food crisis in these areas.

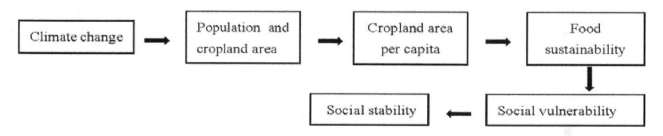

Figure 7. The impacting process of climate change.

4.5. Comparison between Climate Change Phases and the Extent of Sustainable Cropland Use in Shandong

First, it seems that a higher extent of sustainable cropland use occurred in the cold period, while a lower extent of sustainable cropland use occurred in the warm period in Shandong over the past 300 years. In the two cold periods (1620s–1710s and 1800s–1860s), the population density in Shandong was relatively lower. At this time, the labor/household densities of the majority of cities or counties during the 17th–18th century were below 10 Ding/km^2, which reached above 10 Hu/km^2

at the beginning of the 20th century during the warm period. In particular, the labor densities of the Northwest and Southwest agricultural areas increased from above 10 Ding/km^2 during the 17th–18th century to reach 20 Hu/km^2 in many regions until the beginning of 20th century. The cropland area per capita during the 17th–18th century was higher above 0.4 ha, which was eight times higher than the modern warning line of cropland area per capita put forward by the FAO. From the 19th century to the beginning of the 20th century, the cropland area per capita in the majority of counties in middle Shandong and the Jiaodong Peninsula decreased to below 0.1 ha, which is two times higher than the modern warning line. This meant that the strength of social vulnerability and the possibility of social turbulence resulting from a threat to food security increased. Similarly, the pressure index on cropland in the majority of counties during the 17th–18th century was below 0.8, which means that there was no pressure on cropland. Until the 19th century and the beginning of the 20th century, two high-value centers of pressure on cropland appeared in middle Shandong and the Jiaodong Peninsula. In these places, the pressure index on cropland in some cities or counties reached above 1.6, meaning that there was the possibility of a food crisis in these areas.

Second, the turning point of the 1680s from dry to relatively wet in North China seems to be attributed to the decrease in the extent of sustainable cropland use in Shandong, although this was not very distinctive. During the 17th–18th century, the population density in Shandong was relatively lower, the cropland area per capita was higher (above 0.4 ha), and the pressure index on cropland in majority of counties was below 0.8. This means that there was no pressure on cropland during the period around the 1680s. It is likely that the effect of dry/wet change on the sustainability of cropland use was not as obvious as that of temperature change.

5. Discussion

5.1. Impacting Process of Climate Change and the Sustainability of Cropland Use

Many researchers have analyzed the relationships between climate change or extreme climatic events with refugees and social stability in North China (Ye, et al., 2004; Xiao, 2011; Fang et al., 2007; Ye and Fang, 2013; Xiao et al., 2013). These case studies all showed that the impact of historical climate change on social stability in this area was often influenced by a failure in food production. In this present study, the intermediate elements of population, cropland area per capita, and pressure on sustainable cropland use were emphasized as a means of obtaining a better understanding of the impacting and responding processes of climate change.

It appears that a warm climate was beneficial in driving an increase in population and agricultural development, which finally resulted in pressure on the sustainability of cropland use. In the cold periods, the population density in Shandong was relatively lower and the cropland area per capita was higher. During these periods, there was no pressure on sustainable cropland use. In the warm periods of the 20th century, the population density in Shandong increased more quickly, while cropland area increased slowly. This led to a decrease in cropland area per capita, which was followed by a decrease in the sustainability of cropland use. This would intensify the social conflict and increase the probability of social revolts. Therefore, the impacting process of climate change can be depicted as shown in Figure 7.

In addition, the turning point of the 1680s from dry to wet in North China seemed to contribute to the decrease in the extent of sustainable cropland use in Shandong, although this was not very distinctive.

5.2. Special View of this Research and Its Scientific Value

Although some deviation still exists in the reconstructed results of population and cropland area per capita, this paper provides a perspective of food security and social vulnerability in the research area. To a certain extent, it makes up for the deficiency of paying more attention to the comparison of a series of climate change and social results while neglecting some intermediate links in the impact process and social vulnerability in this research area. By combining social vulnerability and human adaptation actions with climate-related hazards and physical exposure, it aims to accurately evaluate the impact of historical climate change or risk on human society. It also provides references to modern sustainable cropland use.

6. Conclusions

By analyzing historical documents, regression analysis, model of pressure index on cropland, and geographic information system (GIS), this paper reconstructed spatial patterns of labor/household density, cropland area per capita, and pressure index on cropland at the county level in Shandong Province during the 17th–20th century. Following this, we analyzed the impacting process of climate change and the sustainability of cropland use during the different periods. The conclusions of this study are as follows:

There was a distinct spatial difference in labor/household density which showed the effect of land suitability for reclamation on the population distribution. There was a greater proportion of the population in the agricultural area of Northwest and Southwest Shandong as well as the Jiaolai Plain. The population density of Shandong Province over the past 300 years has been increasing, especially in the three agricultural areas.

The spatial distribution of cropland area per capita in Shandong over the past 300 years has been relatively uniform. From the 19th century to the beginning of the 20th century, cropland area per capita decreased extensively and social vulnerability was strengthened. There was likely a threat to food security in Binzhou, Linyi, Jining, Laiwu, Taian, Yantai, Weihai, Jinan, Dezhou, Qingzhou.

The pressure index on cropland also showed a similar spatial distribution to cropland area per capita, but its threshold was lower than that in modern times. During the 19th century and the beginning of the 20th century, two high-value centers of pressure on cropland appeared in Middle Shandong and the Jiaodong Peninsula.

A warm climate was beneficial to driving an increase in population and agricultural development, which finally resulted in increasing pressure on the sustainability of cropland use. This increase in pressure on cropland was also related to the growth of the population as well as the spatial differences in land quality between plain and hill areas. The impacting process of climate change was sketched as following the flowchart shown in Figure 7: climate change—population and cropland area—cropland area per capita—food sustainability and society vulnerability—social stability.

Acknowledgments: This work was supported by the National Natural Science Foundation of China (Grant No. 41471156), the National Key Research and Development Program of China (Grant No. 2016YFA0602500), and the Strategic Priority Research Program from Chinese Academy of Sciences (XDA05080102).

Author Contributions: Yu Ye wrote this text and produced the figures. Xueqiong Wei, Xiuqi Fang and Yikai Li listed the references and help to edit the text.

References

1.   McCarthy, J.J.; Canziani, O.F.; Leary, N.A.; Dokken, D.J.; White, K.S. *Climate Change 2001: Impacts, Adaptation and Vulnerability, Contribution Of Working Group II to the Third Assessment Report of the Intergovernmental Panel on Climate change*; Cambridge University Press: Cambridge, UK, 2001.
2.   Field, C.B.; Barros, V.R.; Dokken, D.J.; Mach, K.J.; Mastrandrea, M.D.; Bilir, T.E.; Chatterjee, M.; Ebi, K.L.; Estrada, Y.O.; Genova, R.C.; et al. *Climate Change 2014: Impacts, Adaptation and Vulnerability, Contribution of Working Group II to the Fifth Assessment Report of the International Panel on Climate change*; Cambridge University Press: Cambridge, UK, 2014.
3.   Fang, X.Q.; Zheng, J.Y.; Ge, Q.S. Historical climate change impact-response processes under the framework of food security in China. *Sci. Geogr. Sin.* **2014**, *34*, 1291–1298.
4.   Richard, S.J.; Wagner, T.S. Climate change and violent conflict in Europe over the last millennium. *Clim. Chang.* **2010**, *99*, 65–79.

5. Haug, G.H.; Gunther, D.; Peterson, L.C.; Sigman, D.M.; Hughen, K.A.; Aeschlimann, A. Climate and the collapse of Maya civilization. *Science* **2003**, *299*, 1731–1735. [CrossRef] [PubMed]

6. Patterson, W.P.; Dietrich, K.A.; Holmden, C. Two millennia of North Atlantic seasonality and implications for Norse colonies. *Proc. Natl. Acad. Sci. USA* **2010**, *107*, 5306–5310. [CrossRef] [PubMed]

7. Hsu, K.J. Sun, climate, hunger and mass migration. *Sci. China Ser. D Earth Sci.* **1998**, *41*, 449–472. [CrossRef]

8. Zhang, D.E.; Li, H.C.; Ku, T.L.; Lu, L.H. On linking climate to Chinese dynastic change: Spatial and temporal variations of monsoonal rain. *Chin. Sci. Bull.* **2010**, *55*, 77–83. [CrossRef]

9. Ge, J.X. *The History of China's Population*; Fudan University Press: Shanghai, China, 2002.

10. Li, B.Z. Climate change and several times of Chinese historical population fluctuation. *Popul. Res.* **1999**, *23*, 15–19.

11. He, B.D. Verification and Evaluation of Ancient and Today's Land Data in China. unpublished work.

12. Bian, L. Folk measure method and essence of farmland area in South China during Ming dynasty and Qing dynasty. *China Agric. Hist.* **1995**, *14*, 49–56.

13. Wan, H. Historical comparison between cropland numbers in China during Ming dynasty and early Qing dynasty. *China Agric. Hist.* **2000**, *19*, 34–40.

14. Shi, Z.X. Analysis on change of population and land in the area of Gan, Ning and Qing during the late Qing dynasty. *China Agric. Hist.* **2000**, *19*, 72–79.

15. Geng, Z.J. Tentative analysis of question of men's and farmland's discount in Shanxi during Qing dynasty. *China Agric. Hist.* **2000**, *19*, 67–71.

16. Zhou, R. Integrated review and new calculation of cropland area during the early Qing dynasty. *Jianghan Tribune* **2001**, *9*, 57–61.

17. Ge, Q.S.; Dai, J.H.; He, F.N.; Zheng, J.Y. Change of the amount of cropland resource and analysis of driving forces in partial provinces in China during the past 300 years. *Prog. Natl. Sci.* **2003**, *13*, 825–832.

18. He, F.N.; Li, S.C.; Zhang, X.Z. The reconstruction of cropland area and its spatial distribution pattern in the mid-northern Song Dynasty. *Acta Geogr. Sin.* **2011**, *66*, 1531–1539.

19. Xie, Y.W.; Wang, X.Q.; Wang, G.S.; Yu, L. Cultivated land distribution simulation based on grid in middle reaches of Heihe River Basin in the historical periods. *Adv. Earth Sci.* **2013**, *28*, 71–78.

20. Tian, Y.C.; Li, J.; Ren, Z.Y. Analysis of cropland change and spatial-temporal pattern inLoess Plateau over the recent 300 years. *J. Arid Land Resour. Environ.* **2012**, *26*, 94–101.

21. Ye, Y.; Fang, X.Q. Expansion of cropland area and formation of the eastern farming-pastoral ecotone in northern China during the twentieth century. *Reg. Environ. Chang.* **2012**, *12*, 923–934. [CrossRef]

22. Ye, Y.; Fang, X.Q.; Ren, Y.Y.; Zhang, X.Z.; Chen, L. Cropland cover change in Northeast China during the past 300 years. *Sci. China Ser. D Earth Sci.* **2009**, *52*, 1172–1182. [CrossRef]

23. Cai, Y.L.; Fu, Z.Q.; Dai, E.F. The minimum area per capita of cultivated land and its implication for the optimization of land resource allocation. *Acta Geogr. Sin.* **2002**, *57*, 127–134.

24. Ren, G.Z.; Zhao, X.G.; Chao, S.J.; Dong, L.L.; Zhao, Y.M. Temporal-spatial analysis of cultivated land pressure in China based on the ecological tension indexes of cultivated land. *J. Arid Land Resour. Environ.* **2008**, *22*, 37–41.

25. Liu, X.T.; Cai, Y.L. Grain security of basic cropland pressure index in Shandong Province. *Popul. Resour. Environ.* **2010**, *20*, 334–337.

26. Ye, Y.; Fang, X.Q.; Ge, Q.S.; Zheng, J.Y. Response and adaptation to climate change indicated by the relationship between revolt and drought-flood in Shandong Province during middle and late Qing Dynasty. *Sci. Geogr. Sin.* **2004**, *24*, 680–686.

27. Fang, X.Q.; Ye, Y.; Zeng, Z.Z. Extreme climate events, migration for cultivation and policies: A case study in the early Qing Dynasty of China. *Sci. China Ser. D Earth Sci.* **2007**, *50*, 411–421. [CrossRef]

28. Ye, Y.; Fang, X.Q.; Khan, M. Migration and reclamation in Northeast China in response to climatic disasters in North China during the past 300 years. *Reg. Environ. Chang.* **2012**, *12*, 193–206. [CrossRef]

29. Xiao, L.B.; Ye, Y.; Wei, B.Y. Revolts frequency during 1644–1911 in North China Plain and its relationship with climate. *Adv. Clim. Res.* **2011**, *2*, 218–224. [CrossRef]

30. Xiao, L.B.; Fang, X.Q.; Ye, Y. Reclamation and revolt: social responses in Eastern Inner Mongolia to flood/drought-induced refugees from the North China Plain 1644–1911. *J. Arid Environ.* **2013**, *88*, 9–16. [CrossRef]

31. Niu, P.H. *Table of Administrative Division Evolution during the Qing Dynasty*; China Cartographic Publishing House: Beijing, China, 1990.

32. Wang, S.W.; Wang, R.S. Reconstruction of temperature series of North China from 1380s to 1980s. *Sci. China Ser. B Chem.* **1991**, *34*, 751–759.

33. Ge, Q.S.; Zheng, J.Y.; Hao, Z.X.; Liu, H.L. General characteristics of climate changes during the past 2000 years in China. *Sci. China Earth Sci.* **2012**, *42*, 934–942. [CrossRef]

34. Ye, Y.; Wei, X.Q.; Li, F.; Fang, X.Q. Reconstruction of cropland cover changes in Shandong Province over the past 300 years. *Sci. Rep.* **2015**, *5*. [CrossRef] [PubMed]

35. Wang, W.M. The perspective of 0.8 Mu warning line of cropland per capita. *Chin. Land* **2001**, *10*, 32–33.

36. Guo, S.Y. Food production in rainfed agricultural region in North China in the Qing Dynasty. *Res. Chin. Econ. Hist.* **1995**, *1*, 22–44.

37. Agriculture Department of the Shandong Province. Agriculture Information Website of Shandong Province. 2014. Available online: www.sdny.gov.cn (accessed on 4 July 2017).

3

Assessing Future Spatio-Temporal Changes in Crop Suitability and Planting Season over West Africa: Using the Concept of Crop-Climate Departure

Temitope S. Egbebiyi *, Chris Lennard, Olivier Crespo, Phillip Mukwenha, Shakirudeen Lawal and Kwesi Quagraine

Climate System Analysis Group (CSAG), Department of Environmental and Geographical Science, University of Cape Town, Private Bag X3, Rondebosch, 7701 Cape Town, South Africa
* Correspondence: EGBTEM001@myuct.ac.za

Abstract: The changing climate is posing significant threats to agriculture, the most vulnerable sector, and the main source of livelihood in West Africa. This study assesses the impact of the climate-departure on the crop suitability and planting month over West Africa. We used 10 CMIP5 Global climate models bias-corrected simulations downscaled by the CORDEX regional climate model, RCA4 to drive the crop suitability model, Ecocrop. We applied the concept of the crop-climate departure (CCD) to evaluate future changes in the crop suitability and planting month for five crop types, cereals, legumes, fruits, root and tuber and horticulture over the historical and future months. Our result shows a reduction (negative linear correlation) and an expansion (positive linear correlation) in the suitable area and crop suitability index value in the Guinea-Savanna and Sahel (southern Sahel) zone, respectively. The horticulture crop was the most negatively affected with a decrease in the suitable area while cereals and legumes benefited from the expansion in suitable areas into the Sahel zone. In general, CCD would likely lead to a delay in the planting season by 2–4 months except for the orange and early planting dates by about 2–3 months for cassava. No projected changes in the planting month are observed for the plantain and pineapple which are annual crops. The study is relevant for a short and long-term adaptation option and planning for future changes in the crop suitability and planting month to improve food security in the region.

Keywords: crop-climate departure; Ecocrop; crop suitability; planting month; CORDEX; West Africa

1. Introduction

The West African region has been identified as one of the hotspots with high susceptibility and vulnerability to the impact of climate change and global warming [1]. For example, the global climate is projected to be above 1.5 °C above the pre-industrial level in the next decade [2]. An increase in temperature between 3 °C and 6 °C coupled with a rise in the rainfall variability is projected into the future over West Africa from the AR5 report [3]. Most countries in West Africa heavily rely on agriculture, which is predominantly rainfed, as an important and significant contributor to their economies. It accounts for over 16% of the Gross Domestic Product (GDP) of the region's economy and employs over 60% of the labour force [4–6]. Additionally, West Africa has accounted for about 60% of the total value of the agricultural production in the continent for about 24 years [7]. However, the region has been identified as a hotspot to climate change impacts in the recent time owing to its reducing yields in the total agricultural production since 2007 in comparison to other sub-regions on the continent [7]. Current trends show that there may be further decreases in yields especially in the face of increasing warming and droughts which may lead to food insecurity over the region [8–10].

Findings from the Intergovernmental Panel on Climate Change (IPCC) fifth Assessment Report (AR5) shows widespread impacts from the changing climate to the historical month across all continents [11]. The report reveals a high exposure to climatic events and a low adaptive capacity of the African continent makes it one of the most vulnerable regions of the world. Agriculture is the most and major economic sector of Africa and has been described as the most vulnerable sector to the climate change impact with a great threat to the farming systems, crop production and food security at any level [7,12–14]. For example, past studies e.g., [15–19] have shown the impact of climate change on crop production and yield in Africa and West Africa in particular using different crop models. Sultan et al. [15] showed the decrease in the mean yield of sorghum cultivars due to the impact of climate change resulting from variation in the rainfall pattern and increasing temperature. Jalloh et al. [17] revealed that the impact of climate change will badly affect the production of major staple crops in West Africa particularly sorghum and groundnut in the Sahel. Moreover, Roudier et al. [6] combining the result of 16 published studies, showed that the projected impact of climate change on the crop yield over most African countries is negative (about 11%) with variations among crops, regions and modelling uncertainties posing the challenge for robust assessment of future yields at the regional scale. Further changes in the climate are expected in Africa over the next decades [1], as projections suggest a threat to food security due to the likely increase in climate variability over the next decades in Sub-Saharan Africa (SSA) [7]. As a result, impacts from the changing climate varies from subsectors among regions and different countries in SSA including West Africa but may be more detrimental to the West African region owing to its high susceptibility and low adaptive capacity with further warming [14,20].

The increase in global warming will lead to a new climate regime with a deviation from historical variability with a variation in the timing of emergence for different regions of the world called the climate departure [21,22]. For instance, [21] found that the mean temperature over West Africa will move outside the bounds of historical variability about two decades earlier before the global mean temperature thus making the region a hotspot of climate departure due to the impact of the global warming. On this premise and its direct consequence on rainfed crop production in West Africa, Egbebiyi et al. [23] explored the climate change induced crop realizations of the climate departing from historical variability, developed and proposed the concept called the crop-climate departure (CCD) in the context of recent climate historical variability and future climate projections.

The study defines CCD "as a departure from historical crop suitability threshold, whether in terms of variability, mean or both, over a location both in space and in time resulting from climate change (whether radical climatic change or not)" This concept was used to characterize crop suitability across the three agro-ecological zones (AEZs) of West Africa. However, the CCD concept was only tested and applied using three weather stations, within the three AEZs of West Africa. Although these stations are a representation of the three AEZs, nevertheless these cannot be generalized for the entire region, hence there is a need to examine how CCD at different climate windows, near the future (2031–2050) till end of the century (2081–2100) will affect crop suitability over the region using the concept of CCD.

Based on our definition and understanding on CCD, the aim of this present study is to examine the impact of CCD from the historical variability on future changes in crop suitability and month of planting over the entire West African region. Section 2 describes the data and methods used. Results from the study are outlined in Section 3. The discussion of the results and concluding remarks and recommendations for the future are in Sections 4 and 5, respectively.

2. Data and Methodology

2.1. Study Area

The West African (shown in Figure 1) region comprises of 15 countries namely Benin, Burkina Faso, Gambia, Ghana, Guinea Bissau, Guinea, Ivory Coast, Liberia, Mali, Mauritania, Niger, Nigeria, Senegal, Sierra Leone and Togo. It is geographically located at latitude 4–20 °N and 16 °W–20 °E and has rainfed agriculture as its mainstay economy. The region can be divided into three Food and Agriculture Organization (FAO) agro ecological zones (AEZs) namely, Guinea (4–8 °N), Savanna (8–12 °N) and the Sahel (12–20 °N) [24,25]. The region also has some localized highlands (Cameroon Mountains, Jos Plateau, and Guinea Highlands) which influence its climate. The climate of the region is mainly controlled by the West African Monsoon (WAM) which accounts for about 70% of the annual rainfall [24,26]. WAM is an important and dynamic characteristic of the West African climate during the summer month [27].WAM is produced from the reversal of the land and ocean differential heating and dictates the seasonal pattern of rainfall over West Africa between latitudes 9° and 20 °N. It is characterized by winds that blow south-westerly during warmer months (June–September) and north-easterly during cooler months (January–March) of the year [25,27]. It is the major system that influences the onset, variability and pattern of rainfall over West Africa [28], [29]. It alternates between wet (April–October) and dry seasons (November–March) as the rainfall belt follows the migration of Inter-Tropical Discontinuity (ITD) [30] and thus affects the rainfall producing systems with an impact on the rainfed agriculture and influences crops suitability and food production in the region.

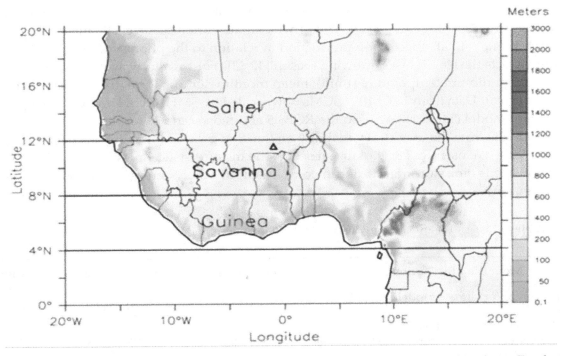

Figure 1. The study area, showing the West African topography and the three Food and Agriculture Organization (FAO) agro-ecological zones, designated as Guinea, Savannah and Sahel, respectively [24,25].

Different crops are grown in various parts of West Africa. Some of the major crops grown in the region are cassava, groundnut, millet, maize, sorghum, yam, plantain, cocoa, rice, wheat [8,26,31,32]. Millet and sorghum account for 64% of the cereal production over the regions in the year 2000 thus making them among the important staple crops in West Africa [26,33]. Cassava is one of the most important staple food crops in terms of production in sub-Saharan West Africa owing to high resilience to drought in the region [26,32,34] This also applies to the Yam production, which account for about

91% of the world's production [26,34,35]. Cereal, maize provides about 20% of the calorie intake in West Africa and is adjudged the most important staple food in the sub-Saharan Africa [26,33]. Other crops such as cocoa and plantain to mention a few contribute significantly to the economy of the region.

2.2. Data

2.2.1. Historical and Future Climate Datasets

For this study, three datasets were used as observations of the present-day climate and the locations where crops are grown as observed from the crop suitability model, Ecocrop output, and modelled simulations of the present and projected crop suitability driven by the observed and projected climate data. The observation dataset was the $0.5° \times 0.5°$ resolution monthly precipitation and minimum and mean temperature gridded dataset for the month of 1901 to 2016 obtained from the Climate Research Unit (CRU TS4.01 version, land only) University of East Anglia [36]. This was used to evaluate the available bias corrected RCMs forced by the 10 CMIP5 global climate models. The bias-corrected climate data were obtained from the Swedish Meteorological and Hydrological Institute, Linköping, Sweden. The modelled climate data were used as inputs into the crop suitability model, Ecocrop [37]. For this study, five different crop types namely; cereals (maize, pearl millet and sorghum), root and tuber (cassava, plantain and yam), legumes (cowpea and groundnut), horticulture (pineapple and tomato) and fruit (mango and orange) were selected based on the FAO 2016 statistics and their economic importance in the region. These different datasets are defined in the sub-sections below.

Temperatures and rainfall are important climate variables used in determining the impacts of climate change at different scales [38,39]. These two climate variables have a significant effect on crop yield [40,41]. While rainfall affects crop production in relation to the photosynthesis and leaf area, the temperature affects the length of the growing season [42,43]. For this study, we used the bias-corrected mean monthly minimum temperature (tmin), mean monthly temperature (tmean) and total monthly precipitation (prec). Data from 10 CMIP5 GCMs downscaled by SMHI-RCA4 are used as input into the crop suitability model (Table 1). We used the RCP8.5 emission scenario for the analysis to investigate the impact of CCD from the historical variability on the crop growth suitability and month of planting over West Africa. We used RCP8.5 because it seems the most realistic emission scenario as seen from the greenhouse gas emission trajectories in comparison to other scenarios and also has the largest simulation ensemble members [44].

Table 1. List of dynamically downscaled Global Climate Models (GCMs) used in the study.

Modelling Institution	Institute ID	Model Name	Resolution
Canadian centre for climate modelling and analysis	CCCMA	CanESM2	$2.8° \times 2.8°$
Centre National de Recherches Meteorolo-Giques/Centre Europeen de Recherche et Formation Avanceesencalcul scientifiqu	CNRMCERFACS	CNRM-CM5	$1.4° \times 1.4°$
Commonwealth Scientific and Industrial Research Organisation in collaboration with the Queensland Climate Change Centre of Excellence	CSIRO-QCCCE	CSIRO-Mk3.6.0	$1.875° \times 1.875°$
NOAA geophysical fluid dynamic laboratory	NOAAGDFL	GFDL_ESM2M	$2.5° \times 2.0°$
UK Met Office Hadley centre	MOHC	HadGEM2-ES	$1.9° \times 1.3°$
EC-EARTH consortium	EC-EARTH	ICHEC	$1.25° \times 1.25°$
Institute Pierre-Simon Laplace	IPSL	IPSL-CM5A-MR	$1.25° \times 1.25°$
Japan agency for Marine-Earth Science and Technology	MIROC	MIROC5	$1.4° \times 1.4°$
Max Planck institute for meteorology	MPI	MPI-ESM-LR	$1.9° \times 1.9°$
Norwegian climate centre	NCC	NorESM1-R	$2.5° \times 1.9°$

2.2.2. Ecocrop—A Crop Suitability Model

The Ecocrop model is a crop suitability model. It uses a crop growth suitability threshold dataset hosted by the FAO [37]. It is a simple mechanistic and empirical model originally developed by Hijmans et al. [37] and based on the FAO-Ecocrop database [45]. It is designed at a monthly scale with the ability to analyse the crop suitability in relation to the climate conditions over a geographical location [37,45]. Ecocrop employs environmental ranges of a crop coupled with numerical assessment of the environmental condition to determine the potential suitable climatic condition for a crop. The suitability rating can be linked to the agricultural yield which is partly dependent on the strength of the climate signal in the agricultural yield [46] The computation of optimal, suboptimal and non-optimal conditions based on these datasets allows for the simulation of the suitability of crops in response to the 12-month climate via t-min, t-mean and prec. [37].

The Ecocrop model evaluates the relative suitability of crops in response to a range of climates including rainfall, temperature and the growing season for optimal crop growth. A suitability index is generated as follows: 0 < 0.20 (not suitable), 0.20 < 0.4 (very marginally suitable), 0.4 < 0.6 (marginally suitable), 0.6 < 0.8 (suitable), and 0.8 < 1.0 (highly suitable) [45,47]. The default Ecocrop parameters were assumed. Although those thresholds may vary with different geographical and/or climatic conditions, previous studies have reported a close correlation between the Ecocrop model and the climate change impact projections from other crop models [45,48–50]. A paucity of data over regions of interest like SSA limits the validation of these processes [51]. Nevertheless, the method contributes to the demand for the regional scale assessment of the crop response to future climate projections.

2.3. Methods

We analyzed 10 CMIP 5 GCMs datasets downscaled by CORDEX RCM, RCA4 to assess the impacts of CCD from the historical variability on crop suitability and planting season over West Africa for five different crop types, cereal (maize, pearl millet and sorghum), fruit (mango and orange), horticulture (pineapple and tomato), legume (cowpea and groundnut) and root and tuber (cassava, plantain and yam). We used the RCA4 simulation output for the monthly minimum and mean temperature and total monthly precipitation as input into Ecocrop, a crop suitability model. Using a 20-year moving average at five year time steps, we computed the Suitability Index Value (SIV) for each crop across the 10 downscaled GCMs over West Africa.

The Ecocrop suitability output were then used to assess the impact of global warming through CCD from the historical variability on the crop suitability and planting season over a month 1951–2100. Across the agro-ecological zones (AEZs) of West Africa. After the simulation, we computed the mean of the best three consecutive suitability index and best three months of planting window within the growing season across each grid point over the region for the historical and future month. Before examining the RCM-projected changes in the future crop suitability and planting season, we evaluated the capability of the models in simulating the crop suitability spatial distribution and planting date/season during the reference month (1981–2000).

We also used the statistical tool to calculate the trend of change across the three windows compared to the historical month. We assessed the trend of change in the crop suitability and month of planting at each global warming levels for each crop using the Theil-Sen estimator or Sen's slope [52,53]. The Theil-Sen slope estimator is an estimation of the average trend rate only and magnitude of the trend. It is a linear slope that is compatible with the Mann-Kendall test and more robust such that it is less sensitive to outliers in the time series as compared to the standard linear regression trend [54]. The Theil-Sen slope method can detect significant trends with the changing rate than the linear trend [55]. Previous studies [56,57] have used this method in calculating trends.

2.3.1. Simulation Approach and Analysis of suitability

Past studies (e.g., [25,58–60] have evaluated the performance of the RCA4 historical data against the CRU dataset in the past climate. Their results showed that there is a good agreement with a strong correlation ($r \geq 0.6$) between the CRU dataset and RCA4 monthly simulated past climate data for both the temperature and precipitation over West Africa. For example, the model replicates the CRU north-south temperature gradient that concurs with previous findings by [58]. Additionally, the RCA4 simulated total monthly rainfall realistically captures the essential features namely, both the zonal pattern and meridional gradient and the rainfall maxima over high topography (i.e., Cameroon Mountains and Guinean Highlands) as observed in CRU which agrees with previous findings by [25,59,60].

The performance of RCA4 in simulating the essential features of West African climate variables, temperature and rainfall, and doubles as the needed input variables for the crop suitability model, Ecocrop makes it suitable and gives confidence in the use of the RCA4 for the crop suitability simulation over the region.

In addition, we compare the Ecocrop simulation over the region with the MIRCA2000 annual harvested area around year 2000 from the global monthly gridded data as described by [61] for six crops, cassava, maize, groundnut, sorghum, millet and plantain available in the MIRCA2000 dataset. The MIRCA2000 dataset provides monthly irrigated and rainfed crops area for 26 crop classes for each month of the year around year 2000 with a spatial resolution about 9.2 km. We compare the spatial agreement between the Ecocrop simulation and MIRCA2000 by using an overlap in the spatial agreement between the two datasets. Although, we admit the short time length of the MIRCA dataset however, it is a useful gridded dataset that has been used to provide information on the crop harvested area across different regions of the world [61] and will be useful to evaluate the simulated Ecocrop spatial suitability distribution at present due to the paucity of the suitability dataset across the globe. To see the overlap and area of agreement in the spatial suitability output of the two datasets, we set the MIRCA2000 annual harvested area dataset as one (1) and the Ecocrop simulated suitable area suitability index value from 0.2 ($SIV \geq 0.2$) as two(2). Where the two datasets agree as three(3). The output shows a good agreement between the Ecocrop and MIRCA2000 data for the examined crops with a strong spatial correlation ($r > 0.7$) (Figure 2). This gives some level of confidence in the use and performance of the Ecocrop simulation over the region.

To assess the impact of CCD from the historical variability on the crop suitability over West Africa, we computed the monthly climatological mean for a 20-year running month, at every five-year timestep for the t-min, t-mean and prec. from 1951–2100. For example, the first 20-year mean computed was 1951–1970, the second 20-year mean was 1956–1975, etc., until the last month 2081–2100. The resulting 12-month values per the 20-year month window was used as an input climatology into the Ecocrop suitability model as developed by the Food and Agriculture Organization, FAO [37] to simulate crop suitability for each downscaled GCM based on the methodologies described in [45].

Ecocrop calculates the crop suitability values in the response climate variables such as a monthly rainfall and temperature datasets and generates an output with a suitability index score from zero (unsuitable) to one (optimal/excellent suitability). It should be noted that this study did not undertake any additional ground-truthing or calibration of the range of climate parameters preferred for either crop and therefore the default EcoCrop parameters were assumed. Suitability index scores were calculated for the range of climate variables reported for the historical baseline (1981–2000) future months, near future (2031–2050), mid-century (2051–2070) and end of century (2081–2100) for the downscaled 10 CMIP5 GCMs that participated in the CORDEX experiment.

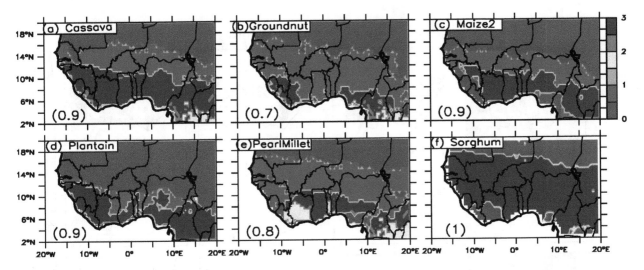

Figure 2. A simulated spatial distribution of the crop harvested area and suitability over West Africa for the year 2000 as simulated by the MIRCA2000 dataset and Ecocrop, respectively. The blue area (represented by 1) are the crop harvested area around the year 2000 as simulated by the MIRCA2000 dataset while the yellow colour represents the suitability index value above0.2 (SIV ≥ 0.2) which is represented by two. The red colour represents the area where the two datasets agree as denoted by three. The number at the left-hand corner represents the spatial correlation (r ≥ 0.7) value between the two datasets. The red colour depicts in Fig. 2a-2f depicts harvested and suitable areas as simulated by MIRCA2000 and Ecocrop from cassava to sorghum respectively. The blue colour depicts MIRCA2000 simulated harvested area only for each crop while yellow means Ecocrop simulated suitable areas for cultivation of each crop in year 2000. The purple colour, 0 depicts non harvested and unsuitable areas as simulated by both MIRCA2000 and Ecocrop for each of the crops around the year 2000.

2.3.2. Assessing the Robustness of Climate Change

We use two conditions (model agreement and statistical significance) to evaluate the robustness of the projected climate change for the three future months. For the model agreement, at least 80% of the simulation must agree on the sign of change. For the statistical significance, at least 80% of the simulations must indicate that the influence of the climate change is statistically significant, at 95% confidence level using a *t* test with regards to the baseline month, 1981–2000. When these two conditions are met then we consider the climate change signal to be significant. [30,44,62,63] have all used the methods to test and indicate the robustness of the climate change signals.

3. Result

3.1. Crop Suitability in the Historical Climate over West Africa

The RCA4 simulated crop suitability from the observed climatology inputs (RCA4-Ecocrop) shows a decreasing mean suitability from south to north over West Africa (north-south suitability gradient). The spatial suitability representation reveals unsuitable or very marginal suitability to the north in the Sahel from lat. 14 °N with a low Suitability Index Value (SIV) value between 0.0–0.4 and a higher suitability to the south in the Guinea-Savanna AEZ with a high SIV (0.6–1.0) sandwiched by an ash/silver suitability line called the Marginal Suitability Line (MSL) with an SIV between 0.41–0.59. In general, the MSL are observed around lat.14 °N in the Sahel AEZ (northern Sahel) for the simulation across the region except for the one observed around lat. 12 °N, the boundary between the Sahel and Savanna AEZ. The RCA4 simulation of all crop types examined, legumes (cowpea and groundnut), root and tuber (cassava, plantain, Yam, white yam), cereals (maize, pearl millet and sorghum) and fruit and horticultural crops (mango, orange, pineapple and tomato) shows that all the crops are very suitable to the south of the MSL but with no or low suitability to the north (Figures 3–6, column 1).

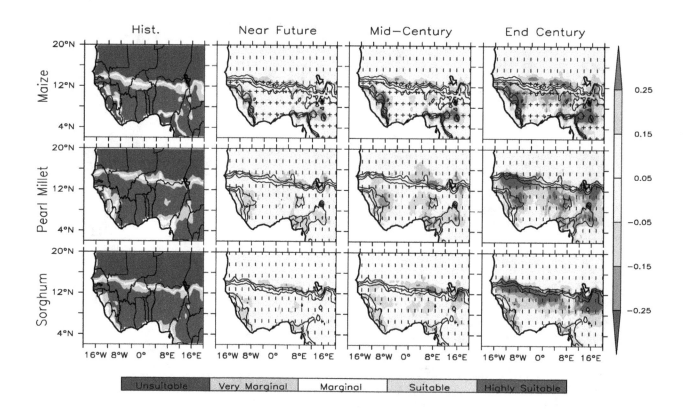

Figure 3. Simulated spatial suitability distribution for the cereal crops, maize pearl millet and sorghum over West Africa for the historical month (1981–2000) (column 1) and the projected change in the crop suitability for the near future month (2031–2050), mid-century (2051–2070) and end of century (2081–2100) (column 2–4, respectively). The vertical strip (|) indicates where at least 80% of the model simulations agrees on the projected sign of change while the horizontal strip (−) indicates where at least 80% of the model simulations agree that the projected change is statistically significant at 99% confidence level. The cross (+) indicates where the two conditions are met, meaning that the change is robust.

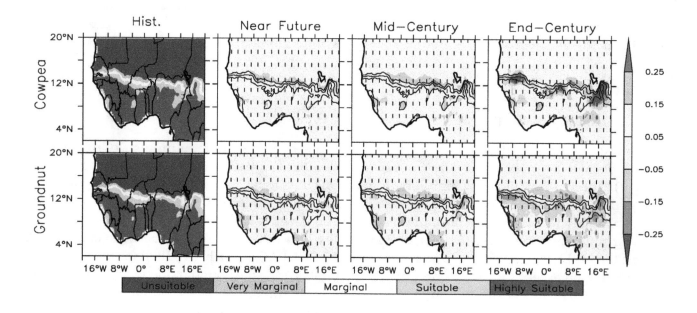

Figure 4. Same as Figure 3 but for the legume crops, cowpea and groundnut.

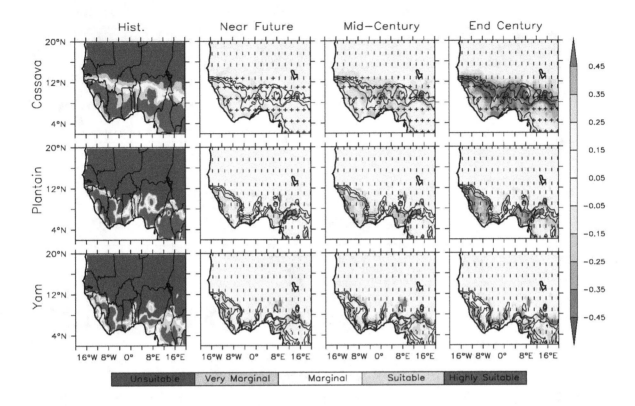

Figure 5. Same as Figure 3 but for the root and tuber crops, cassava, plantain and yam.

Along the coastal areas, legumes and root and tuber crops are suitable along the south-west coast of Senegal to the south-west coast of Cameroon. High SIV are observed for the root and tuber crops, plantain and Yam in the north central part of Nigeria in the Savanna. It is worth mentioning because the surrounding areas are observed to be unsuitable for the cultivation of both crops. For cereals, pearl millet is suitable along the west coast of Senegal and from the south coast Ivory Coast to the south-west coast of Cameroon. Maize is suitable from the south coast of Ivory Coast to the south-west coast of Nigeria. Fruit and horticultural crops are all suitable along the south coast of the Ivory Coast to the south-west coast of Nigeria. Mango and pineapple are suitable along the west coast of Senegal to Gambia while orange and tomato are only suitable along the west coast of Gambia.

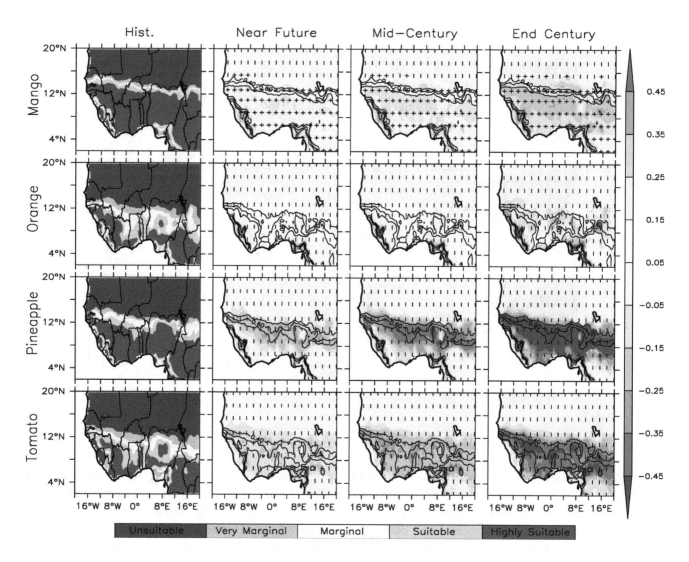

Figure 6. Same as Figure 3 but for the fruit crops, mango and orange and horticultural crops, pineapple and tomato.

RCA4 was also used in simulating the best planting months (PM) from the range of month in a planting window within the Length of Growing Season (LGS) over West Africa for the historical climate (Figures 7–10, column 1). LGS provides information on the start and end of the growing season and can also assist in the simulation process of identifying the best PM within a possible planting window in a growing season over a given location. The simulated planting month represents the first month of the best three months of the planting window. For example, a simulation of April means April–June is the three best PM and varies with crop types across the three AEZs of the region. For the legumes, our simulation shows January–July as the planting windows for cowpea and groundnut over the region (Figure 7, column 1). Jan (January–March) and Feb (February–April) as the best PM for cowpea and groundnut, respectively in the central Guinea and Savanna AEZs except over Sierra Leone, Liberia and the south coast of Nigeria. The month of Feb (Feb-April) was simulated as the best three planting months in the western and eastern Savanna-Sahel AEZs for cowpea, while it was Mar (March–May) over the same area and month for the groundnut. Along the coastal areas, July is simulated as the PM along the southwest coast from southern Sierra Leone to Liberia and the south coast of Nigeria and April along the southwest coast of northern Sierra Leone. For the groundnut, April is the PM along the west coast of Guinea, while May is the PM along the west coast of Sierra Leone and northern Liberia. August and March are the PM at the south coast of Liberia and Nigeria, respectively. The months of December and January are the PMs along the south coast of Ivory Coast to Ghana for the cowpea and groundnut, respectively.

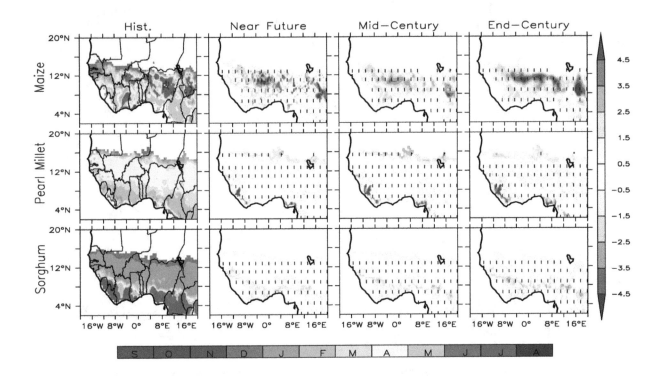

Figure 7. Simulated month of planting for cereals, maize, pearl millet and sorghum over West Africa for the historical month (1981–2000) (column 1) and the projected change in the crop planting month for the near future month (2031–2050), mid-century (2051–2070) and end of century (2081–2100) (column 2–4 respectively). The planting is simulated from September to August. The vertical strip (|) indicates where at least 80% of the model simulations agrees on the projected sign of change while the horizontal strip (−) indicates where at least 80% of the model simulations agree that the projected change is statistically significant at 99% confidence level. The cross (+) indicates where the two conditions are met, meaning that the change is robust.

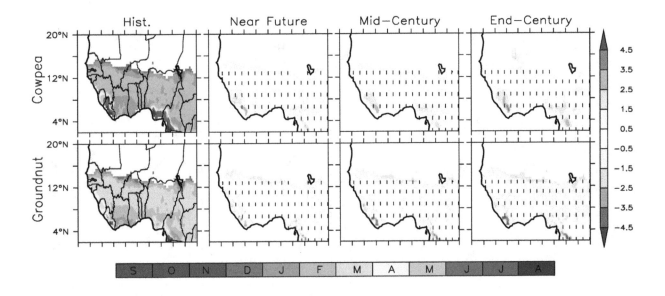

Figure 8. Same as Figure 7 but for the legumes, cowpea and groundnut.

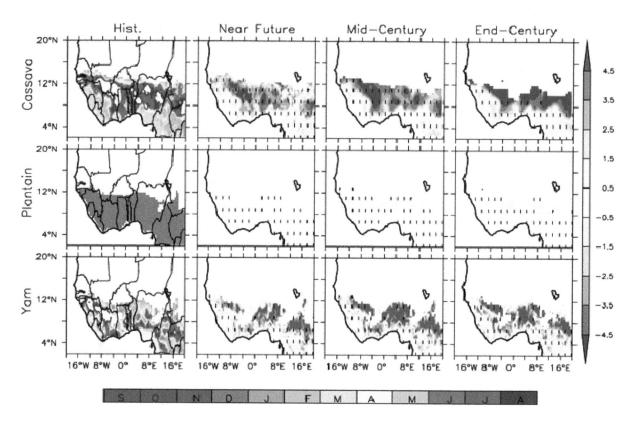

Figure 9. Same as Figure 7 but for the root and tubers, cassava, plantain and yam.

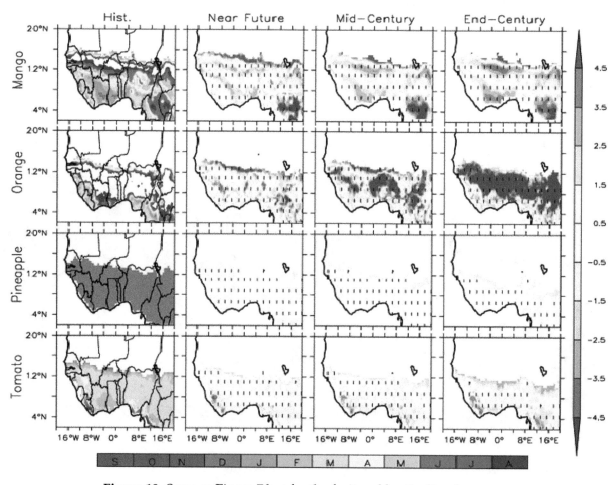

Figure 10. Same as Figure 7 but for the fruit and horticultural crops.

The root and tuber plantain is an annual crop that can be planted at any month of the year (Figure 9). The simulated PM is an overlay of the simulation of other months in the year as the crop may be planted in the suitable zones, Guinea and Savanna at any month/month of the year. For cassava, our simulation shows March (March–May) as the best PM generally over the region (Guinea-Savanna AEZs) except along the south-east coast of Ivory Coast to Ghana with PM in July, northern Guinea to Gambia and south east Senegal and from the boundary of Benin Republic to north west Nigeria in April.

The Ecocrop simulation for Yam shows June as the best PM in the central Guinea zone from the south-east Ghana to the south-east of Nigeria and in the north central part of Nigeria as well as Togo in the central Savanna zone. The month of February is observed as the best PM from the south-east Mali to the north-western part of Nigeria in the Savanna. Over the coastal area, April is observed as the PM along the south west coast from Sierra Leone to Liberia and the south coast of Nigeria and June along the west coast of Guinea and from the south coast of Ivory Coast to the south-west coast of Ghana in the past climate.

Our simulation for cereal crops shows February as the PM for pearl millet in Guinea, March and April are the best PMs in the south Savanna and northern savanna and Sahel AEZs, respectively although with exception. For example, in central savanna, from the northern Benin Republic to the north-western Nigeria for millet, the PM is April while in the north-eastern Nigeria in the Sahel it is March compared to April in the Sahel zone. However, for the pearl millet, the PM is April in the western Sahel along the south-west coast of Senegal, June along the west coast of Guinea and January along the south coast of Ivory Coast to the south-west coast of Nigeria. For maize, the PM is simulated to be in May (May–July) in the Guinea and southern Savanna zone of West Africa while it is in December (December–February) in the northern Savanna into the Sahel zone. For sorghum, June is simulated as the PM over Sierra Leone to Liberia and its coastal areas as well as the south coast of Ivory coast and Nigeria while it is May in the central south of Ivory coast and southern Ghana. The crop is simulated to be best cultivated in January in the Savanna-Sahel zones and best in December in the northern Sahel.

The Ecocrop simulation of the best PM for the horticultural crops (Figure 10) in the past climate shows tomato is mainly planted in March over the regions except from the south-east Ivory coast to south-west Ghana and around 14 °N in the Sahel where the best PM in February, along the west and south coast of Liberia and Nigeria, respectively where the best PM is July. Pineapple is an annual crop and it shows similar characteristics as plantain as mentioned above, which can be planted at any month of the year. For the fruit crop, orange shows February as the best PM over Sierra Leone to Liberia and along the west coast from Guinea to Liberia and the south coast of Nigeria. June is observed as the best PM in the south of Ivory coast to Ghana and Nigeria as well as the south coast of the Ivory and Ghana. June is also simulated as the best PM from Guinea Bissau to north-east Nigeria around lat 14 °N in the Sahel. The Ecocrop simulation for mango in the past climate shows February as the PM from the Guinea to southern savanna AEZ, April, May in the northern savanna AEZ and June as the best PM in the southern Sahel AEZ. Along the coastal areas, March is simulated and observed as the best PM from the west coast of Guinea to Liberia and south coast of Nigeria but February over the south coast of Ivory coast and Ghana.

Nevertheless, the evaluation simulations demonstrate that (RCA4-Ecocrop) captures the spatial variation in the suitability with different crops across the three AEZs of West Africa in the present-day climate and can serve as a baseline for evaluating the changes in crop suitability under global warming at different time windows over the region. The model also captures the spatial distribution of the best planting month within a growing season for crops over the region which varies with different months of the year.

3.2. Projected Changes in Tmin, Tmean and Precip over West Africa

An increasing clear trend of warming is projected across West Africa in the future, with predictions of increases of the t-min and t-mean of approximately 1–4.5 °C (Figure 11, Row 2 and 3 respectively). The mean and minimum monthly temperature (t-mean and t-min) is predicted to increase by 1.5–2 °C in the Guinea-Savanna of the regions, about 2–2.5 °C in the Sahel and increases of about 1 °C predicted for the south-west coastal area in the near future month (2031–2050). By mid- century, the t-mean is projected to increase by 2.5 °C and 3.0 °C over the Guinea-Savanna and Sahel, respectively and 3.0 °C increase over the Guinea and 3.5 °C over the Savanna-Sahel for t-min. At the end of the century, a 4.0 °C temperature increase is projected over the Guinea-Savanna zone except the western area and 3.5 increase over the Sahel for t-mean. The projected change in the minimum temperature by the end of the century showed a different pattern over the region as the Guinea zone, southern Guinea-coastal area, is warmer than the Sahel. The projection shows an increase up to 4.5 °C in the southern Guinea (coastal area) and 4 °C inland. A similar characteristic is also observed over the Sahel as the southern Sahel (12–14 °N) is projected as warmer (4.0 °C) than north of 14 °N (3.5 °C) in the Sahel zone. The savanna zone is however different to the Guinea and Sahel as the temperature increases northward over the zone, i.e., southern Savanna (3.5 °C) is lower to the northern Savanna (4.0 °C) except for the western part of the Savanna zone, which is much cooler than the rest with an increase of 2.5 °C. Our findings are consistent with the findings by [30].

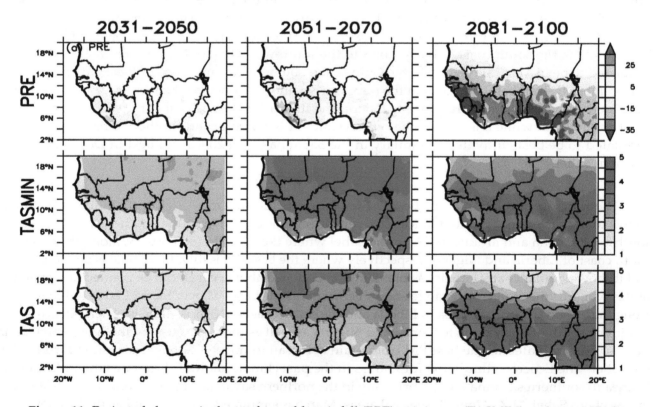

Figure 11. Projected changes in the total monthly rainfall (PRE), minimum (TASMIN) and mean (TAS) monthly temperature over West Africa as simulated by RCA4 for the near future month (2031–2050), mid-century (2051–2070) and the end of the century (2081–2100).

With respect to the predicted effects of climate change on rainfall, it is not a major change in the mean monthly precipitation that is projected over the region except in the south-west Guinea zone extending to the southern Guinea in the Savanna and the south coast of Nigeria (Figure 11, Row 1). The projected increase of about 10 mm extends from the south-west coastal area of Sierra-Leone to Liberia to the south-west coast of Ghana and south coast Nigeria in the near future (2031–2050) compared to the historical climate. By mid-century (2051–2070), the projected change of about 10 mm is expected in the western part of the region from the Guinea zone to the southern Sahel zone and the north-central part of Nigeria. Over the coastal area, an increase up to 25 mm is projected along the south-west coast of Sierra-Leone to Liberia and 10 mm over the south coast of Nigeria. The projection shows that no change is expected in the eastern part of the region by the mid-century. In contrast, by the end of the century, the projected change in the rainfall will be characterized by a decrease in the monthly precipitation across the entire region compared to the baseline. A gradient decrease in the rainfall is projected from south to north with a reduction up to 35 mm over the Guinea-Savanna and about 25 mm over the Sahel. Over the coastal area, a decrease above 35 mm is projected along the west coast of Gambia to northern Liberia and south coast of Nigeria. Our findings are in agreement with [30].

3.3. Impact of CCD on Future Crop Suitability over West Africa

Projected changes in the future crop suitability for all crop types varies across the three future climate windows, from the near future period (2031–2051) to the end of the century (2081–2100) (Figures 3–6, column 2–4). The variation in the impact for the crops can may be linked to the difference in crops response to the different climate window as described in Table 2 below for the three-climate window/period. For the near future, our simulation projects a general no change in the suitability for cereals south of 12 °N except the south coast of Nigeria (Figure 3, column 1). However, a project decreases of about 0.1 SIV is expected in the south coast of Nigeria for all the cereal crop, over Guinea for pearl millet, from Sierra Leone to Liberia for sorghum and from eastern Guinea to Liberia in the western Guinea-Savanna zone. In contrast, an increase in SIV up to 0.2 is expected in the southern Sahel zone for cereals. No suitability change is projected for legumes (Figure 4, column 2) except an increase in SIV of about 0.1 in the southern Sahel (12–14 °N) and up to 0.2 in the central savanna zone, (Figure 6). On the other hand, a projected decrease of 0.1 in SIV is expected along the west coast of Sierra Leone and the south coast of Nigeria for groundnut.

The projected increase in SIV provides an increase in the suitable area for the cultivation of both crops. This is so because a 0.2 increase in SIV for the marginally suitable (SIV, 0.4–0.6) areas in the southern Sahel results in the area becoming suitable (SIV, 0.6–0.8) for both crops. The projected decrease in the SIV values along the coastal areas and over Sierra Leone also does not affect the area negatively as the area remains suitable for these crops. For the root and tuber crop (Figure 5, column 2), a projected decrease of about 0.1 SIV is expected for cassava and up to 0.2 in southern Nigeria and along the west coast of Guinea to Liberia for plantain and yam extending to south of Ivory Coast for plantain. A similar magnitude decrease is also expected in the western Guinea-Savanna zone from Guinea to the western Ivory Coast. For the horticulture and fruit crops (Figure 5, column 1), a 0.1 projected decrease in SIV is expected south of 12 °N and the savanna zone for tomato and pineapple, respectively while up to 0.2 SIV decrease is expected in the south coast of Nigeria for mango and orange. However, a projected increase up to 0.2 SIV is expected in the southern Sahel for mango. The projected suitability changes are robust (i.e., at least 80% of the simulation that the climate change is statistically significant at 95% confidence level) for cassava, maize and mango in the near future month (2031–2050) while the changes are consistent for the other nine crops (i.e., at least 80% of the model agree to the sign of change).

Table 2. Projected changes in crop suitability over West African AEZs at different future window periods.

Crops	Near Future (2031–2050)			Mid-Century (2051–2070)			End-Century (2081–2100)		
	Guinea	Savanna	Sahel	Guinea	Savanna	Sahel	Guinea	Savanna	Sahel
Cassava	No change remains suitable	No change remains suitable	No change remains unsuitable	A 0.2 SIV decrease but still suitable	A 0.2 SIV decrease but still suitable	Same as GWL1.5	About 0.4 SIV decrease still suitable	About 0.4 SIV decrease but still suitable	Same as GWL1.5
Plantain	About 0.1 SIV decrease but still suitable	A 0.1 & 0.2 SIV decrease and increase in west and central respectively	No change unsuitable	About 0.2 decrease in SIV but still suitable	A 0.2 SIV decrease and increase in west and central respectively	No change remains unsuitable	About 0.4 SIV decrease may become marginally suitable	About 0.4 and 0.2 SIV decrease to the west and central respectively	No change unsuitable
Yam	Suitable, but not along the coastal area	Suitable, but in the west & central Savana. Only suitable	No change, unsuitable	Same as in GWL1.5	Same as GWL1.5	Same as GWL1.5	Same as in GWL1.5	Same as GWL1.5	Same as GWL1.5
Maize	No change but 0.1 decrease SIV northern Cameroon	Suitable, but not along the west coast of Guinea to Sierra Leone	About 0.2 SIV increase, now suitable over the southern Sahel	No change in suitability	About 0.1 SIV decrease but still suitable	Same as GWL1.5	No change in suitability	About 0.2 decrease in SIV but still suitable	Same as GWL1.5 but SIV increase up to 0.3
Pearl millet	No change but very marginal suitability in the south coast Nigeria and north Liberia	No change but about 0.1 SIV decrease in eastern Guinea	About 0.2 SIV increase make northern Sahel suitable	Same as GWL1.5	Same as GWL1.5	Same as GWL1.5	Same as GWL1.5	About 0.3 decrease in SIV but still suitable	A 0.4 SIV decrease in west Sahel but still suitable
Sorghum	No change in suitability	No change in suitability	About 0.2 SIV increase make northern Sahel suitable	No change in suitability	About 0.1 decrease in SIV but still suitable	About 0.1 SIV increase makes Sahel suitable	About 0.1 decrease in SIV but still suitable	About 0.2 SIV decrease west respectively	Above 0.2 SIV decrease but still suitable

Table 2. *Cont.*

Crops	Near Future (2031–2050)			Mid-Century (2051–2070)			End-Century (2081–2100)		
	Guinea	Savanna	Sahel	Guinea	Savanna	Sahel	Guinea	Savanna	Sahel
Mango	No change in suitability	No change in suitability	No change in suitability	About 0.1 decrease in SIV but still suitable	About 0.1 decrease in SIV but still suitable	About 0.1 increase in SIV but still unsuitable	About 0.2 SIV decrease but still suitable	About 0.2 SIV increase but still unsuitable	About 0.2 SIV increase but still unsuitable
Orange	About 0.1 SIV increase	About 0.1 SIV increase	No change in suitability	Same as GWL1.5	Same as GWL1.5	Same as GWL1.5	About 0.2 SIV decrease but still suitable	About 0.2 SIV decrease but still suitable	Same as in GWL1.5
Pineapple	No change in suitability	About 0.2 SIV decrease but still suitable	No change in suitability	About 0.1 decrease but still suitable	About 0.3 decrease but still suitable	Same as GWL1.5	About 0.4 decrease but still suitable	About 0.4 SIV decrease but still suitable	Same as GWL1.5
Tomato	About 0.1 decrease but still suitable	About 0.1 decrease but still suitable	About 0.2 SIV increase make southern Sahel suitable	About 0.3 decrease but still suitable	About 0.3 SIV decrease but still suitable	Same as GWL1.5	About 0.4 decrease but still suitable	About 0.4 SIV decrease but still suitable	Same as GWL1.5
Cowpea	No change in suitability	No change in suitability	About 0.2 SIV increase makes southern Sahel suitable	Same as GWL1.5	Same as GWL1.5	Same as GWL1.5	Same as GWL1.5	About 0.1 decrease in SIV but still suitable	Same as GWL1.5
Groundnut	No change in suitability	No change in suitability		Same as GWL1.5	Same as GWL1.5	Same as GWL1.5	Same as GWL1.5	About 0.1 decrease in SIV but still suitable	Same as GWL1.5

The Ecocrop suitability simulation by mid-century (2051–2070) shows a projected increase in the magnitude of change of SIV and spatial suitability distribution of suitable areas compared to the past climate for the different crop types. The projected spatial suitability distribution for mid-century shows a similar spatial pattern as the near future period (2031–2050) with an increase in the suitability spatial extent and the magnitude of change in SIV. For cereals (Figure 3, column 3), the projected change is like the spatial suitability pattern as the near future period except for the spatial extension in the suitable area further north in the central Sahel zone for pearl millet. In contrast, a decrease in the suitable area in the western Nigeria for pearl millet and north-west Nigeria for maize and sorghum.

The legume (Figure 4, column 3) crops show a similar projected suitability spatial pattern as the near future period except a projected decrease in SIV of about 0.1 and 0.2 of the suitable area is expected in the south-west Chad Republic in the eastern Sahel zone for the groundnut and cowpea, respectively. For the root and tubers (Figure 5 column 3), a decrease of about 0.2 SIV is projected for both the plantain and yam but with a similar spatial suitability pattern as shown for the near future period. However, for cassava about 0.2 decrease SIV is projected over the guinea-Savanna zone but the area remains suitable. For the fruit and horticulture crops (Figure 6, column 3), there are no changes in the projected spatial suitability pattern as observed in the near future period by mid-century. However, there is an increase in the magnitude of change of SIV from 0.1 to 0.2 and 0.2 to 0.3 for the tomato and pineapple, respectively. All the projected suitability changes are statistically significant at 95% confidence level for cassava, maize and mango and are consistent for the other nine crops (i.e., at least 80% of the model agree to the sign of change) by mid-century (2051–2070).

The projected increase in global warming will lead to increasing the magnitude in the projected change for the crop SIV and spatial suitability distribution across different crop types by the end of the century (2081–2100). Cereal (Figure 3, column 4) as projected will be severely affected as more areas becomes less suitable by the end of the century. For legume (Figure 4, column 4), the Savanna zone will be less suitable with a decrease of about 0.1 in SIV while a decrease of about 0.2 SIV is expected along the eastern Sahel zone for groundnut as well as the south coast of Nigeria. Cowpea as projected will be more affected with a decrease of about 0.2 SIV in the northern savanna in the southern Chad Republic and Nigeria with its boundary with south-east Niger Republic in the Sahel and south-west Mali in the western Sahel zones. A decrease up to 0.2 in SIV is expected in the southern Sahel for cereal except maize with an increase of about 0.2 in the central southern Sahel zone.

The root and tubers (Figure 5, column 4), show a similar spatial pattern for the decrease in the suitable area as the near future period and mid-century but with an increase in the SIV magnitude of about 0.2, 0.3 and 0.4 for yam, plantain and cassava, respectively. The fruit and horticulture crops (Figure 6, column 4) shows further reduction in the suitable area compared to the near future period with an increase up to 0.4 SIV for the horticulture crop. The Guinea-Savanna will become less suitable with a decrease of 0.1 and 0.2 SIV for orange and mango, respectively. All the projected suitability changes are statistically significant at 95% confidence level for cassava, maize and mango and are consistent for the other nine crops (i.e., at least 80% of the model agree to the sign of change) by the end of the century (2081–2100).

3.4. Impact of CCD on Crop Planting Month over West Africa

At all the three future climate windows, the Ecocrop projected change on the planting month varies for different crop types across the different AEZs of West Africa (Figures 7–10). The impact of CCD resulted in an early or late/delay in the PM for different crops and increases in magnitude across the three zones as described in Table 3 below. It is worth stating that the change in PM describes a change in the best three planting months under the three future windows.

Table 3. Projected changes in time of planting (crop planting months) over West African AEZs at different global warming levels.

Crops	Near Future (2031–2050)			Mid-Century (2051–2070)			End-Century (2081–2100)		
	Guinea	Savanna	Sahel	Guinea	Savanna	Sahel	Guinea	Savanna	Sahel
Cassava	Delayed planting for one month	Early planting by four months	Not applicable	Same as GWL1.5	Same as GWL1.5 but for more area	No planting date	Same as GWL1.5	Same as GWL1.5	No planting date
Plantain	No change in planting date	No change in planting date	No change in planting date	No change in planting date	No change in planting date	No change in planting date	No change in planting date	No change in planting date	No change in planting date
Yam	On month delayed planting	No change in planting date	No change in planting date	Same as GWL1.5	Same as GWL1.5	No change in planting date	Same as GWL1.5	Same as GWL1.5	No change in planting date
Maize	Three months delayed planting	Four months early and delay planting in east and west respectively	No change in planting date	Same as GWL1.5	Same as GWL1.5	No change in planting date	Same as GWL1.5	Same as GWL1.5	No change in planting date
Pearl millet	One-month delayed planting	Two months delayed planting	Two months delayed planting	Same as GWL1.5	Same as GWL1.5	Same as GWL1.5	Same as GWL1.5	Same as GWL1.5	Same as GWL1.5
Sorghum	No change in planting date	No change in planting date	One-month delay in southern Sahel zone	No change in planting date	No change in planting date	No change in planting date	No change in planting date	No change in planting date	No change in planting date
Mango	Delayed planting for two months	Early planting by four months		Same as GWL1.5	Same as GWL1.5 but for more area	No planting, date	Same as GWL1.5	Same as GWL1.5	No planting date

Table 3. *Cont.*

Crops	Near Future (2031–2050)			Mid-Century (2051–2070)			End-Century (2081–2100)		
	Guinea	Savanna	Sahel	Guinea	Savanna	Sahel	Guinea	Savanna	Sahel
Orange	One-month delayed planting	No change in planting date	No change in planting date	Same as GWL1.5	Same as GWL1.5	No change in planting date	Same as GWL1.5	Same as GWL1.5	No change in planting date
Pineapple	No change in planting date	No change in planting date	No change in planting date	No change in planting date	No change in planting date	No change in planting date	No change in planting date	No change in planting date	No change in planting date
Tomato	One-month delayed planting	No change in planting date	No change in planting date	Same as GWL1.5	Same as GWL1.5	No change in planting date	Two months delayed planting	One-month early planting	No change in planting date
Cowpea	One-month delayed planting	Two months delayed planting	Two months delayed planting	Same as GWL1.5	Same as GWL1.5	Same as GWL1.5	Same as GWL1.5	Same as GWL1.5	Same as GWL1.5
Groundnut	No change in planting date	No change in planting date	No change in planting date	No change in planting date	No change in planting date	No change in planting date	No change in planting date	No change in planting date	No change in planting date

In the near future, cereals crops, pearl millet and sorghum are projected to experience a one-month delay over the region and up to 0.2 along the west coast of Sierra Leone to Liberia and the south coast of Nigeria (Figure 7, column 2). In contrast, the two-month delayed planting is expected over the Savanna-Sahel zone for maize. For the legumes crops, cowpea and groundnut (Figure 8 column 2, see Table 4) no projected change in the PM compared to the past climate is expected over the regions except about one-month delay (i.e., from June to July) in planting over Sierra-Leone and Liberia in the Guinea zone and the southern Sahel zone from Senegal to Chad Republic compared to the planting month (June) over the area. For the root and tuber (Figure 9 column 2), about three to four months early (February/March) the planting is projected for cassava in the near future as compared to June/July, the PM from the historical climate across the region except the north-east Nigeria and the coastal areas (Figure 9, Table 3). No change in the PM is expected in the near future over the coastal areas but about three months delay in planting is projected in the north-eastern part of Nigeria. No change in the PM is projected for plantain, an annual crop which can be planted anytime of the year while a 3–4 months delay is expected for yam except in western Guinea-Savanna and the south coast of Nigeria. For fruits and horticulture (Figure 10, column 2), no projected change in the planting month is expected for tomato and pineapple except a two-month delay over Liberia.

Early planting between one to two months is expected in the Guinea-Savanna zone and about three-months delay in the planting of orange in the southern Sahel zone. About two-months and up to four-months delay in planting is projected for mango in the Guinea-Savanna zone and the northern Sahel zone, respectively while a two-month early planting of the crop is expected in the southern Sahel zone. The projected change is consistent for all crops as 80% of the simulation agree to the sign of change.

Table 4. Trends in the projected change in suitability over West Africa for the near future, mid and end of the century periods for different crops.

Crops/Period	2031–2050	2051–2070	2081–2100
Cassava	1.053	1.141	1.497
Cowpea	1.000	1.000	1.002
Groundnut	1.000	1.001	1.030
Maize	1.007	1.021	1.082
Mango	1.013	1.046	1.137
Orange	0.981	0.974	1.089
Pearl millet	1.007	1.022	1.057
Pineapple	1.061	1.216	1.580
Plantain	1.017	1.025	1.215
Sorghum	1.007	1.018	1.032
Tomato	1.219	1.421	1.997
Yam	0.873	0.784	0.779

By mid (2051–2070) and end of the century (2081–2100), most crop types show a similar spatial pattern in the planting month as observed in the near future but with an increase in the magnitude of the delay or early planting period (Figures 7–10, column 3–4). For example, cereal crops show a similar spatial pattern as projected for the near future for sorghum and pearl millet by mid-century (Figure 7, column 3) and the end of the century (Figure 7, column 4) except over Liberia and south coast of Nigeria for pearl millet. These areas are expected to experience a 2–3-month delay in planting. Legume crops, cowpea and groundnut show similar characteristics of no projected change in the PM as the near future period but for an increase in the magnitude a delay period in the south coast of Nigeria and southern Liberia. A delay in planting from one to two months is expected from Sierra-Leone to Liberia and over the south coast of Nigeria for cowpea by mid-century (Figure 8, column 3) and up to three months by the end of century (Figure 8, column 4). A two-month delay in the PM is projected over southern Liberia and a one-month delay in the southern Sahel zone by mid and end of the century (Figure 8, column 3–4). For the root and tuber crops (Figure 9, column 3–4), about a four-month delay in planting is projected over the Savanna zone except the western area of the zone and in the central Guinea zone by mid-century for yam (Figure 9, column 3).

A similar pattern is projected by the end of the century for crop (Figure 9, column 4). No change in the planting period is projected for plantain because it is annual crop over these two periods. For cassava a month delay planting by mid-century and up to two-months by the end of the century is projected in the western Guinea-Savanna zone while an early planting is expected in other parts of the Savanna zone and north of the Guinea zone over the two-climate change period. For the fruit and horticulture crops, there is no change in the PM for pineapple being an annual crop. A one-month PM delay is projected for tomato over the region and up to two-months over Liberia by mid and end of the century.

However, a projected two-month early planting is expected by the end of the century in the southern Sahel zone. For fruit crops, a four-month early planting compared to the historical climate is projected in the Guinea-Savanna zone with a delay of about three-months in the south Sahel by mid-century. By the end of the century, an early planting of about four-month early compared to the historical climate is projected over the region for orange. Similarly, a two-month early planting is projected for mango in the southern Sahel zone for mid and up to three-months by the end of the century. In contrast, a delay in planting of about two-three months is expected over the Guinea-Savanna zone and up to four-months in the northern Sahel zone. All the projected changes are consistent for all crops as 80% of the simulation agree to the sign of change over the two climate periods.

3.5. Trends in Projected Crop Suitability and Crop Planting over West Africa

We used the Theil-Sen slope to evaluate the trend in the projected suitability and month of planting for the crop types for the near future, mid and end of the century over West Africa (Tables 4 and 5). The trend describes the rate of increase and decrease of the suitable area and SIV with increasing warming over the three-window month. In general, all the crop types show an increasing trend in the projected change in the crop suitability compared to the past climate from the near future to the end of the century when compared to the past climate except for yam (Table 4).

The projected change in the suitability index value of suitable areas for tomato showed the highest trend value from 1.219 in the near future month to 1.997 by the end of the century. Compared to other crops, our analysis showed that there was a decreasing trend (from 0.873, the near future to 0.779, end of the century) for yam in the projected suitability change with increasing warming across each time of month from the near future to the end of century over West Africa. Additionally,, there was decrease in the trend between the near future month and mid-century month in the projected change for orange but later increased at end of the century. Moreover, there was no trend in the projected change in the suitability for cowpea over the near future month and mid-century but there was increase in the trend of the projected change for the crop by the end of the century.

Table 5. Trends in the projected change in the month of planting over West Africa for the near future, mid and end of the century periods for the different crops.

Crops/Period	2031–2050	2051–2070	2081–2100
Cassava	1.125	1.171	0.974
Cowpea	0.972	0.957	0.887
Groundnut	0.969	0.952	0.857
Maize	1.000	0.990	0.950
Mango	1.000	0.976	0.909
Orange	1.000	1.111	1.930
Pearl millet	0.980	0.959	0.912
Pineapple	1.000	1.000	1.000
Plantain	1.000	1.000	1.000
Sorghum	1.000	1.000	0.944
Tomato	0.938	0.900	0.851
Yam	1.000	0.924	0.909

Our Theil-Sen slope trend analysis shows a general decreasing trend in the projected change in the planting month compared to the past climate for the different crop types except for orange which gives an increasing trend pattern of the projected planting for all the crop (Table 5). Our trend analysis test show there was no change in the projected change in planting for plantain, pineapple (1.000) and for sorghum for the near future and mid-century month (1.000).

4. Discussion

4.1. Crop Type Sensitivity to CCD and Impact on Food Security

Horticulture, cereals, root and tubers (hereafter HCRT) crops, respectively will be the most impacted by the climate change/departure impact from the historical variability in West Africa. All the five different crop types show a different response to the impact of the global warming induced CCD across the examined three-window month, near the future to the end of the century in West Africa. The variability in the response of the different crop types to CCD is very cardinal to the agricultural production and food security in the region. HCRT are the most negatively affected with decreasing suitability across the three AEZs of West Africa due to the impact of the climate change compared to the legumes and fruit crops. In terms of sensitivity, the HCRT crop suitability show a negative linear relationship with increasing global warming over the region except for cereals with a positive linear relationship in the southern Sahel zone. The negative linear relationship is observed notably over the Guinea-Savanna zone for the HCRT resulting in a decrease in the crop suitable area with increasing warming across the three months examined. The projected negative linear relationship due to an increase in global warming may result in a decrease in the yield of these crop types over West Africa due to a decrease in the crop suitable land [6,64]. For example, previous studies (e.g., Lobell et al. [65], Sultan et al. [15]) have revealed that the impact of climate change will result in a decrease in the yield of cereals by 20% in the near future month over West Africa. Additionally, the result is in line with the findings of [32] that there will be a decline in the suitability and suitable cultivated areas for cassava due to a result of the temperature increases but the crop will remain suitable over the region. In addition, our result also agrees with [66] findings that increasing warming will lead to a decrease in the availability of the suitable land for the cultivation of horticulture with a direct implication on the horticultural production. This agrees with [14] that the variability in the climate will lead to a reduction in the yield quantity of pineapple in Ghana which is one of the key producers of pineapple, which may be linked to the decrease in the suitable areas and SIV as projected in this study.

The projected impacts of CCD on crop suitability will further compound the challenge of food security in West Africa. This is in line with past findings that climate variability and change in the coming decades will further threaten food security in sub-Saharan Africa notably West Africa, a region that plays a major role in the agricultural production [1,7]. West Africa for about 24 years mainly accounts for about 60% of the total value of agricultural outputs within Africa [7]. However, the story has not been the same since 2007 due to instability in the agricultural production over the region and this has been a source of concern [7]. As a result, the projected decrease in crop suitability due to a reduction in the suitable area for crop cultivation coupled with the projected delay in the month of planting will both strongly have a negative impact on the crop yield and agricultural production. This may further plunge the plan for food security in the region into a mirage.

4.2. Impact of CCD on Spatial Suitability Distribution

The impact of CCD will lead to a projected variability in the spatial suitability distribution across the three AEZs for the three future months and different crop types. The magnitude of deviation due to the increase in warming may influence the suitability over the zones as well as crop sensitivity to the projected change in the climate. The crop growth and yield are directly proportional to the climate-crop threshold i.e., climate suitability/threshold [67]. It is important to note that each crop has their climatic or suitability threshold for healthy growth, development and optimal yield and that future changes/departure in the climate generally has a reaching impact on the yield of the crop. This is further buttressed by our finding that CCD may lead to future constraint in the available cultivated area in the Guinea and southern Savanna zones of West Africa. On the other hand, it tends to provide an opportunity in the northern Savanna extending to the southern Sahel. The projected spatial constraint in the suitability and cultivated area will strongly affect the crop production and yield over West Africa. The Guinea-Savanna zone provides and significantly contributes to the agricultural production over the region and a large proportion in the continent [7]. For example, about four of the five different crop types (except the legumes) examined in the study is and will be significantly affected with the projected decrease in SIV and reduction in the cultivated area of the crops. This projected decrease in SIV and the reduction in the spatial distribution of suitable areas for cultivation of major crops such as cassava and the horticulture crops such as pineapple pose a great challenge to the economy of most countries and further raises the challenge of food security in the region. The challenge of food security arising from the projected decrease in the crop suitable area may compound the climatic stress over the region due to the increase in food production to meet the present food demand but with the projected and limited available land for cultivation are not realistic and may become a mirage with the projected increase in the population over the region by mid-century, 2050 [68,69].

On the other hand, crop suitability due to CCD from the historical variability is projected and will lead to an increase in SIV and more suitable area notably in the Southern Sahel. The increase in suitable areas provides an opportunity for more suitable areas in the region for the cultivation of cereals, legumes and mango in the southern Sahel zone (12–14 °N), plantain and yam in the Savanna zone as well as the legume crops in the central savanna zone of West Africa. The projected increase into the Sahel agrees with the previous finding for maize in the Sahel zone with CCD. This shows that the crop spatial suitability distribution and productivity are highly sensitive to variations in the climate such that a departure of the future African climate from the recent range of historical variability will have the most devastating effect on agriculture over the continent [70–72].

4.3. Implication for Socio-Economic Development and Strategy Policy

The above result provides a basis for developing the policy and strategy to reduce future crop loss due to a lack of suitable land and risks of food security over West Africa. At the same time, it advocates for a more proactive response to increase resilience and adaptive options via the urgency and timing of adaptation. For instance, the analysis of crop suitability indicates that a greater proportion of suitable land areas in the West African region may become less suitable or unsuitable in the future from CCD due to global warming, which may enhance a decrease in the crop yield and agricultural production of some crop. On the other hand, the analysis showed an expansion of the suitable area into the Sahel for the cereal and legume crops with CCD, which provide future opportunities for more suitable areas for the cultivation of one of the most staple crops, maize. This will have both positive and negative impacts on regional development and economic activities (e.g., regional trade and international relation in terms of exports and importing goods). The increasing population also implies that the demand for food will be on the increase. However, the projected change in suitability also suggests that a well-planned land use change (through the urgency of adaptation to the CCD) could help reduce the impacts of CCD on the crop yield and food security in the region. Hence, there is a need for the formulation of a strategic policy that can accommodate or encourage such a land-use change. A strategic policy is also required more importantly for the new opportunities such as an expansion into the Sahel for maize and the other crops that may arise out of the impact of CCD over the region. Hence, the results can guide policymakers on how to prioritize their adaptation plan in terms of the urgency of response and redefine mitigation measures to the future impact of CCD on the crop suitability and planting season over West Africa.

5. Conclusions

Summary and Conclusions

In investigating the impact of CCD on the crop suitability and planting month over the entire West African region, we analyzed 10 CMIP 5 GCM datasets downscaled by CORDEX RCM, RCA4 for five different crop types, cereal (maize, pearl millet and sorghum), fruit (mango and orange), horticulture (pineapple and tomato), legume (cowpea and groundnut) and root and tuber (cassava, plantain and yam). The summary from our study are as follows:

We suggest that projected changes in the temperature may lead to an increase between 1–4.5 °C for the minimum and mean temperature over West Africa from the near future to the end of the century. A change of about 10 mm is projected over the western Guinea-Savanna zone and no major changes in other parts of the region and up to 25 mm along the coastal areas (west coast of Sierra-Leone to south-west Ghana and the south coast of Nigeria) for the near future and mid-century. A projected decrease up to 25 mm is expected over the region and up to 35 mm over the coastal area (from the west coast of Gambia to north Liberia) by the end of the century.

Addressing our main objective, the Ecocrop simulated spatial suitability distribution of the crops shows higher suitability are to the south of 14 °N while a lower suitability is to the north. The marginal suitability line (around 12–14 °N) shows the transition between the higher and lower suitability of the crop. Results show that the horticulture crops, pineapple and tomato, respectively are the most negatively affected by the impact of CCD from the historical variability over the region. There is a projected constraint showing a negative linear correlation with increasing warming in the cultivation of most different crop types except for cowpea in the Guinea-Savanna AEZs (south of 14 °N) by the end of the century due to an increasing reduction in the suitable area and crops suitability index value due to the climate departure although most of the crop remains suitable. The impact of CCD will provide opportunities for more suitable areas in the southern Sahel zone for cereals, mango and legumes crops showing a positive linear correlation with increasing warming thus creating more land for cultivation, which can in turn increase the yield and production of the crops. Generally, a projected delay of 1–4 months is expected for most of the crop types with CCD except for orange and cassava as well as maize in the Savanna zone. No projected changes are observed for plantain and pineapple, mainly because they are annual crops.

Statistically, we demonstrated that over 80% of the simulations agree with the sign of the projected change for all the crop types due to the CCD and the changes are statistically significant at 95% confidence interval for maize, cassava and mango. Additionally, we showed there is an increasing trend in the projected crop suitability for all crops except yam with a decreasing trend due to CCD from the historical variability while a decreasing trend is projected for the future change in the month of planting of the crops.

Despite our analysis, the results of this study can be improved and applied to reduce the future impact of crop suitability and risks of food security over West Africa in many ways. For instance, future studies may investigate the impact of CCD on the crop suitability and planting season over the region using more RCMs with different forcing GCMs other than only RCA4. This may help resolve the challenge of uncertainty in the future simulation of the crop suitability and planting season. In addition, the results of the study will be more robust and improve our knowledge on the impact of CCD and its influence on the crop suitability and planting season over West Africa. Further studies on how to reduce the uncertainty will improve the credibility and application of the results. Nevertheless, the present work shows the impact of CCD on the crop suitability and planting season using GCMs downscaled with RCMs. This establishes a premise for future work in advancing our knowledge into how CCD influences the crop suitability and planting season in West Africa.

In conclusion, the application of the concept of CCD in this study has demonstrated future changes in how the crop suitability and planting season can be analyzed. The application of CCD established the impact of climate change on crop suitability over West Africa and further identified spatial variability in the future suitability showing that horticulture, cereal, root and tubers crops will be most negatively affected by the impact of CCD in West Africa. It also identifies the three best planting months in a growing season and the changes in the planting time is about four month delay in the planting season for most crops but early planting for cassava, orange and maize but only in the savanna zone. The application of CCD aims to underpin future works to advance the study of future changes in crop suitability and planting in any region of the world. This type of analysis is important for adaptation options and planning for future changes in the crop suitability and planting period to improve food security.

Author Contributions: T.S.E. was responsible for conceptualization, developing the initial content of the manuscript, including literature search, data analysis and writing of the manuscript. O.C. and C.L. are the supervisors for the research and provided guidance in terms of the article structure, data analysis and finalization of the manuscript. P.M. provided guidance with data computation, S.L. assisted with review and editing of the manuscript, K.Q. assisted with some data processing, analysis and editing of the manuscript.

Acknowledgments: This study was supported with bursaries from the National Research Foundation (NRF, South Africa), Alliance Centre for Climate and Earth Systems Science (ACCESS, South Africa) and the JW Jagger Centenary Scholarship and Sari Johnson scholarship from the Postgraduate Funding Office, University of Cape Town, South Africa and IMPALA project.

References

1.   IPCC. Summary for Policymakers. In *Climate Change 2013: The Physical Science Basis. Contribution of Working Group I to the Fifth Assessment Report of the Intergovernmental Panel on Climate Change*; Stocker, T.F., Qin, D., Plattner, G.-K., Tignor, M., Allen, S.K., Eds.; Cambridge University Press: Cambridge, UK, 2013.

2.   Kirtman, B.; Power, S.B.; Adedoyin, A.J.; Boer, G.J.; Bojariu, R.; Camilloni, I.; Doblas-Reyes, F.; Fiore, A.M.; Kimoto, M.; Meehl, G.; et al. Near-term Climate Change: Projections and Predictability. In *Climate Change 2013: The Physical Science Basis. Contribution of Working Group I to the Fifth Assessment Report of the Intergovernmental Panel on Climate Change*; Stocker, T.F., Qin, D., Eds.; Cambridge University Press: Cambridge, UK; New York, NY, USA, 2013; Chapter 11; pp. 953–1028.

3.   Riede, J.O.; Posada, R.; Fink, A.H.; Kaspar, F. What's on the 5th IPCC Report for West Africa? In *Adaptation to Climate Change and Variability in Rural West Africa*; Yaro, J.A., Hessellberg, J., Eds.; Springer: Cham, Switzerland, 2013; Volume 19, pp. 7–24.

4.   Benhin, J.K. South African crop farming and climate change: An economic assessment of impacts. *Glob. Environ. Chang.* **2008**, *18*, 666–678. [CrossRef]

5.   Schlenker, W.; Lobell, D.B. Robust negative impacts of climate change on African agriculture. *Environ. Res. Lett.* **2010**, *5*, 014010. [CrossRef]

6.   Roudier, P.; Sultan, B.; Quirion, P.; Berg, A. The impact of future climate change on West African crop yields: What does the recent literature say? *Glob. Environ. Chang.* **2011**, *21*, 1073–1083. [CrossRef]

7.   OECD/FAO. *OECD-FAO Agricultural Outlook 2016–2025: Special Focus on Sub-Sharan Africa*; OECD Publishing: Paris, France, 2016.

8. Nelson, G.C.; Rosegrant, M.W.; Koo, J.; Robertson, R.; Sulser, T.; Zhu, T.; Magalhaes, M. Climate change: Impact on agriculture and costs of adaptation. *Intl. Food Policy Res. Inst.* **2009**, *21*.

9. Nelson, G.C.; Van Der Mensbrugghe, D.; Ahammad, H.; Blanc, E.; Calvin, K.; Hasegawa, T.; Havlik, P.; Heyhoe, E.; Kyle, P.; Lotze-Campen, H.; et al. Agriculture and climate change in global scenarios: Why don't the models agree. *Agric. Econ. (UK)* **2014**, *45*, 85–101. [CrossRef]

10. Ray, D.K.; Foley, J.A. Increasing global crop harvest frequency: Recent trends and future directions. *Environ. Res. Lett.* **2013**, *8*, 044041. [CrossRef]

11. IPCC. Summary for policymakers. In *Climate Change 2014: Impacts, Adaptation, and Vulnerability. Part A: Global and Sectoral Aspects. Contribution of Working Group II to the Fifth Assessment Report of the Intergovernmental Panel on Climate Change*; Field, C.B., Barros, V.R., Dokken, D.J., Mach, K.J., Mastrandrea, M.D., Bilir, T.E., Chatterjee, M., Ebi, K.L., Estrada, Y.O., Genova, R.C., et al., Eds.; Cambridge University Press: Cambridge, UK; New York, NY, USA, 2014; pp. 1–32.

12. Rurinda, J.; Mapfumo, P.; Van Wijk, M.T.; Mtambanengwe, F.; Rufino, M.C.; Chikowo, R.; Giller, K.E. Sources of vulnerability to a variable and changing climate among smallholder households in Zimbabwe: A participatory analysis. *Clim. Risk Manag.* **2014**, *3*, 65–78. [CrossRef]

13. Challinor, A.; Wheeler, T.; Garforth, C.; Craufurd, P.; Kassam, A. Assessing the vulnerability of food crop systems in Africa to climate change. *Clim. Chang.* **2007**, *83*, 381–399. [CrossRef]

14. Williams, P.A.; Crespo, O.; Abu, M. Assessing vulnerability of horticultural smallholders' to climate variability in Ghana: Applying the livelihood vulnerability approach. *Environ. Dev. Sustain.* **2018**, 1–22. [CrossRef]

15. Sultan, B.; Guan, K.; Kouressy, M.; Biasutti, M.; Piani, C.; Hammer, G.L.; McLean, G.; Lobell, D.B. Robust features of future climate change impacts on sorghum yields in West Africa. *Environ. Res. Lett.* **2014**, *9*, 104006. [CrossRef]

16. Parkes, B.; Defrance, D.; Sultan, B.; Ciais, P.; Wang, X. *Projected Changes in Crop Yield Mean and Variability Over West Africa in a World 1.5 K Warmer Than the Pre-Industrial Era*; Copernicus Publications: Gottingen, Germany, 2018; Volume 9, pp. 119–134.

17. Jalloh, A.; Nelson, G.C.; Thomas, T.S.; Roy-Macauley, H. *West African Agriculture and Climate Change: A Comprehensive Analysis*; International Food Policy Research Institute: Washington, DC, USA, 2013; 444p.

18. Ramirez-Villegas, J.; Thornton, P.K. *Climate Change Impacts on African Crop Production*; CCAFS Working Paper no. 119; CGIAR Research Program on Climate Change, Agriculture and Food Security (CCAFS): Copenhagen, Denmark; 127p, Available online: www.ccafs.cgiar.org (accessed on 24 March 2017).

19. Thornton, P.K.; Jones, P.G.; Ericksen, P.J.; Challinor, A.J. Agriculture and food systems in sub-Saharan Africa in a 4°C+ world. *Philos. Trans. R. Soc. A Math. Phys. Eng. Sci.* **2011**, *369*, 117–136. [CrossRef] [PubMed]

20. Adger, W.N. Social Capital, Collective Action, and Adaptation to Climate Change. *Econ. Geogr.* **2003**, *79*, 387–404. [CrossRef]

21. Mora, C.; Frazier, A.G.; Longman, R.J.; Dacks, R.S.; Walton, M.M.; Tong, E.J.; Sanchez, J.J.; Kaiser, L.R.; Stender, Y.O.; Anderson, J.M.; et al. The projected timing of climate departure from recent variability. *Nature* **2013**, *502*, 183–187. [CrossRef] [PubMed]

22. Hawkins, E.; Sutton, R. Time of emergence of climate signals. *Geophys. Res. Lett.* **2012**, *39*, L01702. [CrossRef]

23. Egbebiyi, T.S.; Crespo, O.; Lennard, C. Defining Crop–climate Departure in West Africa: Improved Understanding of the Timing of Future Changes in Crop Suitability. *Climate* **2019**, *7*, 101. [CrossRef]

24. Abiodun, B.J.; Adeyewa, Z.D.; Oguntunde, P.G.; Salami, A.T.; Ajayi, V.O. Modeling the impacts of reforestation on future climate in West Africa. *Theor. Appl. Climatol.* **2012**, *110*, 77–96. [CrossRef]

25. Egbebiyi, T.S. Future Changes in Extreme Rainfall Events and African Easterly Waves Over West Africa. MSc. Thesis, University of Cape Town, Cape Town, South Africa, May 2016.

26. Sultan, B.; Gaetani, M. Agriculture in West Africa in the Twenty-First Century: Climate Change and Impacts Scenarios, and Potential for Adaptation. *Front. Plant Sci.* **2016**, *7*, 1262. [CrossRef] [PubMed]

27. Janicot, S.; Caniaux, G.; Chauvin, F.; De Coëtlogon, G.; Fontaine, B.; Hall, N.; Kiladis, G.; Lafore, J.-P.; Lavaysse, C.; Lavender, S.L.; et al. Intraseasonal variability of the West African monsoon. *Atmos. Sci. Lett.* **2011**, *12*, 58–66. [CrossRef]

28. Omotosho, J.B.; Abiodun, B.J. A numerical study of moisture build-up and rainfall over West Africa. *Meteorol. Appl.* **2007**, *14*, 209–225. [CrossRef]

29. Nicholson, S.E. The West African Sahel: A Review of Recent Studies on the Rainfall Regime and Its Interannual Variability. *ISRN Meteorol.* **2013**, *2013*, 453521. [CrossRef]

30. Klutse, N.A.B.; Ajayi, V.O.; Gbobaniyi, E.O.; Egbebiyi, T.S.; Kouadio, K.; Nkrumah, F.; Quagraine, K.A.;

Olusegun, C.; Diasso, U.; Abiodun, B.J.; et al. Potential impact of 1.5 °C and 2 °C global warming on consecutive dry and wet days over West Africa. *Environ. Res. Lett.* **2018**, *13*, 055013. [CrossRef]

31. Paeth, H.; Capo-Chichi, A.; Endlicher, W. Climate Change and Food Security in Tropical West Africa—A Dynamic-Statistical Modelling Approach. *Erdkunde* **2008**, *2*, 101–115. [CrossRef]

32. Jarvis, A.; Ramírez-Villegas, J.; Campo, B.V.H.; Navarro-Racines, C. Is Cassava the Answer to African Climate Change Adaptation? *Trop. Plant Biol.* **2012**, *5*, 9–29. [CrossRef]

33. FAOSTAT. Statistical Yearbook of 2012: Europe and Central Asia; 2012. Available online: http://www.fao.org/3/a-i3621e.pdf (accessed on 1 December 2018).

34. Srivastava, A.K.; Gaiser, T.; Ewert, F. Climate change impact and potential adaptation strategies under alternate climate scenarios for yam production in the sub-humid savannah zone of West Africa. *Mitig. Adapt. Strateg. Glob. Chang.* **2016**, *21*, 955–968. [CrossRef]

35. FAOSTAT. FAO Statistical Yearbook 2014, Africa Food and Agriculture; 2014. Available online: http://www.fao.org/3/a-i3590e.pdf (accessed on 1 December 2018).

36. Harris, I.; Jones, P.D.; Osborn, T.J.; Lister, D.H. Updated high-resolution grids of monthly climatic observations—The CRU TS3.10 Dataset. *Int. J. Climatol.* **2014**, *34*, 623–642. [CrossRef]

37. Hijmans, R.J.; Guarino, L.; Cruz, M.; Rojas, E. Computer tools for spatial analysis of plant genetic resources data: 1. *DIVA-GIS. Plant Genet. Resour. News.* **2001**, *127*, 15–19.

38. Cong, R.-G.; Brady, M. The interdependence between rainfall and temperature: Copula analyses. *Sci. World J.* **2012**, *2012*, 405675. [CrossRef]

39. IPCC. *Meeting Report of the Intergovernmental Panel on Climate Change Expert Meeting on Climate Change, Food, and Agriculture;* Mastrandrea, M.D., Mach, K.J., Barros, V.R., Bilir, T.E., Dokken, D.J., Edenhofer, O., Field, C.B., Hiraishi, T., Kadner, S., Krug, T., et al., Eds.; World Meteorological Organization: Geneva, Switzerland, 2015; 68p.

40. Medori, M.; Michelini, L.; Nogues, I.; Loreto, F.; Calfapietra, C. The impact of root temperature on photosynthesis and isoprene emission in three different plant species. *Sci. World J.* **2012**, *2012*, 525827. [CrossRef]

41. Abbate, P.E.; Dardanelli, J.L.; Cantarero, M.G.; Maturano, M.; Melchiori, R.J.M.; Suero, E.E. Climatic and water availability effects on water-use efficiency in wheat. *Crop Sci.* **2004**, *44*, 474–483. [CrossRef]

42. Olesen, J.E.; Bindi, M. Consequences of climate change for European agricultural productivity, land use and policy. *Eur. J. Agron.* **2002**, *16*, 239–262. [CrossRef]

43. Cantelaube, P.; Terres, J.-M. Seasonal weather forecasts for crop yield modelling in Europe. *Tellus Ser. A Dyn. Meteorol. Oceanogr.* **2005**, *57*, 476–487. [CrossRef]

44. Abiodun, J.B.; Makhanya, N.; Petja, B.; Abatan, A.A.; Oguntunde, G.P. Future projection of droughts over major river basins in Southern Africa at specific global warming levels. *Theor. Appl. Climatol.* **2018**, *137*, 1785–1799. [CrossRef]

45. Ramírez-Villegas, J.; Jarvis, A.; Läderach, P. Empirical approaches for assessing impacts of climate change on agriculture: The EcoCrop model and a case study with grain sorghum. *Agric. For. Meteorol.* **2013**, *170*, 67–78. [CrossRef]

46. Ramírez-Villegas, J.; Lau, C.; Kohler, A.K.; Jarvis, A.; Arnell, N.; Osborne, T.M.; Hooker, J. *Climate Analogues: Finding Tomorrow's Agriculture Today;* CGIAR Research Program on Climate Change, Agriculture and Food Security (CCAFS): Frederiksberg, Denmark, 2011.

47. Hunter, R.; Crespo, O. *Large Scale Crop Suitability Assessment Under Future Climate Using the Ecocrop Model: The Case of Six Provinces in Angola's Planalto Region;* Springer: Cham, Switzerland, 2018.

48. Rippke, U.; Ramirez-Villegas, J.; Jarvis, A.; Vermeulen, S.J.; Parker, L.; Mer, F.; Diekkrüger, B.; Challinor, A.J.; Howden, M.; Howden, S. Timescales of transformational climate change adaptation in sub-Saharan African agriculture. *Nat. Clim. Chang.* **2016**, *6*, 605–609. [CrossRef]

49. Challinor, A.J.; Watson, J.; Lobell, D.B.; Howden, S.M.; Smith, D.R.; Chhetri, N.; Challinor, A.; Howden, S. A meta-analysis of crop yield under climate change and adaptation. *Nat. Clim. Chang.* **2014**, *4*, 287–291. [CrossRef]

50. Vermeulen, S.J.; Challinor, A.J.; Thornton, P.K.; Campbell, B.M.; Eriyagama, N.; Vervoort, J.M.; Kinyangi, J.; Jarvis, A.; Läderach, P.; Ramirez-Villegas, J.; et al. Addressing uncertainty in adaptation planning for agriculture. *Proc. Natl. Acad. Sci. USA* **2013**, *110*, 8357–8362. [CrossRef]

51. White, J.W.; Hoogenboom, G.; Kimball, B.A.; Wall, G.W. Methodologies for simulating impacts of climate change on crop production. *Field Crop Res.* **2011**, *124*, 357–368. [CrossRef]

52. Theil, H. A rank-invariant method of linear and polynomial. *Mathematics* **1950**, *392*, 387.

53. Sen, P.K. Estimates of the Regression Coefficient Based on Kendall's Tau. *J. Am. Stat. Assoc.* **1968**, *63*, 1379–1389. [CrossRef]

54. Wilcox, R.R. Simulations on the Theil-Sen regression estimator with right-censored data. *Stat. Probab. Lett.* **2003**, *39*, 43–47. [CrossRef]

55. Ohlson, J.A.; Kim, S. Linear valuation without OLS: The Theil-Sen estimation approach. *Rev. Acc. Stud.* **2015**, *20*, 395–435. [CrossRef]

56. Wilcox, R.R. A note on the Theil-Sen regression estimator when the regresser is random and the error term is heteroscedastic. *Biom. J.* **1998**, *40*, 261–268. [CrossRef]

57. Peng, H.; Wang, S.; Wang, X. Consistency and asymptotic distribution of the Theil-Sen estimator. *J. Stat. Plan. Inference* **2008**, *138*, 1836–1850. [CrossRef]

58. Gbobaniyi, E.; Sarr, A.; Sylla, M.B.; Diallo, I.; Lennard, C.; Dosio, A.; Dhiédiou, A.; Kamga, A.; Klutse, N.A.B.; Hewitson, B.; et al. Climatology, annual cycle and interannual variability of precipitation and temperature in CORDEX simulations over West Africa. *Int. J. Climatol.* **2014**, *34*, 2241–2257. [CrossRef]

59. Klutse, N.A.B.; Sylla, M.B.; Diallo, I.; Sarr, A.; Dosio, A.; Diedhiou, A.; Kamga, A.; Lamptey, B.; Ali, A.; Gbobaniyi, E.O.; et al. Daily characteristics of West African summer monsoon precipitation in CORDEX simulations. *Theor. Appl. Climatol.* **2016**, *123*, 369–386. [CrossRef]

60. Abiodun, B.J.; Adegoke, J.; Abatan, A.A.; Ibe, C.A.; Egbebiyi, T.; Engelbrecht, F.; Pinto, I. Potential impacts of climate change on extreme precipitation over four African coastal cities. *Clim. Chang.* **2017**, *143*, 399–413. [CrossRef]

61. Portmann, F.T.; Siebert, S.; Döll, P. MIRCA2000-Global monthly irrigated and rainfed crop areas around the year 2000: A new high-resolution data set for agricultural and hydrological modeling. *Glob. Biogeochem. Cycles* **2010**, 1–24. [CrossRef]

62. Nikulin, G.; Lennard, C.; Dosio, A.; Kjellström, E.; Chen, Y.; Hänsler, A.; Kupiainen, M.; Laprise, R.; Mariotti, L. Cathrine Fox Maule The effects of 1.5 and 2 degrees of global warming on Africa in the CORDEX The effects of 1.5 and 2 degrees of global warming on Africa in the CORDEX ensemble Manuscript version: Accepted Manuscript. *Environ. Res. Lett.* **2018**, *13*, 065003.

63. Maure, G.A.; Pinto, I.; Ndebele-Murisa, M.R.; Muthige, M.; Lennard, C.; Nikulin, G.; Dosio, A.; Meque, A.O. The southern African climate under 1.5 °C and 2 °C of global warming as simulated by CORDEX regional climate models. *Environ. Res. Lett.* **2018**, *13*, 065002. [CrossRef]

64. Ahmed, K.F.; Wang, G.; Yu, M.; Koo, J.; You, L. Potential impact of climate change on cereal crop yield in West Africa. *Clim. Chang.* **2015**, *133*, 321–334. [CrossRef]

65. Lobell, D.B.; Burke, M.B.; Tebaldi, C.; Mastrandrea, M.D.; Falcon, W.P.; Naylor, R.L. Prioritizing Climate Change Adaptation Needs for Food Security in 2030 Region. *Science* **2008**, *319*, 607–610. [CrossRef]

66. Malhotra, S.K. Horticultural crops and climate change: A review. *Indian J. Agric. Sci.* **2017**, *87*, 12–22.

67. Luo, Q. Temperature thresholds and crop production: A review. *Clim. Chang.* **2011**, *109*, 583–598. [CrossRef]

68. UNDP. *The 2030 Agenda for Sustainable Development*; A/RES/70/1; UNDP: New York, NY, USA, 2015; Volume 16301, pp. 13–14.

69. FAO. *The State of Food Security and Nutrition in the World 2018. Building Climate Resilience for Food Security and Nutrition*; Licence: CC BY-NC-SA 3.0 IGO; FAO: Rome, Italy, 2018.

70. Lobell, D.B.; Gourdji, S.M. The Influence of Climate Change on Global Crop Productivity. *Plant Physiol.* **2012**, *160*, 1686–1697. [CrossRef]

71. Taylor, K.E.; Stouffer, R.J.; Meehl, G.A. An overview of CMIP5 and the experiment design. *Bull. Am. Meteorol. Soc.* **2012**, *93*, 485–498. [CrossRef]

72. Zhang, X.; Cai, X. Climate change impacts on global agricultural water deficit. *Geophys. Res. Lett.* **2013**, *40*, 1111–1117. [CrossRef]

4

The Impact of Climatic Change Adaptation on Agricultural Productivity in Central Chile: A Stochastic Production Frontier Approach

Lisandro Roco [1,*], Boris Bravo-Ureta [2,3], Alejandra Engler [3,4] 🆔 and Roberto Jara-Rojas [3,4]

[1] Department of Economics and Institute of Applied Regional Economics (IDEAR),
    Universidad Católica del Norte, Antofagasta 1240000, Chile
[2] Department of Agricultural and Resource Economics, University of Connecticut, Storrs 06269, CT, USA;
    boris.bravoureta@uconn.edu
[3] Department of Agricultural Economics, Universidad de Talca, Talca 3460000, Chile;
    mengler@utalca.cl (A.E.); rjara@utalca.cl (R.J.-R.)
[4] Center for Socioeconomic Impact of Environmental Policies (CESIEP), Talca 3460000, Chile
* Correspondence: lisandro.roco@ucn.cl

Abstract: Adaptation to climate change is imperative to sustain and promote agricultural productivity growth, and site-specific empirical evidence is needed to facilitate policy making. Therefore, this study analyses the impact of climate change adaptation on productivity for annual crops in Central Chile using a stochastic production frontier approach. The data come from a random sample of 265 farms located in four municipalities with different agro-climatic conditions. To measure climate change adaptation, a set of 14 practices was used in three different specifications: binary variable, count and index; representing decision, intensity and quality of adaptation, respectively. The aforementioned alternative variables were used in three different stochastic production frontier models. Results suggest that the use of adaptive practices had a significant and positive effect on productivity; the practice with the highest impact on productivity was irrigation improvement. Empirical results demonstrate the relevance of climate change adaptation on farmers' productivity and enrich the discussion regarding the need to implement adaptation measures.

Keywords: climate change; adaptation; agricultural systems; productivity; technical efficiency; Chile

---

1. Introduction

Agriculture represents a relevant economic sector for the analysis of climate change, given that it is situated at the interface between ecosystems and society, and it is highly affected by changes in environmental conditions [1,2]. Climate change is affecting food prices, food security, land use [3] and raising uncertainty for crop managers [4]. According to Kahil [5], the severity of climate change impact depends on the degree of adaptation at the farm level, farmers' investment decisions and policy choices, and these factors are interrelated. Thus, it is necessary to recognize the effect that limitations in natural resources will have on agriculture to build resilience to climate change at the farm level [6].

On the other hand, as natural resources available for food production become more constraining, crop productivity is essential for fostering the growth and welfare of the agricultural sector [7]. To relax these constraints, farmers have been modifying their practices to cope with climatic variability for centuries; however, climate change is now threatening their livelihoods with increasing unpredictability, including frequent and intense weather extremes such as droughts, floods and frosts [8]. According to Zilberman et al. [9], adaptation is the response of economic agents and societies to major shocks such as climate change. Adaptation practices are adjustments intended to enhance resilience or reduce vulnerability to observed or expected changes in climate [10]. Nelson et al. [11] claim that adaptation is imperative for three reasons: (i) many future environmental risks are now more apparent and predictable than ever; (ii) even where risks are not quantifiable, environmental changes may be very significant; and (iii) environmental change, although often the outcome of multiple drivers, has indisputable human causes. Changes in food production affect all consumers; however, it is producers that need to adapt to insure adequate supplies and who bear the costs involved in improving efficiency [12].

There is a wide range of methodological approaches that have been developed over the years to generate multiple measures of productivity and efficiency [13]. A relevant measure of productivity for management recommendations is technical efficiency (TE) [14]. This indicator evaluates the difference between frontier or maximum attainable output and observed output given an input bundle and technology. Given that TE is an important component in overall productivity, the development and implementation of public policies can be more effective if the TE of any given farming system is known [15]. Several studies have investigated factors associated with agricultural productivity across the globe, but the literature linking TE with climate change adaptation is scanty. One exception is the study by Mukherjee et al. [16], which finds that heat stress in the southeastern U.S. has a significant and negative impact on milk production, while adaptation through a fairly simple cooling technology has a positive and significant effect on efficiency. In addition, in the same analysis, when climate change is factored into the production function (frontier) specification, the resulting estimates are more accurate, because they avoid possible parameter bias stemming from the omitted variable problem.

It is thus important to model the full range of interactions that might exist between productivity and climate change [17]. Most of the scientific information related to climate change and its effects on agriculture comes from case studies in developed countries. In developing countries, where there are high levels of uncertainty and vulnerability to climate change, there is need to target policy instruments to adapt the productive systems, particularly considering the lack of articulation between climate change adaptation and agricultural policy [18].

In this work, we investigate whether adaptive practices can increase productivity in different agricultural production systems based on annual crops in Central Chile. Major adaptation practices in farming systems include: conserving soil, using water efficiently, planting trees, changing planting dates and using improved varieties [19–23]. It is expected that farmers who are more aware of and better adapted to climate change will be able to make more efficient use of their resources and thus cope with any adversities. This study adds valuable information for agricultural policy design, as it provides evidence of the impact of alternative adaptation strategies to climate change. Additionally, farmer and agricultural system characteristics are linked to productivity to inform agricultural policy.

The rest of the paper is organized as follows: Section 2 gives a description of the study area, the methodological approach and the empirical models; Section 3 presents and discusses the empirical results; and Section 4 summarizes and concludes.

2. Materials and Methods

2.1. Study Area and Data

The study area covers 8,958 farms in four municipalities of the Maule Region, in Central Chile, a Mediterranean transition zone between the arid north and the rainy south. Projections for the study area comprise a decrease in precipitation of up to 40% and a rise in temperatures between 2 °C and 4 °C in the next 40 years [24,25]. This region is a major contributor to the agricultural output of the country and, despite rapid technological progress in recent years, the cultivation of annual crops, fruits and vegetables is not changing fast enough to counteract the predicted adverse effects of climate change [26,27]. Specific adverse effects expected in the near future concern losses in the quality of the environment for agricultural production [28].

The four municipalities selected for the study were: Pencahue, San Clemente, Cauquenes and Parral. Pencahue and Cauquenes are dryland areas; San Clemente is primarily composed of irrigated land near the Andes Mountains; and Parral is in the central irrigated valley. San Clemente has a total of 226,826 hectares (ha) dedicated largely to the production of forage, cereals and seeds. Cauquenes and Parral have 128,017 and 125,630 ha, respectively, with a significant area devoted to vineyards, cereals and forage. Pencahue is the smallest municipality, with 65,118 ha dedicated mostly to vineyards, orchards and cereals [26]. Table 1 presents some key characteristics of the four municipalities and the main cropping systems for each one.

Table 1. General information for the study area.

Municipality	Area	Rainfall (mm/Year)	Farms	Farms Interviewed	Main Crop System (%)				
					Wheat and Oat	Spring Crops [a]	Spring Vegetable [b]	Rice	Others Crops [c]
Pencahue	Irrigated dryland	709	1129	40	12.5	35.0	52.5	0.0	0.0
Cauquenes	Non-irrigated dryland	670	3026	81	97.5	2.5	0.0	0.0	0.0
San Clemente	Irrigated Andean foothill	920	2990	89	40.4	42.6	12.4	0.0	4.6
Parral	Irrigated central valley	900	1813	89	54.5	7.3	1.8	36.4	0.0
	Total		8958	265	56.6	77.4	12.5	7.5	1.5

[a] Spring crops are: maize, beans and potatoes. [b] Spring vegetables are: peas, onion, tomato, melon, watermelon, cucumber and squash. [c] Other crops are: tobacco and cabbage.

During August and November of 2011, a random survey was conducted that involved 274 interviews, representing 3.06% of the farmers in the study area. This survey targeted farmers that specialized in annual crops. The surveys with missing information were excluded from the analysis, leaving 265 valid surveys. Previous work in the study area inquired about the perception of and adaptation to climate change [24,26]; however, this article goes further by linking adaptation to climate change and productivity at the farm level.

Table 2 shows a description of the variables used in the study. The mean crop production value is US$66,383 (MM$31.2 where MM$ is equivalent to millions of Chilean pesos; and the prevailing exchange rate was 470 Chilean pesos per U.S. dollar when the data were collected). Farms range in size from 0.5–595 hectares, with a mean of 55.5 hectares. The average cultivated land area is 17.1 hectares. The mean value of purchased inputs (seeds, fertilizers, pesticides and hired machinery) is MM$11.4, and the mean investment in labor for crop production is MM$2.2. Crop diversification is measured

using a variant of the Herfindahl index (H) calculated for each farm as: $H = \left(1 - \sum_{i=1}^{n} \left(\frac{c_i}{T}\right)^2\right) \times 100$,

where c_i is the area under the i-th crop and T is the total cropped area [29]. The H index for the sample is 23.7%, ranging from 0–96.4%.

The average age for farmers is 55.5 years, while the average level of formal education is 7.2 years. The majority (82.6%) of farmers claimed that agriculture is their main income source, accounting, on average, for 62.1% of their total income. Eighty-one farms are in dryland areas. Meteorological information from the Internet and mass media (radio, TV and newspaper) is used by 93.2% of the farmers, and 52.4% of them participate in farmer associations. The mean distance from the farms to the city of Talca, the regional capital, is 77.4 kilometers.

Table 2. Description of the variables used in the stochastic production frontier (SPF) and inefficiency models. MM\$, millions of Chilean pesos.

Variable	Name		Unit	Definition	Mean	SD
		Production Function Variables				
y	Agricultural production		MM\$	Crop production value in Chilean pesos [a]	31.2	14.0
L	Cultivated land		Ha	Hectares with crops	17.1	53.3
C	Capital		MM\$	Value of seeds, fertilizers, pesticides and machinery contracted in Chilean pesos	11.4	51.7
W	Labor		MM\$	Value of family and hired labor	2.2	6.8
D	Dryland		%	Dummy variable = 1 if the farm is located in a dryland area and 0 otherwise	30.6	46.2
H	Diversification		%	Crop diversification index	23.7	27.5
A_1	Climate change adaptation	Decision	%	Dummy variable = 1 if there are at least one practice adopted and 0 otherwise	56.6	49.7
A_2		Intensity	Number	Number of climate change adaptation practices adopted in the farm	1.8	2.2
A_3		Quality	%	Index of adaptation based on experts' opinion	12.6	15.4
		Inefficiency Model Variables				
z_1	Age		Years	Age of the head of the farm in years	55.5	14.1
z_2	Schooling		Years	Years of schooling of the head of the farm	7.2	4.1
z_3	Dependence		%	Dummy variable = 1 if agriculture is the main source of income for the household and 0 otherwise	82.6	37.9
z_4	Specialization		%	Percent of total income that corresponds to income from crops	62.1	32.0
z_5	Use of meteorological information		%	Dummy variable = 1 if the farmer is a user of meteorological information and 0 otherwise	93.2	25.2
z_5	Membership		%	Dummy variable = 1 if the farmer is a member of an association and 0 otherwise	52.4	50.0
z_7	Farm size		Ha	Total farm size in hectares	56.4	122.3
z_8	Distance to market		Km	Distance to the regional capital city in kilometers	77.4	43.8

[a] Four hundred seventy Chilean pesos = US\$1 for the study period.

2.2. Practices Considered for Climate Change Adaptation

In recent studies, adaptive practices are identified as investment in technologies such as irrigation, the use of drought- and heat-tolerant and early-maturing varieties [19,30] and the adoption of strategies such as changing planting and harvesting dates, crop diversification, agroforestry and soil and water conservation practices [20–22]. Tambo and Abdoulaye [23] highlight the relevance of adaptation and its intensity regarding climate change. The authors just mentioned use as a first hurdle the decision to adopt a drought-resistant variety of maize and then intensity as the degree to which they will invest in adaptation measured as the area cultivated with the resistant variety.

A panel of experts was consulted to determine the most appropriate climate change adaptation strategies for the farming systems of Central Chile. This expert panel was composed of 14 national experts in agricultural systems and climate change. These experts were asked to assign a score from 0–3, where 0 is no impact and 3 is high impact, to 14 practices according to the importance of each practice for adaptation. These practices, described in Table 3, fall into three main categories: (1) water and soil conservation practices (WSC); (2) changes in cropping schedule and varieties (Cr); and (3) improvement of irrigation systems (I). These practices have been used previously in the literature [19,20,31]. We used this list of practices in the producers' survey to learn about what practices are being used by them. In several quantitative studies, the adaptation to climate change has been measured as the adoption of strategies, practices and technologies to increase the capacity of a farm to cope with changing climate and variability ([19–23] and others), and in most studies, the adaptation variable is defined as a binary decision. To carry out a more comprehensive analysis of adaptation, we include alternative measures of adaptation, from a simple binary variable to a more complex adaptation quality index. Each measure accounts for different interpretations of adaptation described as follows:

- Binary decision: a dichotomous variable indicating that at least one practice was adopted (A_1). In this case, the aim is to analyze the impact of being able to carry out a basic strategy.
- Intensity: measured as the number of practices or technologies adopted on the farm (A_2). Compared to A_1, this measure analyzes the impact of passing the first hurdle, i.e., the decision to adapt.
- Quality: an index calculated as the sum of adaptation practices weighted by the experts' score (A_3). The objective here is to estimate the impact of adopted practices that are more effective to face climate change. The weights were estimated by normalizing the average scores (0–3) given by the panel of experts to each practice, to generate a scale. The quality adaptation index (A_3) was constructed considering the sum of all the practices on a given farm multiplied by the weight assigned by experts (W_{ij}), divided by the sum of all weights (W_i). The formula used is as follows: $A_{3_i} = \left[\frac{\sum_{j=1}^{14} W_{ij}}{\sum_{j=1}^{14} W_j} \right] \times 100$, where i are the farms (from 1–265) and j are the practices (from 1–14).) The value of A_3 ranges from 0–100% where 100% implies that the practice presents the highest valuation assigned by the experts.

The number of farmers who have decided to adopt at least one of the practices is 150, representing 56.6% of the sample. The intensity in the number of practices adopted by farmers ranges from 0–11, with a mean of 1.8. The quality of adaptation (average index) is 12.6%, ranging from 0–79.3% (as can be seen in Table 2).

Table 3. Climate change adaptation practices according to the recommendation by experts.

Practice	Type [a]	Weight %	Farmers (n = 265)	
			No. of Respondents	% of Total
Incorporation of crop varieties resistant to droughts	Cr	85.7	2	0.7
Use of drip and sprinkler	I	83.3	31	11.7
Incorporation of crops resistant to high temperatures	Cr	80.9	2	0.7
Changes in planting and harvesting dates	Cr	78.6	110	41.5
Afforestation	WSC	76.2	5	1.9
Zero tillage	WSC	69.0	3	1.1
Use of water accumulation systems	I	66.7	38	14.3
Use of green manure	WSC	66.0	33	12.4
Use of mulching	WSC	61.9	24	9.0
Use of cover crops	WSC	61.9	16	6.0
Other WSC practices	WSC	61.9	16	6.0
Use of hoses and pumps for irrigation	I	59.5	52	19.6
Implementation of infiltration trenches	WSC	57.1	19	7.1
Cleaning of canals	WSC	54.8	60	22.6

[a] Cr: changes in crops, I: improvement of irrigation systems, WSC: water and soil conservation practices.

2.3. Analytical Framework and Empirical Model

The stochastic production frontier (SPF) model developed by Battese and Coelli [32] was used to estimate the following Cobb–Douglas frontier:

$$lny_i = \beta_0 + \beta_1 lnL_i + \beta_2 lnC_i + \beta_3 lnW_i + \beta_4 D_i + \beta_5 H_i + \beta_6 A_i + (v_i - u_i) \qquad (1)$$

where y_i is the value of agricultural production of the i-th farm, including the value of the output marketed, as well as the value of home consumption; L is the number of hectares assigned to annual crops by the farmer; C represents capital and is the sum of seeds, fertilizers, pesticides purchased and machinery contracted; W is the value of family and hired labor; D is a dichotomous variable that indicates if a farm is located in a dryland area and is thus expected to have lower production; H is the crop diversification index used to control for the intensity of agricultural activity and land use on the farm; A is the climate change adaptation measured as explained in Section 2.2; βs are the parameters to be estimated; and $v - u = \varepsilon$ is the composed error term.

The term v is a two-sided random error with a normal distribution ($v \sim N [0, \sigma_v^2]$) that captures the stochastic effect of factors beyond the farmer's control and statistical noise. The term u is a one-sided

$(u \geq 0)$ component that captures the TE of the producer; in other words, u measures the gap between observed production and its maximum value given by the frontier. This error can follow various statistical distributions including half-normal, exponential or gamma [33–35]. A high value of u implies a high degree of technical inefficiency; conversely, a value of zero implies that the farm is completely efficient. According to Battese and Coelli [32], the TE of the i-th farm is given by:

$$TE_i = exp(-u_i) \tag{2}$$

where u is the efficiency term specified in (1). TE for each farm is calculated using the conditional mean of $exp(-u)$, given the composed error term for the stochastic frontier model [36]. The maximum-likelihood method developed by Battese and Coelli [32] allows for a one-step estimation of u and v, and u can be expressed in terms of a set of explanatory variables Z_{nj} as:

$$u_j = \delta_0 + \sum_{n=1}^{k} \delta_n Z_{nj} + e_j \tag{3}$$

where δ_n are unknown parameters to be estimated.

The variables that affect technical inefficiency in our study (Table 2) are related to human capital (age, schooling, dependence, specialization and the use of meteorological information); social capital variables (membership in associations or organizations); and structural factors (distance to regional capital and farm size).

The adoption of climate change adaptation practices is a choice variable and, as in studies related to soil conservation adoption and credit access (e.g., [37–39]), might be correlated with the error term in Equation (1). Instrumental variables are commonly used to address endogeneity biases, and the Durbin–Wu–Hausman test (DWH) [40] is often the approach employed to statistically evaluate if this is indeed a problem. This test is based on the difference between the ordinary least square (OLS) and instrumental variables estimators [41]. The idea of the DWH test is to check whether the dissimilarity across these estimators is significantly different from zero given the data from the available sample. Under the null hypothesis that the error terms are uncorrelated with all the regressors against the alternative that they are correlated with at least some of the regressors, an F-test is performed [42]. The instrumental variables approach has been used in several recent studies of agricultural production analysis [43–47].

Therefore, to resolve the potential endogeneity of the variables A_1, A_2 and A_3, an instrumental variable approach was used to obtain their predicted values in a first-step regression, where A_1', A_2', and A_3' are the predicted values for A_1, A_2 and A_3, respectively. In the first step regression, the predicted values were generated as follows: A_1' was estimated using a logistic regression model; A_2' was assumed to have a zero-inflated negative binomial distribution; and for A_3', a truncated regression was applied. The models used to estimate the first step are shown in the Tables A2–A4, respectively.

To identify possible differences in TE across various technologies, we performed a Student's t-test comparing the mean of the expected TE for producers that did and did not adopt the following: (a) at least one irrigation improvement, (b) change in planting and harvesting schedule, and (c) at least two conservation practices. This simple procedure allows one to compare two independent groups by testing the null hypothesis of equal means.

3. Results and Discussion

3.1. Production Frontiers

Table 4 shows the estimations of the three SPF models. The parameter gamma is significant at the 1% level for the three models, with values of 0.42 for the Intensity model and 0.54 for the Decision and Quality models. In addition, the null hypothesis that sigma is equal to zero is rejected, confirming that the stochastic model is superior to the model that would result from using OLS. The presence of endogeneity is confirmed according to the DWH test implemented (as detailed in Table A1).

For the three models, the parameter for L, C and W are positive and statistically significant at the 1% level presenting also similar values across models. Capital (C) represents the most important production factor, with estimated coefficients around 0.60. Other studies reveal that capital is also important in the production function, with estimated parameters between 0.3 and 0.5 [39,48,49]. The size of the area under cultivation has an estimated parameter close between 0.23 and 0.29, consistent with those reported in other studies [50–52]. The lowest values are related to labor, L, around 0.11, consistent with the results from Rahman et al. [53] and Mariano et al. [52].

As expected, D is significant and negative, indicating that farms located in areas with lower quality soils and without irrigation are relatively less productive. Various agricultural production studies have shown that less-favored areas in terms of soil fertility or irrigation have lower productivity levels [52,54] and that this condition tends to be associated with high levels of inefficiency [48,49].

On the other hand, it is expected that crop diversification helps farmers to increase output, *ceteris paribus*, by allowing the continuous and more intensive use of the available soil and labor, and other resources. Crop diversification is one of the strategies used by farmers to minimize agricultural risk and to stabilize income [55]. Based on the H index, our results are consistent with expectations, revealing that higher diversification is positively associated with productivity. The Herfindahl index has been used in several studies to measure crop concentration or diversification [29,56]. Manjunatha et al. [57] incorporated this index in a production function for crops in India; Rahman [51] used it as a variable explaining crop efficiency in Bangladesh, demonstrating that crop diversification is associated with high levels of TE; and Kassali et al. [55] established a positive relation between crop diversification and efficiency among farmers in Nigeria.

The adoption of climate change adaptation technologies, for the three specifications (A_1, A_2, A_3) resulted in a positive and significant effect on productivity, evidencing the importance of adaptation in farming. As envisioned by Sauer et al. [58], over the next two decades, there will be pressing need for new agricultural responses in the face of population and economic growth, and these responses include increases in irrigated area and in water use intensity. Adaptation measures will need to play an increasingly important role to equilibrate food supply and demand in a global context [11,17].

The sum of the coefficients associated with L, C and W (partial elasticities of production) is close to one, an indication of nearly constant returns to scale for all models. This finding is consistent with those of Nyemeck et al. [48], Karagiannis and Sarris [50], Sauer and Park [59], and Reddy and Bantilan [49], but differs from that of Jaime and Salazar [60], who found increasing returns to scale in a sample of Chilean wheat farmers.

3.2. Technical Efficiency

Table 4 (bottom) shows that the average values of TE for the three models accounting for endogeneity are 67.8% (Decision), 76.4% (Intensity) and 72.3% (Quality). The mean TEs for models of decision are statistically the same. Table 5 shows that the range of TE for the 30% most efficient farms (the last three intervals) ranges from 53.9% to 74.1%. The average TE value is consistent with other studies done in Latin America using SPF models. Solís et al. [39] reported an average TE of 78%, and Bravo-Ureta et al. [14] reported a value of 70%. Table 5 also reveals high correlation coefficients between TE levels across the various models with values exceeding 0.95. In addition, Table 5 shows that the estimated TE values tend to be higher for models acknowledging endogeneity, indicating the relevance of considering this issue in the analysis.

Now we go back to Table 4 to examine the results concerning the Inefficiency Model. According to Gorton and Davidova [61], variables affecting farm efficiency can be divided into agency and structural factors. Agency factors, such as age, experience, education, specialization and training (i.e., human and social capital), represent the capacity of individuals to act independently and to make their own free choices. By contrast, structural factors, such as access to markets and credit, land tenure and farm size, influence or limit an agent in his or her decisions.

Table 4. Cobb–Douglas parameters for the stochastic production frontiers estimated considering endogeneity and three different specifications to measure climate change adaptation.

Variables	Climate Change Adaptation Measurement		
	Decision	Intensity	Quality
Constant (β_0)	4.1356 (0.9463) ***	4.7996 (0.9253) ***	4.7690 (0.9894) ***
Land (β_1)	0.2284 (0.0849) ***	0.2876 (0.0850) ***	0.2726 (0.0877) ***
Capital (β_2)	0.6184 (0.0739) ***	0.5950 (0.0710) ***	0.6041 (0.0779) ***
Labor (β_3)	0.1224 (0.0278) ***	0.1044 (0.0276) ***	0.1140 (0.0275) ***
Dryland (β_4)	−0.3485 (0.1303) ***	−0.4280 (0.1222) ***	−0.3882 (0.1350) ***
Diversification (β_5)	0.5670 (0.1312) ***	0.5933 (0.1373) ***	0.6074 (0.1361) ***
Climate change adaptation (β_6)	0.1092 (0.3012) ***	0.1656 (0.0546) ***	0.0075 (0.0052) *
Inefficiency Model			
Constant (δ_0)	0.2005 (0.6762)	0.3462 (0.6594)	−0.3082 (0.6554) ***
Age (δ_1)	0.0124 (0.0083) *	0.0171 (0.0080) **	0.0212 (0.0084) ***
Schooling (δ_2)	0.0200 (0.0175)	0.0147 (0.0270)	0.0107 (0.0296)
Dependence (δ_3)	−0.7099 (0.1738) ***	−0.8436 (0.1797) ***	−0.7310 (0.1800) ***
Specialization (δ_4)	−0.0085 (0.0034) ***	−0.0099 (0.0034) ***	−0.0112 (0.0031) ***
Use of meteorological information (δ_5)	−0.6258 (0.2770) **	−0.8279 (0.2463) ***	−0.7480 (0.2556) ***
Membership (δ_6)	0.2027 (0.1698)	0.2533 (0.1742) *	0.1915 (0.1884)
Farm size (δ_7)	−0.0036 (0.0008) ***	−0.0028 (0.0029)	−0.0035 (0.0026) *
Distance to market (δ_8)	0.0085 (0.0033) ***	0.0038 (0.0031) *	0.0057 (0.0031) **
Returns to scale	0.9692	0.9870	0.9907
Maximum Likelihood Function	−209.18	−209.60	−212.76
Sigma^2	0.4209 (0.0731) ***	0.4203 (0.0693) ***	0.4828 (0.0747) ***
Gamma	0.5363 (0.1043) ***	0.4247 (0.1111) ***	0.5411 (0.0989) ***
TE	67.8	76.4	72.3
TE difference with models without correcting endogeneity	ns	***	***

Climate change adaptation (A) is estimated through a logit regression (A_1') in the model for Decision, a zero-inflated negative binomial regression (A_2') in the model for Intensity and using a truncated regression (A_3') in the model for Quality (see the Appendix A). Numbers in parentheses are standard errors. * $p < 0.1$; ** $p < 0.05$; *** $p < 0.01$; ns: not significant. Estimations using Frontier Version 4.1 and STATA 11.1.

Most of the literature on TE uses human capital as the main source for explaining inefficiency [61]. Studies show that the relation of the age of farmers and TE levels varies according to geographic region and context. A negative and significant relation was described by Jaime and Salazar [60] for Chilean farmers; similar results were found by Mariano et al. [52] for rice producers in The Philippines and by Bozoğlu and Ceyhan [62] for vegetable farms in Turkey. Conversely, a positive relation is described by other authors [51,54,63]. In our study, the positive sign for age indicates that older farmers are less efficient.

It is expected that schooling has a negative effect on inefficiency levels, as noted by Jaime and Salazar [60], because education improves access to information, facilitates learning and the adoption of new processes and promotes forward-looking attitudes. Other studies support this conclusion [39,48,51,54,63–65]. However, in our study, schooling, measured by the number of years of formal instruction, has a negative, though not significant relationship with TE.

Our study found that the farmers who depend on agriculture as a primary source of income tend to be more efficient than those who do not. Similarly, Jaime and Salazar [60] report that the degree of dependence of Chilean wheat farmers on agriculture has a significant and positive relation with efficiency. Along this same line, Melo-Becerra and Orozco-Gallo [66] found that Colombian households that are dedicated exclusively to agricultural production are more efficient.

A similar relationship was found between specialization and TE; producers who specialize in crop production are more efficient than those who do not. Karagiannis et al. [67] showed that TE depends on specialization for both organic and conventional milk farms. Guesmi et al. [68], using the proportion of vineyard revenue to total agricultural revenue as a measure for specialization, also observed a positive relation between specialization and TE.

The use of meteorological information also shows a positive and significant relation with TE; farmers with access to meteorological information can be more alert about changes in weather and, in this way, minimize negative effects on productivity at the farm level. It is to be expected that access to information can have a positive effect on farm management and on the adoption of technologies related to farm productivity improvements. The use of meteorological information can represent a way to reduce uncertainty in productive operations. However, Lemos et al. [69] and Roco et al. [26] argue that the use of forecasts in decision-making is not straightforward and that much work is required to narrow the gap between producers and users of this kind of information.

Table 5. Distribution of TE and the correlation matrix for fitted models.

Interval TE		Farms in Interval (%)					
		Not-Correcting Endogeneity			Correcting Endogeneity		
		Decision	Intensity	Quality	Decision	Intensity	Quality
0–29		2.6	3.0	3.0	6.4	2.6	3.4
30–39		9.1	5.3	5.3	7.9	3.0	4.5
40–49		7.2	6.8	6.4	6.4	4.9	6.0
50–59		10.6	6.4	6.4	9.1	6.0	6.4
60–69		16.6	13.3	13.7	10.9	9.4	12.1
70–79		25.6	23.0	23.0	22.7	16.7	23.4
80–89		23.8	35.8	34.7	30.6	45.7	35.9
>90		4.5	6.4	7.5	6.0	11.7	8.3
Average TE		67.5	71.3	71.5	67.8	76.4	72.3
Correlation Matrix for TE Values							
Not-correcting for endogeneity	Decision	1	-	-	-	-	-
	Intensity	0.9872	1	-	-	-	-
	Quality	0.9876	0.9999	1	-	-	-
Correcting for endogeneity	Decision	0.9666	0.9779	0.9766	1	-	-
	Intensity	0.9532	0.9842	0.9841	0.9569	1	-
	Quality	0.9874	0.9967	0.9969	0.9741	0.9839	1

Social capital is another important factor to be considered in efficiency analyses. Membership in farmers' organizations can help to reduce inefficiency. Dios et al. [70] relate technical efficiency to innovation among farmers in Spain. Jaime and Salazar [60] note that in the Bío Bío Region in Chile, producers with higher levels of participation in organizations had higher levels of efficiency. Similar results were found by Nyemeck et al. [48] among producers in Cameroon. While in general, we found a positive relation between membership in organizations and TE levels, our results are not conclusive.

Intra- and inter-organizational arrangements are relevant for farm efficiency [61]. Our analysis, reveals a positive association between farm size and TE levels. There is evidence supporting the notion that large farms have higher levels of efficiency, due to advantages derived from economies of scale [49,53,54,60,63,66,71,72]. Considering the high percentage of small farms in the area under study, 28.6% according to ODEPA, which is the Chilean National Service for Agricultural Policy (the acronym stands for *Oficina de Estudios y Políticas Agrarias*) [72], this factor is likely a barrier to improve productivity levels in the region.

As expected, our results indicate that distance from the regional capital city has a negative and significant effect on TE levels. Proximity to markets, extension agencies and information coming from the regional capital tend to enhance farmers' TE. Tan et al. [54] claim that distance to a major city has a negative effect on TE levels for rice producers in China. Nyemeck et al. [48] highlight the importance of accessibility and find that TE is higher for farmers located near main roads.

In fact, Henderson et al. [73] found a strong and statistically-significant relationship between market participation and performance for crop-livestock smallholders in Sub-Saharan Africa.

3.3. Efficiency and Climate Change Adaptation

The analysis of efficiency in agriculture has been widely used to propose improvements in the management of farm systems. Areal et al. [74] argue that if the information received by policy makers concerning farm efficiency levels is harmonized with policy aims, policy measures may be targeted to support the targeted farms. This deserves further consideration given that the literature that links efficiency and climate change adaptation is limited.

Various t-tests were performed to relate efficiency levels and climate change adaptation (Table 6). We found a positive relation between TE and adopting at least one irrigation technology, i.e., farmers that adopt irrigation improvements exhibit a higher TE. In this regard, Yigezu et al. [75] argue that the use of modern irrigation methods yields an improvement of 19% in TE for wheat farmers in Syria.

However, a comparison across municipalities shows considerable geographical variability. In San Clemente, TE and the implementation of at least one irrigation alternative is evident regardless of the crops involved. For Pencahue, no differences are found between groups probably because most of the farmers in the sample (62.5%) have adopted at least one irrigation technology. In Cauquenes and Parral, we also find no significant difference and this is probably due to the low number of adopters. These results demonstrate the importance of climate change adaptation through the improvement of irrigation at the farm level to increase resource use efficiency. Kahil et al. [5] argue that water management policies, such as irrigation subsidies and efficient water markets, are key to face climate change in agriculture. Policy measures include supply enhancements to remove the threat of immediate water scarcity along with demand management measures and improved governance [76].

In general, changes in planting and harvesting dates show no relation with TE levels; however, in San Clemente, where the crops are highly diversified, farmers who have changed their planting calendars appear to have higher efficiency. Thus, it appears that this strategy that a priori could be expected to play a significant role for climate change adaptation, does not have a clear direct effect on efficiency. Additional information is required, to understand in a deeper way, the effects of a climate change practices portfolio on productivity and efficiency of agricultural systems.

The higher TE values detected for the groups who have more intensive adaptation strategies and with higher quality (number of practices and quality index) substantiate the importance of further research focusing on adaptation. It is not only necessary to adapt, but is also relevant to determine what and how much to adapt. Therefore, it is essential to foster effective adaptation and to improve the design of relevant programs to promote the adaptation capacity across farming systems. In Pencahue, 65% of the sample has adopted at least one adaptation practice, and 60% is above 25% in the adaptation index. In contrast, only 3.7% of the sample for Cauquenes has implemented at least one adaptation practice, and none of the farmers interviewed show an adaptation index over 25%. Based on this analysis, it seems clear that climate change adaptation in agriculture requires a complex set of actions including technical and managerial dimensions to reduce vulnerability and improve farmer productivity.

Table 6. t-tests for average TE levels grouped into various categories.

Average TE	Model	Grouping Criteria											
		Adoption of at Least One Irrigation Improvement			Changes in Planting and Harvesting Schedules			Adoption of at Least Two Adaptation Practices			Value of Adaptation Index ≥ 25%		
		Yes	No	Sig	Yes	No	Sig	Yes	No	Sig	Yes	No	Sig
Complete sample	Decision	73.5	65.3	***	64.9	70.0	**	81.4	65.1	***	86.5	65.4	***
	Intensity	80.9	74.4	***	75.1	77.3	ns	86.3	74.5	***	88.8	74.8	***
	Quality	77.5	69.9	***	70.9	73.3	ns	84.4	69.9	***	87.3	70.3	***
	%	54.7			42.6			16.2			11.3		

Table 6. *Cont.*

Average TE	Model	Adoption of at Least One Irrigation Improvement			Changes in Planting and Harvesting Schedules			Adoption of at Least Two Adaptation Practices			Value of Adaptation Index ≥ 25%		
		Yes	No	Sig	Yes	No	Sig	Yes	No	Sig	Yes	No	Sig
Pencahue	Decision	86.1	85.0	ns	85.7	85.7	ns	85.1	86.0	ns	86.0	85.2	ns
	Intensity	88.2	88.5	ns	87.8	88.7	ns	88.2	88.5	ns	88.1	88.5	ns
	Quality	86.7	86.7	ns	86.3	87.0	ns	86.8	86.6	ns	86.7	86.7	ns
	%	62.5			37.5			65.0			60.0		
Cauquenes	Decision	45.2	50.2	ns	47.2	50.9	ns	41.7	49.4	ns	-	-	
	Intensity	62.0	62.8	ns	62.0	63.3	ns	63.6	62.6	ns	-	-	
	Quality	55.7	58.2	ns	56.3	59.0	ns	59.6	57.6	ns	-	-	
	%	21.0			48.1			3.7			0.0		
San Clemente	Decision	82.2	77.0	***	83.3	77.1	***	81.6	78.2	ns	86.3	78.3	**
	Intensity	86.3	81.7	***	87.2	81.8	***	87.6	82.5	**	90.0	82.8	**
	Quality	83.2	77.4	***	84.4	77.6	***	85.0	78.4	**	87.8	78.9	**
	%	31.5			24.7			13.5			4.5		
Parral	Decision	66.5	64.1	ns	64.0	65.8	ns	79.5	64.0	*	93.1	63.5	***
	Intensity	80.0	76.3	ns	76.7	77.6	ns	89.0	76.6	*	94.5	76.4	**
	Quality	75.7	70.9	ns	71.2	71.6	ns	86.1	71.2	*	92.9	71.0	***
	%	20.0			67.3			3.6			3.6		

* $p < 0.1$; ** $p < 0.05$; *** $p < 0.01$, ns: not significant.

4. Concluding Remarks

This study analyzes the impact of climate change adaptation in productivity and efficiency for producers of annual crops in Central Chile. We used three measures of adaptation: a binary choice of adopting at least one adaptation practice or technology; an intensity measure given by the number of practices or technologies adopted; and a quality index measure. A positive association between productivity and climate change adaptation was observed for the three measures. The fitted stochastic production frontier models revealed that climate change adaptation is endogenous. Incorporation of instrumental variables allowed us to check the robustness of our results and improved the TE estimations. The fitted models showed important levels of inefficiency, suggesting the potential for increasing crop production using the current level of inputs and available technology.

Our results also show that factors such as dependence on annual crop production for income and high levels of specialization in production are associated with elevated TE levels. The use of meteorological information is also positively related with TE. In addition, our results indicate that farm size is positively related to efficiency while distance to a major city exhibits a negative relationship.

Farmers who have adopted irrigation technologies have higher TE levels. These results suggest that climate change adaptation is significant for agricultural production, especially for the intensity of climate change adaptation. Our results validate the importance, of incorporating climate change adaptation in agricultural policies designed to promote productivity growth. Our analysis also sheds light on the relevance of using meteorological information by farmers given the positive link between the latter variable and technical efficiency.

The connection between productivity with the implementation of specific farm-level adaptive practices, as well as with actions that ease adoption barriers deserves additional analyses. These analyses are essential to generate information required by policy makers to formulate robust action plans across differing cultural, economic and agricultural environments.

Acknowledgments: This work was supported by a research grant from The Latin American and Caribbean Environmental Economics Program (LACEEP). The authors thank the farmers who courteously answered our survey and the Excellence Program of Interdisciplinary Research: Adaptation of Agriculture to Climate Change (A2C2) of The University of Talca.

Author Contributions: All authors contributed extensively to the work presented in this paper. L.R., B.B-U. and A.E. designed the study. L.R. and R.J-R. implemented the fieldwork. L.R. wrote the paper. L.R. and R.J-R. conducted the analysis. B.B-U. and A.E. contributed extensively to the revision of several drafts of the manuscript.

Appendix A

Table A1. Cobb–Douglas parameters for stochastic production frontiers estimated considering three different specifications for the measurement of climate change adoption and without considering endogeneity.

Variables	Climate Change Adaptation Measurement		
	Decision	Intensity	Quality
Constant (β_0)	4.0218 (0.9857) ***	4.6682 (0.9741) ***	4.6090 (0.9891) ***
Land (β_1)	0.2314 (0.0887) ***	0.2654 (0.0869) ***	0.2602 (0.0857) ***
Capital (β_2)	0.6828 (0.0764) ***	0.6206 (0.0754) ***	0.6255 (0.0758) ***
Labor (β_3)	0.1043 (0.0283) ***	0.1112 (0.0283) ***	0.1110 (0.0270) ***
Dryland (β_4)	−0.4204 (0.1334) ***	−0.3578 (0.1270) ***	−0.3614 (0.1314) ***
Diversification (β_5)	0.5990 (0.1381) ***	0.5957 (0.1357) ***	0.6054 (0.1349) ***
Climate change adaptation (β_6)	0.0331 (0.0735)	0.0035 (0.0017) ***	0.0046 (0.0024) **
Inefficiency Model			
Constant (δ_0)	0.2035 (0.6194)	0.2166 (0.5937)	0.1591 (0.7713)
Age (δ_1)	0.0189 (0.0072) ***	0.0177 (0.0075) ***	0.0185 (0.0096) **
Schooling (δ_2)	0.0097 (0.0250)	0.0129 (0.0259)	0.0130 (0.0262)
Dependence (δ_3)	−0.4878 (0.1697) ***	−0.7657 (0.1776) ***	−0.7480 (0.2117) ***
Specialization (δ_4)	−0.0099 (0.0031) ***	−0.0098 (0.0034) ***	−0.0099 (0.0032) ***
Use of meteorological information (δ_5)	−0.7010 (0.2326) ***	−0.7406 (0.2423) ***	−0.7420 (0.2981) ***
Membership (δ_6)	0.0877 (0.1701)	0.2591 (0.1773) *	0.2663 (0.1970) *
Farm size (δ_7)	−0.0040 (0.0026) *	−0.0035 (0.0010) ***	−0.0034 (0.0009) ***
Distance to market (δ_8)	0.0056 (0.0029) **	0.0053 (0.0029) **	0.0051 (0.0030) **
Returns to scale	1.0185	0.9972	0.9967
MLF	−218.13	−211.14	−211.51
Sigma^2	0.4588 (0.0611) ***	0.4516 (0.0652) ***	0.4493 (0.0725) ***
Gamma	0.5632 (0.0996) ***	0.5222 (0.1044) ***	0.5178 (0.1144) ***
TE	67.52	71.34	71.50
Endogeneity (F value)	4.868 ***	14.266 ***	13.012 ***

Climate change adaptation (A) is measured as: the adoption of at least one practice (A_1) in the model for decision; the number of practices adopted (A_2) in the model for intensity; the number of practices weighted according to experts' opinion (A_3) in the model for quality. Numbers in parentheses are standard errors. * $p < 0.1$; ** $p < 0.05$; *** $p < 0.01$. Estimations using Frontier Version 4.1 and STATA 11.1.

Table A2. Logit regression estimation.

Variable Name	Description	Coefficient
A_1	**Dependent Variable**	
ExpAgIndep	Years of independent experience in agriculture.	−0.0153 * (0.0087)
SanClemente	Dummy variable = 1 if the farm is located in San Clemente and 0 otherwise	−0.9189 *** (0.2886)
TTPropia	Dummy variable = 1 if the farmer is owner and 0 otherwise	0.3590 (0.2821)
Internet	Dummy variable = 1 if the farmer has access to meteorological information principally form the Internet and 0 otherwise	0.9667 *** (0.3290)
Constant		0.5849 ** (0.3012)
	Log-likelihood	−170.73
	N	265
	Pseudo R^2	5.86
	Correctly classified values by Logit (%)	62.2

Numbers in parenthesis are standard errors. * $p < 0.1$; ** $p < 0.05$; *** $p < 0.01$.

Table A3. Zero inflated negative binomial regression estimation.

Variable Name	Description	Coefficient
A_2	**Dependent Variable**	
ExpAgIndep	Years of independent experience in agriculture.	−0.0121 *** (0.0034)
RXP	Dummy variable = 1 if the farmer has adopted any irrigation improvement and the location is in Pencahue municipality and 0 otherwise	0.7731 *** (0.1324)

Table A3. *Cont.*

Variable Name	Description	Coefficient
A_2	**Dependent Variable**	
SupProd	Surface designated to production in hectares	0.0003 (0.0003)
Internet	Dummy variable = 1 if the farmer has access to meteorological information principally form the Internet and 0 otherwise	0.2233 * (0.1329)
Constant		1.0172 *** (0.1362)
	Log-likelihood	−411.76
	N	265
	Correlation of predicted values ($A_1{}'$) with A_1 (%)	53.51

Numbers in parenthesis are standard errors. * $p < 0.1$; *** $p < 0.01$.

Table A4. Truncated linear regression estimation.

Variable Name	Description	Coefficient
A_3	**Dependent Variable**	
ExpAgIndep	Years of independent experience in agriculture.	−0.2518 *** (0.0893)
RXP	Dummy variable = 1 if the farmer has adopted any irrigation improvement and the farm location is Pencahue and 0 otherwise	18.445 *** (4.2773)
SupProd	Surface designated to production in hectares	0.0173 * (0.0105)
Internet	Dummy variable = 1 if the farmer has access to meteorological information principally form the Internet and 0 otherwise	3.4477 (3.1870)
Constant		20.1456 *** (2.8415)
	Log-Likelihood	−574.27
	N	265
	Correlation of predicted values ($A_2{}'$) with A_2 (%)	51.35

Regression was truncated in values with 0 as the lower limit and 100 as the upper limit. Numbers in parenthesis are standard errors. * $p < 0.1$; *** $p < 0.01$.

References

1. Oelesen, J.; Bindi, M. Consequences of climate change for European agricultural productivity, land use and policy. *Eur. J. Agron.* **2002**, *16*, 239–262. [CrossRef]
2. IPCC. *Managing the Risk of Extreme Events and Disasters to Advance Climate Change Adaptation*; Special Report; Cambridge University Press: Cambridge, UK, 2012; p. 582.
3. Lobell, D.B.; Field, C.B. Global scale climate-crop yield relations and the impacts of recent warming. *Environ. Res. Lett.* **2007**, *2*, 1–7. [CrossRef]
4. Pathak, H.; Wassmannn, R. Quantitative evaluation of climatic variability and risk for wheat yield in India. *Clim. Chang.* **2009**, *93*, 157–175. [CrossRef]
5. Kahil, M.T.; Connor, J.D.; Albiac, J. Efficient water management policies for irrigation adaptation to climate change in Southern Europe. *Ecol. Econ.* **2015**, *120*, 226–233. [CrossRef]
6. Jackson, T.M.; Hanjra, M.; Khan, S.; Hafeez, M.M. Building a climate resilient farm: A risk based approach for understanding water, energy and emissions in irrigated agriculture. *Agric. Syst.* **2012**, *104*, 729–745. [CrossRef]
7. Fuglie, K.; Schimmelpfennig, D. Introduction to the special issue on agricultural productivity growth: A closer look at large, developing countries. *J. Product. Anal.* **2010**, *33*, 169–172. [CrossRef]
8. Clements, R.; Haggar, J.; Quezada, A.; Torres, J. *Technologies for Climate Change Adaptation—Agriculture Sector*; Zhu, X., Ed.; UNEP Risø Centre: Roskilde, Denmark, 2011.
9. Zilberman, D.; Zhao, J.; Heirman, A. Adoption versus adaptation, with emphasis on climate change. *Annu. Rev. Resour. Econ.* **2012**, *4*, 27–53. [CrossRef]
10. IPCC. *Climate Change 2007: Impacts, Adaptation and Vulnerability*; Intergovernmental Panel on Climate Change, Fourth Assessment Report; Cambridge University Press: Cambridge, UK, 2007.
11. Nelson, D.; Adger, W.N.; Brown, K. Adaptation to environmental change: contributions of a resilience framework. *Annu. Rev. Resour. Econ.* **2007**, *32*, 395–419. [CrossRef]

12. AGRIMED. *Impactos Productivos en el Sector Silvoagropecuario de Chile Frente a Escenarios de Cambio Climático*; U. de Chile, CONAMA, ODEPA, FIA Report; Universidad de Chile: Santiago, Chile, 2008.

13. Paul, C. Productivity and efficiency measurement in our "New Economy": Determinants, interactions, and policy relevance. *J. Product. Anal.* **2003**, *19*, 161–177. [CrossRef]

14. Bravo-Ureta, B.; Solís, D.; Moreira, V.; Maripani, J.; Thiam, A.; Rivas, T. Technical efficiency in farming: A meta-regression analysis. *J. Product. Anal.* **2007**, *27*, 57–72. [CrossRef]

15. Coelli, T.; Rao, D.S.; O'Donnell, C.; Battesse, G. *An Introduction to Efficiency and Productivity Analysis*, 2nd ed.; Springer: New York, NY, USA, 2005; p. 350.

16. Mukherjee, D.; Bravo-Ureta, B.; de Vries, A. Dairy productivity and climatic conditions: Econometric evidence from South-eastern United States. *Aust. J. Agric. Resour. Econ.* **2013**, *57*, 123–140. [CrossRef]

17. USDA. *Climate Change and Agriculture in the United States: Effects and Adaptation*; United States Department of Agriculture Technical Bulletin 1935; USDA: Washington, DC, USA, 2013; p. 186.

18. Roco, L.; Poblete, D.; Meza, F.; Kerrigan, G. Farmers' options to address water scarcity in a changing climate: Case studies from two basins in Mediterranean Chile. *Environ. Manag.* **2016**, *109*, 958–971. [CrossRef] [PubMed]

19. Deressa, T.T.; Hassan, R.M. Economic impact of climate change on crop production in Ethiopia: Evidence from cross-section measures. *J. Afr. Econ.* **2009**, *18*, 529–554. [CrossRef]

20. Gbetibouo, G.A. *Understanding Farmers' Perceptions and Adaptations to Climate Change and Variability: The Case of the Limpopo Basin, South Africa*; IFPRI Discussion Paper No. 849; International Food Policy Research Institute: Washington, DC, USA, 2009; p. 36.

21. Di Falco, S.; Veronesi, M.; Yesuf, M. Does adaptation to climate change provide food security? A micro-perspective from Ethiopia. *Am. J. Agric. Econ.* **2011**, *93*, 829–846. [CrossRef]

22. Sofoluwe, N.; Tijane, A.; Baruwa, O. Farmers' perception and adaptation to climate change in Osun State, Nigeria. *Afr. J. Agric. Resour. Econ.* **2011**, *6*, 4789–4794.

23. Tambo, J.A.; Abdoulaye, T. Climate change and agricultural technology adoption: The case of drought tolerant maize in rural Nigeria. *Mitig. Adapt. Strateg. Glob. Chang.* **2012**, *17*, 277–292. [CrossRef]

24. Roco, L.A.; Engler, B.; Bravo-Ureta, B.E.; Jara-Rojas, R. Farm level adaptation decisions to face climatic change and variability: Evidence from Central Chile. *Environ. Sci. Policy* **2014**, *44*, 86–96. [CrossRef]

25. Chilean Ministry of Agriculture. Plan de Adaptación al Cambio Climático del Sector Silvoagropecuario. 2012. Available online: http://www.mma.gob.cl/1304/articles-52367_PlanAdaptacionCCS.pdf (accessed on 16 September 2017).

26. Roco, L.A.; Engler, B.; Bravo-Ureta, B.E.; Jara-Rojas, R. Farmers' perception of climate change in Mediterranean Chile. *Reg. Environ. Chang.* **2015**, *15*, 867–879. [CrossRef]

27. FIA. *El Cambio Climático en el Sector Silvoagropecuario de Chile*; Fundación para la Innovación Agraria, Ministerio de Agricultura de Chile: Santiago, Chile, 2010; p. 16.

28. Hannah, L.; Ikegami, M.; Hole, D.G.; Seo, C.; Butchart, S.H.; Peterson, A.T.; Roehrdanz, P.R. Global climate change adaptation priorities for biodiversity and food security. *PLoS ONE* **2013**, *8*, e72590. [CrossRef] [PubMed]

29. Malik, D.P.; Singh, I.J. Crop diversification—An economic analysis. *Indian J. Agric. Res.* **2002**, *36*, 61–64.

30. Moniruzzaman, S. Crop choice as climate change adaptation: Evidence from Bangladesh. *Ecol. Econ.* **2015**, *118*, 90–98. [CrossRef]

31. Bryan, E.; Deressa, T.T.; Gbetibouo, G.A.; Ringler, C. Adaptation to climate change in Ethiopia and South Africa: Options and constraints. *Environ. Sci. Policy* **2009**, *12*, 413–426. [CrossRef]

32. Battese, G.E.; Coelli, T.J. A model for technical inefficiency effects in stochastic frontier production function for panel data. *Empir. Econ.* **1995**, *20*, 325–332. [CrossRef]

33. Aigner, D.J.; Lovell, C.A.K.; Schmidt, P. Formulation and estimation of stochastic frontier production function models. *J. Econom.* **1977**, *6*, 21–37. [CrossRef]

34. Greene, W.H. Maximum likelihood estimation of econometric frontier functions. *J. Econom.* **1980**, *13*, 27–56. [CrossRef]

35. Meeusen, W.; van den Broeck, J. Efficiency estimation from Cobb–Douglas production function with composed error. *Int. Econ. Rev.* **1977**, *18*, 435–444. [CrossRef]

36. Battese, G.E.; Coelli, T.J. Prediction of firm-level technical efficiencies with a generalized frontier production function and panel data. *J. Econom.* **1988**, *38*, 387–399. [CrossRef]

37. Jones, S. A framework for understanding on-farm environmental degradation and constraint to the adoption of soil conservation measures: Case studies from Highland Tanzania and Thailand. *World Dev.* **2002**, *30*, 1607–1620. [CrossRef]

38. Chavas, J.P.; Ragan, P.; Roth, M. Farm household production efficiency: Evidence from The Gambia. *Am. J. Agric. Econ.* **2005**, *87*, 160–179. [CrossRef]

39. Solís, D.; Bravo-Ureta, B.; Quiroga, R. Technical efficiency among peasant farmers participating in natural resource management programs in Central America. *J. Agric. Econ.* **2009**, *60*, 202–219. [CrossRef]

40. Davidson, R.; MacKinnon, J.G. *Estimation and Inference in Econometrics*; Oxford University Press: New York, NY, USA, 1993.

41. Cameron, C.; Trvedi, P. *Microeconometrics Using Stata*, Revised ed.; Stata Press: College Station, TX, USA, 2010; p. 706.

42. Davidson, R.; MacKinnon, J.G. *Econometric Theory and Methods*; Oxford University Press: New York, NY, USA, 2004.

43. Di Falco, S.; Yesuf, M.; Kohlin, G.; Ringler, C. Estimating the impact of climate change on agriculture low-income countries: Household level evidence from the Nile Basin, Ethiopia. *Environ. Resour. Econ.* **2012**, *52*, 457–478. [CrossRef]

44. Weber, J.; Key, N. How much do decoupled payments affect production? An instrumental variable approach with panel data. *Am. J. Agric. Econ.* **2012**, *94*, 52–66. [CrossRef]

45. Stifel, D.; Fafchamps, M.; Minten, B. Taboos, agriculture and poverty. *J. Dev. Stud.* **2011**, *47*, 1455–1481. [CrossRef]

46. Mishra, A.K.; El-Osta, H.S.; Shaik, S. Succession decisions in US family farm business. *J. Agric. Resour. Econ.* **2010**, *35*, 133–152.

47. Kilic, T.; Carletto, C.; Miluka, J.; Savanasto, S. Rural nonfarm income and its impact on agriculture: Evidence from Albania. *Agric. Econ.* **2009**, *40*, 139–160. [CrossRef]

48. Nyemeck, J.B.; Tonyé, J.; Wandi, N.; Nyambi, G.; Akoa, M. Factors affecting the technical efficiency among smallholder farmers in the slash and burn agriculture zone of Cameroon. *Food Policy* **2004**, *29*, 531–545.

49. Reddy, A.A.; Bantilan, M.C. Competitiveness and technical efficiency: Determinants in the groundnut sector of India. *Food Policy* **2012**, *37*, 255–263. [CrossRef]

50. Karagiannis, G.; Sarris, A. Measuring and explaining scale efficiency with the parametric approach: The case of Greek tobacco growers. *Agric. Econ.* **2005**, *33*, 441–451. [CrossRef]

51. Rahman, S. Women's labour contribution to productivity and efficiency in agriculture: Empirical evidence fron Bangladesh. *J. Agric. Econ.* **2010**, *61*, 318–342. [CrossRef]

52. Mariano, M.J.; Villano, R.; Fleming, E. Technical efficiency of rice farms in different agroclimatic zones in the Philipines: An application of a stochastic metafrontier model. *Asian Econ. J.* **2011**, *25*, 245–269. [CrossRef]

53. Rahman, S.; Wiboopongse, A.; Sriboonchitta, S.; Chaovanapoonphol, Y. Production efficiency of jasmine rice producers in Northern and North-eastern Thailand. *J. Agric. Econ.* **2009**, *60*, 419–435. [CrossRef]

54. Tan, S.; Heerink, N.; Kuyvenhoven, A.; Qu, F. Impact of land fragmentation on rice producers' technical efficiency in South-East China. *NJAS Wagening. J. Life Sci.* **2010**, *57*, 117–123. [CrossRef]

55. Kassali, R.; Ayanwale, A.B.; Idowu, E.O.; Williams, S.B. Effect of rural transportation systems on agricultural productivity in Oyo State, Nigeria. *J. Agric. Rural Dev. Trop. Subtrop.* **2012**, *113*, 13–19.

56. Sarris, A.; Savastano, S.; Christiaensen, L. *The Role of Agriculture in Reducing Poverty in Tanzania: A Household Perspective from Rural Kilimanjaro and Ruvuma*; FAO Commodity and Trade Policy Research Working Paper No. 19; FAO: Rome, Italy, 2006; p. 30.

57. Manjunatha, A.V.; Anikc, A.R.; Speelmand, S.; Nuppenaua, E.A. Impact of land fragmentation, farm size, land ownership and crop diversity on profit and efficiency of irrigated farms in India. *Land Use Policy* **2013**, *31*, 397–405. [CrossRef]

58. Sauer, T.; Havlik, P.; Scheneider, U.A.; Schmidt, E.; Kindermann, G.; Obersteiner, M. Agriculture and resource availability in a changing world: The role of irrigation. *Water Resour. Res.* **2010**, *46*, W06503. [CrossRef]

59. Sauer, J.; Park, T. Organic farming in Scandinavia—Productivity and market exit. *Ecol. Econ.* **2009**, *68*, 2243–2254. [CrossRef]

60. Jaime, M.; Salazar, C. Participation in organizations, technical efficiency and territorial differences: A study of small wheat farmers in Chile. *Chil. J. Agric. Res.* **2011**, *71*, 104–113. [CrossRef]

61. Gorton, M.; Davidova, S. Farm productivity and efficiency in the CEE applicant countries: A synthesis of results. *Agric. Econ.* **2004**, *30*, 1–16. [CrossRef]
62. Bozoğlu, M.; Ceyhan, V. Measuring the technical efficiency and exploring the inefficiency determinant of vegetable farms in Samsun province, Turkey. *Agric. Syst.* **2011**, *94*, 649–656. [CrossRef]
63. Külekçi, M. Technical efficiency analysis for oilseed sunflower farms: A case study in Erzurum, Turkey. *J. Sci. Food Agric.* **2010**, *90*, 1508–1512. [CrossRef] [PubMed]
64. Phillips, J. Farmer education and farmer efficiency: A meta-analysis. *Econ. Dev. Cult. Chang.* **1994**, *43*, 149–165. [CrossRef]
65. Phillips, J.; Marble, R. Farmer education and efficiency: A frontier production approach. *Econ. Educ. Rev.* **1996**, *5*, 257–264. [CrossRef]
66. Melo-Becerra, L.A.; Orozco-Gallo, A.J. Technical efficiency for Colombian small crop and livestock farmers: A stochastic metafrontier approach for different production systems. *J. Product. Anal.* **2017**, *47*, 1–16. [CrossRef]
67. Karagiannis, G.; Salhofer, K.; Sinabell, F. *Technical Efficiency of Conventional and Organic Farms: Some Evidence for Milk Production*; OGA Tagungsband: Wien, Austria, 2006.
68. Guesmi, B.; Serra, T.; Kallas, Z.; Roig, M.G. The productive efficiency of organic farming: The case of grape sector in Catalonia. *Span. J. Agric. Res.* **2012**, *10*, 552–566. [CrossRef]
69. Lemos, M.C.; Kirchhoff, C.J.; Ramprasad, V. Narrowing the climate information usability gap. *Nat. Clim. Chang.* **2012**, *2*, 789–794. [CrossRef]
70. Dios, R.; Martínez, J.M.; Vicario, V. Eficiencia versus innovación en explotaciones agrarias. *Estud. Econ. Apl.* **2003**, *21*, 485–501.
71. Yang, Z.; Mugera, A.M.; Zhang, F. Investigating yield variability and inefficiency in rice production: A case study in Central China. *Sustainability* **2016**, *8*, 787. [CrossRef]
72. ODEPA. Caracterización de la Pequeña Agricultura en Chile. 2011. Available online: http://www.odepa.gob.cl/odepaweb/servicios-informacion/publica/Pequena_agricultura_en_Chile.pdf (accessed on 16 September 2017).
73. Henderson, B.; Godde, C.; Medina-Hidalgo, D.; van Wijkb, M.; Silvestri, S.; Douxchamps, S.; Stephenson, E.; Power, B.; Rigolot, C.; Cacho, O.; et al. Closing system-wide yield gaps to increase food production and mitigate GHGs among mixed crop-livestock smallholders in Sub-Saharan Africa. *Agric. Syst.* **2016**, *143*, 106–113. [CrossRef] [PubMed]
74. Areal, F.J.; Tiffin, R.; Balcombe, K.G. Provision of environmental output within a multi-output distance function approach. *Ecol. Econ.* **2012**, *78*, 47–54. [CrossRef]
75. Yigezu, Y.; Ahmed, M.; Shideed, K.; Aw-Hassan, A.; El-Shater, T.; Al-Atwan, S. Implications of a shift in irrigation technology on resource use efficiency: A Syrian case. *Agric. Syst.* **2013**, *118*, 14–22. [CrossRef]
76. Vargherse, S.K.; Veettil, P.C.; Speelman, S.; Buysse, J.; van Huylenbroeck, G. Estimating the causal effect of water scarcity on the groundwater use efficiency of rice farming in South India. *Ecol. Econ.* **2013**, *86*, 55–64. [CrossRef]

5

Climate-Smart Agriculture and Non-Agricultural Livelihood Transformation

Jon Hellin [1],* and Eleanor Fisher [2]

[1]   Sustainable Impact Platform at the International Rice Research Institute (IRRI),
      Metro Manila 1301, Philippines
[2]   School of Agriculture, Policy and Development at the University of Reading, Reading RG6 6AH, UK;
      e.fisher@reading.ac.uk
*   Correspondence: j.hellin@irri.org

Abstract: Agricultural researchers have developed a number of agricultural technologies and practices, known collectively as climate-smart agriculture (CSA), as part of climate change adaptation and mitigation efforts. Development practitioners invest in scaling these to have a wider impact. We use the example of the Western Highlands in Guatemala to illustrate how a focus on the number of farmers adopting CSA can foster a tendency to homogenize farmers, instead of recognizing differentiation within farming populations. Poverty is endemic in the Western Highlands, and inequitable land distribution means that farmers have, on average, access to 0.06 ha per person. For many farmers, agriculture per se does not represent a pathway out of poverty, and they are increasingly reliant on non-agricultural income sources. Ineffective targeting of CSA, hence, ignores small-scale farming households' different capacities for livelihood transformation, which are linked to the opportunities and constraints afforded by different livelihood pathways, agricultural and non-agricultural. Climate-smart interventions will often require a broader and more radical agenda that includes supporting farm households' ability to build non-agricultural-based livelihoods. Climate risk management options that include livelihood transformation of both agricultural and non-agricultural livelihoods will require concerted cross-disciplinary research and development that encompasses a broader set of disciplines than has tended to be the case to date within the context of CSA.

Keywords: climate-smart agriculture; livelihood transformation; Guatemala; climate change

1. Introduction

Climate change will have a detrimental impact on agricultural productivity in many parts of the developing world [1]. Farmers have long adapted to climate variability, but the severity of the predicted changes may be beyond many farmers' current ability to adapt and improve their livelihoods [2,3]. There is an urgent need to work with farmers to develop climate change adaptation, mitigation and transformation strategies. Sustainable development goal (SDG) 13 is on *Climate Action* and, hence, there is much interest in the promotion of climate-smart agricultural practices (CSA). These are practices that contribute to an increase in global food security (and other development goals), an enhancement of farmers' ability to adapt to a changing climate and the mitigation of emissions of greenhouse gases [3,4]. CSA, hence, not only contributes to the realization of SDG 13, but is also intrinsically linked to several other SDGs, for example, SDG 1: *No Poverty* and SDG 2: *Zero Hunger*.

Transformative approaches have gained traction in contemporary policy debates on climate impacts, stimulated, amongst other factors, by the United Nations Sustainable Development Goals (SDGs) and the Intergovernmental Panel on Climate Change (IPCC). CSA can be transformative in terms of its aims to ensure food security via a reorientation of agricultural development in the context of the realities of climate change. For example, recent research on the climate-smart village (CSV) approach [4] highlights the potential of scaling out so as to benefit larger number of farmers. However, there has been limited scaling of the CSV approach. One of the challenges is that the scaling of the CSV approach is premised largely on identifying a portfolio of CSA options and the financial or institutional mechanisms that enhance adoption by farmers, and targeting these at regions with similar agro-ecological conditions [4], with less attention being given to the local context [5,6]. The danger is the a priori belief that CSA is a pathway out of poverty. For many farmers, adaptation to climate change in ways that lead to an escape from poverty, and greater prosperity may not be via CSA [7].

Lipper et al. [3] stress that CSA results in higher resilience and lower risks to food security. While this may be the case, there are farmers for whom agricultural-based livelihoods are so precarious that even "climate-proofing" their agricultural systems represents a higher risk to food security and prosperity than non-agricultural livelihood options. The challenge, therefore, is that at the same time that international calls for transformative approaches are made, current and future rural livelihood conditions are so adverse that, for some, this is a matter of changing to grasp any livelihood opportunity, including adverse coping strategies, without any ability to improve agricultural practices in ways that could be considered synonymous with transformative change in a positive sense. Hence, "climate-smart" may actually mean the need for actions that focus on supporting people in building non-agricultural-based livelihoods [3,4]. If this livelihood transformation is to be positive, it will require concomitant policy and development support to provide enabling conditions for non-agricultural livelihoods to be built. Moreover, this needs to be performed in ways that improve household income and security, i.e. are prosperity-enhancing, thus avoiding recourse to adverse coping strategies. The Western Highlands of Guatemala illustrates this challenging development scenario.

2. Climate-Smart Agriculture in the Western Highlands, Guatemala

Scientific evidence points to negative impacts on agriculture in Guatemala, and other parts of Central America, due to changing temperature and rainfall patterns, e.g., [8]. Inequalities in land distribution have forced many resource-poor farmers to farm steep hillsides, areas that are very susceptible to soil and land degradation. The response has often been the promotion of CSA. Development practitioners are rediscovering technologies and practices that were promoted in the region in the 1980s and 1990s under the guise of soil and water conservation [9]. These included live barriers, stone terraces, cover crops, green manures and agro-forestry. Farmers' uptake of these technologies and practices was disappointing 20–30 years ago [10], largely because, as is the case worldwide, a technology-led approach tends to ignore the needs for institutional enabling factors, which are very important when it comes to farmers' uptake of agricultural technologies [6,11,12].

The promotion of CSA in Guatemala is particularly challenging. The country suffers from extreme rural poverty and food insecurity [13]. Guatemala is ethnically very diverse, and indigenous groups (who make up almost 40% of the total population) live mainly in the Western Highlands. The underpinnings of present-day poverty are rooted in conflict, linked to Guatemala's 36-year civil war, which ended in the mid-1990s, and during which tens of thousands of indigenous people died [14]. This has left a legacy of inequality and continued social tension.

Small-scale farmers practice largely subsistence and some market-oriented agriculture. The most important cultivated food crop is maize, which is intercropped with beans, chilies and squash [15]. Recent research has shown that land availability in the Western Highlands is 0.06 ha per person [13]. This contributes to considerable food insecurity: farm households produce enough maize (the main staple crop) for fewer than seven months of consumption per year, and for household consumption have to purchase maize to make up the deficit. As a consequence, the majority of farmers seek off-farm employment on a temporary basis, while a minority have managed to branch into the production of higher-value vegetable crops for the export market [16].

Donors have invested much in rural development projects [17,18]. One such rural development project in the Western Highlands was implemented from 2013–2018. The Buena Milpa project was supported by the United States Agency for International Development (USAID), through its Global Hunger and Food Security Initiative "Feed the Future". Its main objectives were to reduce poverty, food insecurity and malnutrition, while increasing the sustainability and resilience of maize-based farming systems (The ideas reported here stem from Hellin's involvement as a socio-economist in this project in the Western Highlands of Guatemala). More details are provided in [13,19]. A strong emphasis of the project was the promotion of CSA, and the project worked through a number of non-governmental organizations. During the course of the work, it became clear that more attention needed to be focused on farmers' different capacities to engage in climate risk management. The danger was that poor targeting of CSA would lead to weak farmer uptake, by implication excluding many poor farmers and/or including those farmers for whom farming (and improvements in farm productivity via the use of CSA) would do little to enable them to escape poverty.

That project brought to the fore the need to recognize more explicitly the heterogeneity of farm households and the need to broaden the portfolio of livelihood options available to them. A further challenge was to accommodate the understandable desires of the donor to see an impact on the ground. There was pressure to scale CSA, in terms of enhanced farmer uptake of technologies and practices. Implicitly, this served to dismiss emerging evidence that the role of non-agriculture-based livelihoods needed to be taken into account in decision-making regarding appropriate interventions; the promotion of livelihood improvement through CSA was inappropriate for some categories of farmer. There was a danger that the focus on numbers would distract from whether farmer uptake of CSA, while contributing to an improvement in food security, would still leave farmers trapped in poverty, not to mention the potential for other unanticipated impacts, such as when wealthier farmers are able to capture the benefits of CSA, with the consequence that their wealth grows at the expense of poorer farmers, leading, ultimately, to greater social inequality.

The project in Guatemala clearly demonstrates the importance of priority-setting and factoring in the varied possibilities and local conditions that farmers face when it comes to targeting project interventions. Thornton et al. [5] provide a useful framework for CSA priority-setting that is based on six elements, and is designed to help guide best-bet CSA intervention. There is widespread recognition of the trade-offs when implementing CSA among the three pillars of food security, adaptation and mitigation [5,6]. The example of the Western Highlands illustrates "higher-level" trade-offs between some of the SDGs. These include trade-offs between SDG 13: *Climate Action* and SDG 5: *Gender Equality* together with SDG 10: *Reduced Inequality*.

In short, the Guatemalan project illustrates that a focus on the number of farmers adopting CSA can divert attention from the far more important issue, which is to support farmers' adaptation to climate change, either through making their agriculture-based systems more climate-resilient and/or by expanding their envelope of prosperity-enhancing non-agricultural livelihoods. The latter has been less prevalent in CSA interventions, and this has been at the expense of potentially ensnaring poorer categories of small-scale farmers in an agricultural-based poverty trap.

3. Climate-Smart Agriculture and Poverty Reduction

Farm households can be distinguished based on their asset endowment, e.g., their amount of land, access to key agricultural inputs etc., coupled with characteristics that determine the livelihood strategies available to them. These livelihood strategies, in turn, influence the livelihood incomes that hopefully enable a household to maintain and strengthen its livelihood security. The livelihood pathways available to a farm household are determined by the household's characteristics (e.g., dependency ratio, availability of labour, etc.), along with the interaction between the available assets (financial, natural, social, human) and the enabling or disabling economic, institutional and policy environment. An understanding of these livelihood pathways informs decisions as to where to target CSA and where to develop enabling approaches that facilitate livelihood changes.

There is no doubt that CSA and agricultural interventions can contribute to poverty reduction and enhanced prosperity. Numerous examples abound, e.g., [20–22]. However, the agricultural future is bleak for some farmers struggling with few resources and the additional challenge of climate change. Harris and Orr [23] argue that for rain-fed agriculture, crop production could be a pathway from poverty where smallholders are able to increase farm size or where markets stimulate crop diversification, commercialization and increased farm profitability. The potential to improve productivity is also, of course, important. Nevertheless, as Cavanagh et al. [24] comment, *"the poor and less poor are [...] more capable of diversifying into off-farm and non-farm activities compared to the very poor, whose small land holdings and poor access to capital constrain their ability to diversify away from on-farm income and seasonal off-farm wage labour"*. We certainly found this to be the case in the Western Highlands of Guatemala [13].

Agriculture is not a pathway out of poverty for all farm households. Hence, for certain categories of household, poverty reduction will come from farmers moving out of agriculture and into the non-farm economy. For poor households, non-agricultural livelihood transformation can, of course, represent nothing more than a negative coping strategy. The challenge is to ensure that non-agricultural livelihood options are positive, i.e., prosperity-enhancing. It is a challenge in all parts of the world due to profound rural changes. In many parts of the world, agricultural production will have to increase hugely, along with labour productivity; the latter will lead to fewer people engaging in agriculture [25]. This has already led, in parts of Asia, to what Li [26] refers to as a rural population that is "surplus" to the needs of capital, as many of those dispossessed from their land are also unable to find meaningful employment off-farm. It is also increasingly common in Latin America.

The idea of a "surplus" population mirrors the earlier thesis of "functional dualism", proposed by de Janvry et al. [27] and expanded on by Blaikie [28]. The authors suggest that farmers rely upon returns from market activities to complement their agricultural returns from farming plots of land that are too small to allow for self-sufficiency. Farmers are often obliged to work as part-time wage laborers due to their resulting food insecurity, thus needing to make up shortfalls of staples and cash requirements for household goods, as well as to pay for inputs for the production process itself on their farms. They are increasingly dependent on non-farm sources of income but are unable to find sufficient employment opportunities or capital to migrate (and abandon the agricultural sector) or to depend fully on wage earnings for their subsistence. Returns from subsistence-oriented agricultural activities provide a necessary complement to the low wages that farmers receive in the labour market. In addition, where opportunities for wage labour are primarily in the agricultural sector, poor returns from own-farm agricultural production are reinforced, given that peak demand for agricultural labour may coincide with labour demands on farmers' own land.

The situation in the Western Highlands of Guatemala, as described in the section above, is in keeping with the functional dualism thesis, and has major implications for identifying and targeting appropriate pathways leading to rural poverty reduction. As suggested, for many farmers in the Western Highlands, CSA may not be an attractive option because of labour and land shortages. In the case of many farmers in the Western Highlands of Guatemala, temporary migration in search of non-farm employment has been a traditional coping strategy, with farmers investing the earned off-farm income in their villages and/or diversifying into non-farm agricultural activities, such as setting up a local shop. For many farmers, labour, essential for investment in soil improvement or maintenance of conservation structures, is not available, because they are working off-farm (and households may also have high dependency ratios) [23]. Similarly, another refrain, when CSA practices such as conservation agriculture are promoted, is that farmers should not burn their fields to clear the vegetation prior to planting because of the adverse impact on soil quality, especially biological health. For farmers who have been working off-farm, and for whom labour is scarce, this can be an unattractive recommendation.

CSA is also very problematic when it comes to small-scale farmers with very small landholdings, as is the case in the Western Highlands. Firstly, in the case of cross-slope soil conservation technologies, such as live-vegetation barriers and stone walls, land is taken out of production. In the case of live barriers, however, this can be partly compensated for by using species that make a contribution to the farm household, e.g., edible products for humans and/or animals. Secondly, even if CSA were to lead to significant improvements in agricultural productivity, the increase (while a contribution to food security) would be unlikely to help the farmer escape poverty. *"For most smallholders, however, small farm size and limited access to markets mean that returns from improved technology are too small for crop production alone to lift them above the poverty line"* [23].

This raises the question of how best to support categories of farmers who are being targeted but whose small holdings, household structure and asset endowment may be inappropriate for the measures advocated under the guise of CSA. It may be the case that the CSV approach would be more successful, but in the absence of effective scaling of this approach and comprehensive impact studies, this remains an under-researched area.

4. Non-Agricultural Livelihood Transformation

In rural contexts where small-holder farmers are based, development involves decreasing livelihood vulnerability and increasing incomes, typically through changes in livelihood activities [29]. Ideally, CSA should enable farmers to pursue livelihood pathways that lead to greater prosperity, while also building resilience. Recent thinking has advocated addressing the need to support changes that can be transformative, in the face of climate-related impacts that imply dramatic changes to environmental conditions. There is much research on developing a framework for assessing and comparing different types of interventions that address the key elements of CSA [5,10]. This research is necessary and important; however, in the context of agricultural transformation, the focus needs to broaden to systematically factor in livelihood trajectories outside of agriculture. CSA, to enhance food security and meet the SDGs, will require a longer-term perspective and bolder action that comprehensively targets farmer livelihoods [5].

The reality is that positive, sustainable livelihood pathways within the agricultural sector may not be an option for all types of farmers, i.e., not all households face the agro-ecological and socio-economic conditions necessary to move from one asset threshold and livelihood pathway to another, enabling them to escape poverty, while still remaining in agriculture. Guatemala epitomizes this reality. A report produced for the United States Agency for International Development (USAID) noted that *"given the agricultural foundation and 'capital' that many Western Highland communities continue to hold, [there is a need to] re-assess the productive options available in agriculture or agriculture-based livelihoods; and engage youth (many of whom have written agriculture off as an option) in development of potential integrated economic/environmental/social development initiatives"* [18].

Incorporating non-agricultural livelihood transformation within CSA requires innovative and open thinking on the way forward for CSA. This has been acknowledged by proponents of CSA, e.g., [3,4], but it poses disciplinary challenges and has not led to the type of holistic and transformative changes are that needed. What is required is a broader and more comprehensive understanding of the realities faced by farmers and the changes needed to foster large-scale transformation in their livelihood trajectories [30]. This means that CSA thinking has to involve those from a plethora of disciplines from the natural and social sciences [31]. In the context of CSA, we have a practical example of how transformation also becomes a political issue [32]. The debate around adaptation to environmental change often avoids questioning the socio-economic and political reasons why farmers' livelihoods are so vulnerable [33].

In the context of Guatemala, serious discussion around climate change adaptation, mitigation and transformation will have to contend with politically divisive issues. In a small way, within a project, "politics" can mean challenging the premises that drive inappropriate scaling. Within the bigger picture, it also means taking into account the political economies within which CSA is advocated for small-scale farmers. In the Guatemalan context, this political economy relates to several decades of conflict and on-going socio-economic inequality that structurally disadvantage small-scale farmers in the Western Highlands. While there are, of course, specificities to the political economy of Guatemala, which have shaped its uptake of CSA, in any given context there will be political economy issues underpinning the implementation of CSA that cannot and should not be ignored. The climate change discourse has tended to focus on the adaptation and mitigation of greenhouse gas emissions, rather than *"problems of unevenly distributed power relations, networks of control and influence, and rampant injustices of the 'system"* [34].

Clearly, a disregard for issues of power and inequality is not tenable if CSA is to provide a viable mechanism for livelihood transformation and a contribution to the SDGs. Indeed, there is growing evidence of CSA proponents adopting a more "radical" agenda, factoring in more readily political and institutional issues and ensuring that CSA debate and implementation does not remain largely a discourse among "elite development and research agencies" [35]. Such recognition of the political realities of small-holder agricultural development are important if CSA is to have continued longevity and relevance within international agendas on climate change action and the SDGs.

5. Conclusions

Climate adaptation requires transformative change. The CSA approach needs to move even more squarely beyond a focus on resilience of food systems to encompass systematic thinking and action with respect to the resilience of farm households. This poses a real challenge, because CSA has tended to overlook targeting issues related to socio-economic differentiation within small-scale farming populations, although recent CSA initiatives have more readily included analyses of the institutional dimensions of CSA. Greater acknowledgement of institutional issues, and indeed the politics, of CSA interventions within rural planning are to be welcomed. CSA has nevertheless, in practice, tended to exclude systematic consideration of support for non-agricultural livelihood transformation that is positive for farm households in marginal contexts, such as the Western Highlands of Guatemala.

In some cases, CSA can lead to the triple win of increased productivity, adaptation and mitigation, but this is not the case for all types of farmers. We argue that more systematic attention be directed at climate risk management that moves beyond the more conventional adaptation and mitigation discourse, towards an approach that includes livelihood transformation from a broader perspective, i.e., one that does not just focus on rural–agricultural transformation, but also identifies (and embraces) where agriculture per se is not a pathway out of poverty and where support for positive non-agricultural livelihood trajectories are needed for small-scale farmers. This requires more disciplines working together, and, perhaps, meeting the challenge of addressing entrenched power balances, both within communities of scientists and in the small-scale farming populations that are the subject of CSA interventions.

Author Contributions: Conceptualisation, J.H. and E.F.; writing—original draft preparation, J.H. and E.F.; writing—review and editing, J.H. and E.F.

Acknowledgments: The authors would also like to thank two anonymous reviewers who provided invaluable comments on earlier versions of this paper.

References

1. Vermeulen, S.J.; Challinor, A.J.; Thornton, P.K.; Campbell, B.M.; Eriyagama, N.; Vervoort, J.M.; Kinyangi, J.; Jarvis, A.; Läderach, P.; Ramirez-Villegas, J.; et al. Addressing uncertainty in adaptation planning for agriculture. *Proc. Natl. Acad. Sci. USA* **2013**, *110*, 8357–8362. [CrossRef]
2. Adger, W.N.; Huq, S.; Brown, K.; Conway, D.; Hulme, M. Adaptation to climate change in the developing world. *Prog. Dev. Stud.* **2003**, *3*, 179–195. [CrossRef]
3. Lipper, L.; Thornton, P.; Campbell, B.M.; Baedeker, T.; Braimoh, A.; Bwalya, M.; Caron, P.; Cattaneo, A.; Garrity, D.; Henry, K.; et al. Climate-smart agriculture for food security. *Nat. Clim. Chang.* **2014**, *4*, 1068–1072. [CrossRef]
4. Aggarwal, P.; Jarvis, A.; Campbell, B.; Zougmoré, R.; Khatri-chhetri, A.; Vermeulen, S.; Loboguerrero, A.M.; Sebastian, S.; Kinyangi, J.; Bonilla-Findji, O.; et al. The climate-smart village approach: Framework of an

integrative strategy. *Ecol. Soc.* **2018**, *23*, 15. [CrossRef]

5. Thornton, P.K.; Friedmann, M.; Kilcline, K.; Keating, B.; Nangia, V.; West, P.C.; Howden, M.; Cairns, J.; Baethgen, W.; Claessens, L.; et al. A framework for priority-setting in climate smart agriculture research. *Agric. Syst.* **2018**, *167*, 161–175. [CrossRef]

6. Totin, E.; Segnon, A.C.; Schut, M.; Affognon, H.; Zougmoré, R.B.; Rosenstock, T.; Thornton, P.K. Institutional perspectives of climate-smart agriculture: A systematic literature review. *Sustainability* **2018**, *10*, 1990. [CrossRef]

7. Hellin, J.; Fisher, E. Building pathways out of poverty through climate smart agriculture and effective targeting. *Dev. Pract.* **2018**, *28*, 974–979. [CrossRef]

8. Lobell, D.B.; Burke, M.B.; Tebaldi, C.; Mastrandrea, M.D.; Falcon, W.P.; Naylor, R.L. Prioritizing Climate Change Adaptation Needs for Food Security in 2030. *Science* **2008**, *319*, 607–610. [CrossRef]

9. Partey, S.T.; Zougmoré, R.B.; Ouédraogo, M.; Campbell, B.M. Developing climate-smart agriculture to face climate variability in West Africa: Challenges and lessons learnt. *J. Clean. Prod.* **2018**, *187*, 285–295. [CrossRef]

10. Hellin, J.; Haigh, M.J. Better land husbandry in Honduras: Towards the new paradigm in conserving soil, water and productivity. *Land Degrad. Dev.* **2002**, *13*, 233–250. [CrossRef]

11. Hellin, J.; López Ridaura, S. Soil and water conservation on Central American hillsides: If more technologies is the answer, what is the question? *AIMS Agric. Food* **2016**, *1*, 194–207. [CrossRef]

12. Chambers, R.; Pacey, A.; Thrupp, L. (Eds.) *Farmer First: Farmer Innovation and Agricultural Research*; Intermediate Technology Publications: London, UK, 1989.

13. Hellin, J.; Cox, R.; López-Ridaura, S. Maize Diversity, Market Access, and Poverty Reduction in the Western Highlands of Guatemala. *Mt. Res. Dev.* **2017**, *37*, 188–197. [CrossRef]

14. Steinberg, M.; Taylor, M. Guatemala's Altos de Chiantla: Changes on the high frontier. *Mt. Res. Dev.* **2008**, *28*, 255–262. [CrossRef]

15. Isakson, S.R. The agrarian question, food sovereignty, and the on-farm conservation of agrobiodiversity in the Guatemalan highlands. *J. Peasant Stud.* **2009**, *36*, 725–759. [CrossRef]

16. Hamilton, S.; Fischer, E.F. Non-traditional agricultural exports in highland Guatemala: Understandings of Risk and Perceptions of Change. *Lat. Am. Res. Rev.* **2003**, *38*, 82–110. [CrossRef]

17. Copeland, N. 'Guatemala Will Never Change': Radical Pessimism and the Politics of Personal Interest in the Western Highlands. *J. Lat. Am. Stud.* **2011**, *43*, 485–515. [CrossRef]

18. Democracy International. *Legacies of Exclusion: Social Conflict and Violence in Communities and Homes in Guatemala's Western Highlands*; Democracy International: Bethesda, MD, USA, 2015; ISBN 3019611660.

19. Hellin, J.; Ratner, B.D.; Meinzen-Dick, R.; Lopez-Ridaura, S. Increasing social-ecological resilience within small-scale agriculture in conflict-affected Guatemala. *Ecol. Soc.* **2018**, *23*, 5. [CrossRef]

20. Kassie, M.; Shiferaw, B.; Muricho, G. Agricultural Technology, Crop Income, and Poverty Alleviation in Uganda. *World Dev.* **2011**, *39*, 1784–1795. [CrossRef]

21. Verkaart, S.; Munyua, B.G.; Mausch, K.; Michler, J.D. Welfare impacts of improved chickpea adoption: A pathway for rural development in Ethiopia? *Food Policy* **2017**, *66*, 50–61. [CrossRef]

22. Dinesh, D.; Frid-Nielsen, S.; Norman, J.; Mutamba, M.; Loboguerrero Rodriguez, A.; Campbell, B. *Is Climate-Smart Agricultre Effective? A Review of Case Studies*; CCAFS Working Paper; CCAFS: Wageningen, The Netherlands, 2015.

23. Harris, D.; Orr, A. Is rainfed agriculture really a pathway from poverty? *Agric. Syst.* **2014**, *123*, 84–96. [CrossRef]

24. Cavanagh, C.J.; Chemarum, A.K.; Vedeld, P.O.; Petursson, J.G. Old wine, new bottles? Investigating the differential adoption of 'climate-smart' agricultural practices in western Kenya. *J. Rural Stud.* **2017**, *56*, 114–123. [CrossRef]

25. Collier, P.; Dercon, S. African Agriculture in 50 Years: Smallholders in a Rapidly Changing World? *World Dev.* **2014**, *63*, 92–101. [CrossRef]

26. Li, T.M. To Make Live or Let Die? Rural Dispossession and the Protection of Surplus Populations. *Antipode* **2009**, *41*, 66–93. [CrossRef]

27. de Janvry, A.; Sadoulet, E.; Young, L.W. Land and labour in Latin American agriculture from the 1950s to the 1980s. *J. Peasant Stud.* **1989**, *16*, 396–424. [CrossRef]

28. Blaikie, P. Explanation and policy in land degradation and rehabilitation for developing countries. *Land Degrad. Dev.* **1989**, *1*, 23–37. [CrossRef]

29. Dorward, A. Integrating Contested Aspirations, Processes and Policy: Development as Hanging In, Stepping

Up and Stepping Out. *Dev. Policy Rev.* **2009**, *27*, 131–146. [CrossRef]

30. O'Brien, K. Responding to environmental change: A new age for human geography? *Prog. Hum. Geogr.* **2010**, *35*, 542–549. [CrossRef]

31. Reid, W.V.; Chen, D.; Goldfarb, L.; Hackmann, H.; Lee, Y.T.; Mokhele, K.; Ostrom, E.; Raivio, K.; Rockström, J.; Schellnhuber, H.J.; et al. Environment and development. Earth system science for global sustainability: Grand challenges. *Science* **2010**, *330*, 916–917. [CrossRef]

32. Castree, N. Geography and Global Change Science: Relationships Necessary, Absent, and Possible. *Geogr. Res.* **2015**, *53*, 1–15. [CrossRef]

33. O'Brien, K. From adaptation to deliberate transformation. *Prog. Hum. Geogr.* **2012**, *36*, 667–676. [CrossRef]

34. O'Brien, K. Global environmental change III: Closing the gap between knowledge and action. *Prog. Hum. Geogr.* **2013**, *37*, 587–596. [CrossRef]

35. Chandra, A.; McNamara, K.E.; Dargusch, P. Climate-smart agriculture: Perspectives and framings. *Clim. Policy* **2018**, *18*, 526–541. [CrossRef]

6

Diamondback Moth, *Plutella xylostella* (L.) in Southern Africa: Research Trends, Challenges and Insights on Sustainable Management Options

Honest Machekano [1], Brighton M. Mvumi [2] and Casper Nyamukondiwa [1,*]

[1] Department of Biological and Biotechnological Sciences, Botswana International University of Science and Technology, P. Bag 16, Palapye, Gaborone 0267, Botswana; honest.machekano@studentmail.biust.ac.bw
[2] Department of Soil Science and Agricultural Engineering, Faculty of Agriculture, University of Zimbabwe, P.O. Box MP 167, Mt. Pleasant, 00263 Harare, Zimbabwe; mvumibm@agric.uz.ac.zw
* Correspondence: nyamukondiwac@biust.ac.bw

Academic Editor: Suren N. Kulshreshtha

Abstract: The diamondback moth (DBM), *Plutella xylostella*, is a global economic pest of brassicas whose pest status has been exacerbated by climate change and variability. Southern African small-scale farmers are battling to cope with increasing pressure from the pest due to limited exposure to sustainable control options. The current paper critically analysed literature with a climate change and sustainability lens. The results show that research in Southern Africa (SA) remains largely constrained despite the region's long acquaintance with the insect pest. Dependency on broad-spectrum insecticides, the absence of insecticide resistance management strategies, climate change, little research attention, poor regional research collaboration and coordination, and lack of clear policy support frameworks, are the core limitations to effective DBM management. Advances in Integrated Pest Management (IPM) technologies and climate-smart agriculture (CSA) techniques for sustainable pest management have not benefitted small-scale horticultural farmers despite the farmers' high vulnerability to crop losses due to pest attack. IPM adoption was mainly limited by lack of locally-developed packages, lack of stakeholders' concept appreciation, limited alternatives to chemical control, knowledge paucity on biocontrol, climate mismatch between biocontrol agents' origin and release sites, and poor research expertise and funding. We discuss these challenges in light of climate change and variability impacts on small-scale farmers in SA and recommend climate-smart, holistic, and sustainable homegrown IPM options propelled through IPM-Farmer Field School approaches for widespread and sustainable adoption.

Keywords: small-scale farmers; pest management; brassicas; farmer-extension-researcher networking; insecticide misuse

1. Introduction

Brassica vegetables, like cabbage (*Brassica oleracea* var. *capitata*) and cauliflower (*B. oleracea* var. *botrytis*), and open leaf kales, like rape (*Brassica napus*) and covo (*Brassica carinata*), are the popular staple relish and most widely grown leafy vegetables in the tropical and subtropical regions of Southern Africa (SA), cutting across a wide range of cultures and agro-ecologies [1–5]. These vegetables are grown throughout the year [6] and form the fastest growing agricultural subsector that contributes significantly to national and regional incomes [6,7]. With the persistent droughts, extreme temperatures, and flooding challenges faced in field crop production due to climate change [8,9], irrigable small vegetable plots remain comparatively reliable as an attractive source of food and income for rural households, who make up over 80% of the farming community in SA [6,10] and whose farming systems are more vulnerable to effects of climate change [9,11]. On the other hand, African urban areas face high food demand because of rapid rural to urban migration, which has grown from 53 million to 400 million between 1960 and 2010, with a potential to increase to 600 million by 2030 [11]. As a result, high unemployment and low per capita income in the highly populated urban areas have created an ever increasing demand for food [11]; hence the need for horticultural expansion in rural, urban, and peri-urban agriculture (UPA) to meet fresh vegetable food demand, supplement incomes, and meet nutritional needs [10–12].

Despite doubling as a household income generating enterprise, brassicas also serve as an important inexpensive source of vitamins and minerals [7,10]. Due to the simplicity with which they can be grown, numerous small-scale farmers make a living out of brassica production, relying on the proximal urban markets [10,12–14]. Similarly farmers distant from the city typically rely on alternative markets [14]. However South Africa still exports brassicas (especially cabbage) to some of its regional neighbours, including Zambia (0.2%), Mozambique (3.3%), Angola (3.4%), Namibia (5%), Swaziland (6.5%), Botswana (31.4%), and Lesotho (46.3%) [15], thus lending credence to the theory of high demand against a production deficit in the region. The global demand for organically produced vegetables [16] has also significantly opened new lucrative markets for these African economies with the potential to substantially increase their Gross Domestic Product (GDP) if the required quality standards are met. This, however, is challenged by the scourge of the diamondback moth (DBM), *Plutella xylostella* (L.) (Lepidoptera: Plutellidae), a cosmopolitan insect pest of brassicas [17,18].

The DBM is the major, ubiquitous, and year-round insect pest hindering the economic production of brassica crops in SA [17–19]. Small-scale farmers are facing difficulty coping with DBM damage-induced losses and management challenges [3,19–24]. The economic importance of DBM is derived from its exceptional pest status that originates from its genetic diversity, high and year-round abundance, high reproductive potential, high genetic elasticity, cosmopolitan distribution, multivoltinity, and continuous suppression of the pest's natural enemies by synthetic pesticides [5,18,25] and possible survival failures by efficient natural enemies in the pest's new invasion areas [26]. Global losses of leafy vegetables attributed to damage and control costs of DBM alone were estimated to be around US\$ 4–5 billion [27]. Partitioning crop losses in SA under small-scale farmer conditions has not been explored in detail. However, in Kenya, an estimated 31% loss has been reported [28]. If uncontrolled, losses of up to 100% are possible [5,29], as has been reported in Botswana ([30]; and from personal observation during fieldwork in 2014 and 2015. There is little knowledge on the actual loss data of brassicas due to DBM in SA countries. However, cases of abandoning brassicas and changing production timing (i.e., concentrating only on winter production) as a means of infestation avoidance have been widely recorded [5,23,25].

Temperature is a critical climatic factor, which influences insect biological activities such as survival, reproduction, growth, development, geographical distribution, and fitness [31–33]. An increase in temperature reduces the time taken to acquire the number of degree-days required to complete the *P. xylostella* life cycle, thus decreasing its generation time and increasing the number of generations per year [27,32–34]. An increase in average temperature with global change may imply reduced overwintering time or a total absence of diapause for some economic insects [32,33], with consequent implications on pest management and food security. In addition, global warming has the potential to impair the potential of *P. xylostella* biological control if an increase in temperature disrupts the life cycle synchronisation of the host and its parasitoids [34]. Recent modelling data has predicted a decrease in ecological niches for some insects with climate change [35], and, similarly, invertebrate biocontrol agents are not an exception. Previous work reported broad lethal temperature limits for adult DBMs [36–39]: the minimum body temperature that 0% of the moths could survive, known as lower lethal temperature (LLT$_0$), was $-16.5\ °C$. The maximum body temperature that 0% of the moths could survive, known as upper lethal temperature (ULT$_0$), was $42.6\ °C$. The minimum body temperature for 25% moth survival (LLT$_{25}$) was $-15.2\ °C$, while the maximum body temperature for 25% moth survival (ULT$_{25}$) was $41.8\ °C$ [37,38]. However thermal tolerance for its major parasitoids has not been fully studied [34,36,38]. Unless the thermal tolerance of the major parasitoids matches that of the host DBM and evidence is presented that these traits may have coevolved, parasitoid efficacy in the face of climate change may be compromised [26]. Without coevolution of thermal tolerance, DBM challenge may likely intensify due to conducive climatic conditions [37–41] that may stimulate increased pest activity (feeding, breeding, and migration) [26,38]. Therefore, without efficient control mechanisms, the DBM problem could continue to increase despite the intensive pesticide use, which to-date may have been short-term, ineffective, unsustainable and expensive [18].

In this paper, we review the status of DBM management in the context of practice in SA. Specifically, we examine the past and current DBM pest status, management practices by farmers, DBM research, and development linkages among member countries (or the lack thereof) in SA with special reference to small-scale farmers who are the most affected. We also analyse the perspectives of researchers, farmers, and agricultural extension agents regarding DBM management and identify challenges and principal areas that require cooperation. We propose research on sustainable climate-smart agriculture and the selection of compatible integrated pest management (IPM) components that provide effective management of the DBM under small-scale farmers in SA, in the context of current and projected climate change scenarios.

2. Horticulture and DBM in Southern Africa

Due to socio-economic challenges and high unemployment rates in SA [11], horticulture is fast transforming into an intensive production and high-income-generating enterprise [39,40]. However, despite large expansion in land committed to horticulture in the region, returns per unit land area are still minimal [41], mainly due to pest related losses and, in some cases, high production costs. In addition, small-scale brassica farming systems are dominated by low scale cultivation of non-rotated monocrops with heavy dependence on family labour and locally available inputs [4,13]. Due to this perception, the management of the DBM (and other pests) is an in-built farming practice based on prophylactic pesticide use with the intention to 'eliminate' rather than to 'manage' the pest; and therefore economic threshold levels based on insect pest monitoring and scouting are not observed [5,13,40].

Consumer perception is another driver to intensive pesticide use. Urban consumers are biased to aesthetically damage-free vegetables and their demand for such produce cannot be ignored as a driver to intensive insecticide use by the farmers [14,42]. For small-scale farmers, the market is typified by vegetable vendors under make-shift stalls in urban and peri-urban roadsides. These vendors are an important market link between small-scale farmers and the urban market as they not only determine the market price for different levels of pest damage but are also directly linked to consumers [2]. This vendor market, just like urban supermarkets, has the capacity to influence price and quality; it triggers the excessive use of pesticides as farmers compete to produce and supply shiny, damage-free, 'quality' brassicas to satisfy the 'market standards'. Research in SA, however, has not contextualized these and other market forces in the light of acceptable damage levels on leafy vegetables, especially with reference to DBM attack. Reports indicate that market rejection and strong legislative frameworks influence the chemical application behaviour of farmers [43], forcing them to change chemical use patterns as fear of market loss supersede concern for public health [44].

DBM damage substantially hinders production and marketing of brassicas in SA [17,18,25]. Farmers' perceptions and practices on the management of this pest in the region are not yet fully understood [39]. Research to date has been survey-based [5,20,22,40] and generalised on both insect pests and diseases for all horticultural crops. This approach generalised and limited the information that could be generated regarding a specific pest. One of the main features of climate change, amongst others, is the rise in global mean temperatures and prolonged hot weather conditions [9]. In SA, temperature is projected to increase by 1–3 °C by 2050 [45–47] and its effects are likely to be more pronounced in the drier tropics than the humid subtropics [8,9]. In laboratory experiments, DBM showed activity over a broad temperature range, measured as LLTs and ULTs [36,38]. This may mean that, under the currently projected climate change in SA, DBM pest status is likely to increase, exacerbating already failing management practices [18,25,48]. Field population peaks, determined by both pheromone trap catches and crop infestation scouting, were observed in the warmer austral summer [19,22,49]. Regardless of the population source, high temperatures were shown to hasten development and thus shorten life cycles in *P. xylostella* [36,37]. However, temperature may differentially affect organisms, such that different insect pests (hosts) and their associated natural enemies may develop at different rates and thus affect host-prey/parasitoids synchronisation [26,33,50,51]. Extreme temperatures eliminate natural enemies that are susceptible to very high/low temperatures, whereas divergence from thermal preferences also disrupts the temporal and spatial synchronization of host/parasitoid phenologies, resulting in a high risk of challenging pest (host) outbreaks [34,35]. An increase in atmospheric carbon dioxide levels associated with global climate change may also reduce the efficacy of biological control agents against DBM by precluding or reducing the production of plants' secondary metabolites, which are necessary for the recruitment of natural enemies as part of the plants' natural defence mechanisms [52,53]. This and the misalignment of host-natural enemy life cycles may affect the natural enemy's efficacy and thus jeopardise the future of biological control programs [26]. There is a scarcity of published literature on climate-related coevolution of DBM and its natural enemies for optimising the efficacy of biological control. Nevertheless, some researchers have recommended that IPM programmes aimed at improving efficacy under global climate change should develop resilient agro-ecosystems, which incorporate populations' evolutionary potential and buffers against climate change effects [26,51,54].

3. Why is Southern Africa Hard Hit by the DBM Scourge?

3.1. Vulnerability to Effects of Climate Change

Sub-Saharan Africa will continue to be the area most hard-hit by climate change effects, due to increased mean temperature and increased rainfall variability [9]. With a record of 0.5 °C regional temperature increase, [9,35,45,46] predicts a projected increase in temperature of 3–4 °C by 2080, reduced rainfall, and increased degree days, aridity index, and evapotranspiration gradients. These factors will increase stress on already debilitating horticultural ecosystems, especially pest management, through changed pest dynamics, spatio-temporal distribution and increased pressure. Insecticide resistance associated with high temperatures has been recorded in different species [55], including variations in *P. xylostella* susceptibility to some organophosphates [55,56]. Therefore, under current climatic projections in SA [9,38,57], it is highly unlikely that DBM populations will decline due to the physiological stress associated with high or low temperature scenarios [36]. Sub-Saharan Africa's majority of rural small-scale farmers remains at the core of food production, but their production ecosystems are the most prone to climate change effects [9,45,57,58]. Using a prediction model [45], between 8% and 22% field crop losses have already been reported in sub Saharan Africa.

3.2. Farmers' Behaviour and Insecticide Use

Details and comprehensive data on farmers' behaviour relating to pesticide usage on DBM in SA are lacking [18]. However, survey baseline results show that between 75% and 100% of farmers in SA totally rely on chemical insecticides (Table 1). By global standards, these farmers use the greatest variety of chemicals, highest application rates, and the highest application frequency [5,43]. Frequency of application ranges from once every three weeks to three times a week [5,20,43,59].

At any given time, brassica farmers possess at least two to six different insecticides [42] and up to five different insecticides have been mixed in a single sprayer tank without technical recommendation or manufacturer instructions [60]. This might result in unknown phytotoxicity and unwanted (and seldom known) chemical reactions into compounds, which are possibly more hazardous and persistent in the environment [61]. Such hazardous compounds, even when geographically concentrated in pattern, could create significant exposure to the environment and the public through non-occupational exposure, where individuals not directly involved with chemical use get exposed to the chemical hazards through a contaminated environment [61,62]. Magauzi et al. [63] and Macharia et al. [28] detailed pesticide-related illnesses in Zimbabwe and Kenya respectively, and it has been reported that various symptoms related to pesticide poisoning have significantly increased as most small-scale farmers misuse chemicals and do not use personal protective equipment (PPE) [64]. Moreover farmers tend to ignore or take for granted certain levels of illnesses from synthetic chemicals, which they feel do not warrant medical attention, as an expected normal part of farmwork [60]. Consequently, there is scant information on the details of health effects and costs related to pesticide exposure, as most cases go unreported [28,60,61]. However, Magauzi et al. [63] reported high organophosphate levels in young horticultural farmworkers' blood and also recorded 24.1% abnormal cholinesterase activity in 50% of the sprayers (occupational exposure) and 49% of workers entering previously sprayed fields (non-occupational exposure) in Zimbabwe. Khoza et al. [64] reported similar results with both organophosphates and organochlorines and further reported chronic illnesses that were often misdiagnosed and mistreated in health centres; possibly due to rampant pesticide incorrect use [65]. Similar results were also recently reported in Kenya [28], as supported by reports of high proportion of small-scale horticulture farmers using insecticides in Africa (Table 1).

Table 1. Proportion of small-scale horticultural farmers using synthetic insecticides in Southern Africa and other parts of Africa.

Country	Farmers Using Pesticides (%)	Reference
Southern Africa		
Mozambique	100	[5]
Botswana	98	[20]
Zimbabwe	No data	
Zambia	75	[40]
Malawi	75	[40]
Other selected African countries		
Tanzania	98	[60]
Cameron	90	[66]
Ghana	85	[67]
Kenya	No data	

Occupational exposure is exacerbated by inefficient chemical use by small-scale farmers [42,43,60]. This ranges from using inappropriate chemicals, incorrect dosages, and wrong application timing and targeting, to non-calibrated or poorly maintained and defective (often leaky) application equipment [1,42,68]. Mvumi et al. [65] also reported the first three problems on synthetic grain protectants in Zimbabwe. Leakages were observed to lead to about 29 mL of dermal exposure per person per hour [60], depending on leakage rates, which might be currently higher due to cheap and faulty spraying equipment from non-reputable manufacturers flooding the market. Other forms of inefficient use resulting in exposure include the choice of extremely hazardous chemicals (Class 1a and 1b by WHO standards) [1,20,40,42,61], the use of banned chemicals [42,60,61], applying chemicals using twig/leaf bunches or home-made grass brushes/brooms, making homemade 'insecticide cocktails', and tongue-testing to assess concentrations [43]. Due to economic challenges, farmers sometimes often procure pesticides from unlicensed and unscrupulous dealers, thus increasing the risk of exposure and the chances of fraud and adulteration [60,65]. Reports from Zimbabwe indicate a failure to adhere to safety withdrawal periods, presumably due to market pressure; inefficient chemical use (only 35%–50% of sprayed chemical reach the target organism) [67,68]; application of the wrong pesticides (e.g., fungicides on insects); and abuse associated with the need to clear last seasons' expired pesticides [13,67]. This uncontrolled misuse and overuse of insecticide was reported to have significantly contributed to the increased resistance and suppression of potential biological control agents [17,18,25,30].

3.3. Lack of Insecticide Resistance Research

The DBM has shown resistance to 91 active ingredients of agricultural chemicals worldwide, including 12 strains of *Bacillus thuringiensis* (*Bt*), between 1953 and 2014 [48,69–71]. Compared to other parts of the world, DBM insecticide resistance in SA is relatively low (see Figure 1). Farmers tend to rely on their personal observation of insecticide efficacy failures to detect resistance. Following resistance 'detection', farmers usually continue using the same active ingredients at higher frequency, higher dosages, or in cocktails with other 'powerful' chemicals, which exacerbates the situation [5,42,43,60]. Despite the widespread use of hard chemicals to combat DBM in SA, we have not found any published comprehensive study on DBM resistance to commercially registered pesticides in this region (Figure 1). Management options and extension recommendations have been based on reports from the relatively advanced economies (China, Brazil, India, Australia, Nicaragua, Pakistan and USA) (Figure 1). However, resistance is highly geographical and highly correlated to insect strain as regards chemical exposure history, hence 'foreign' recommendations may not be directly applicable to the spatially heterogenous nature of the SA small-scale farming communities.

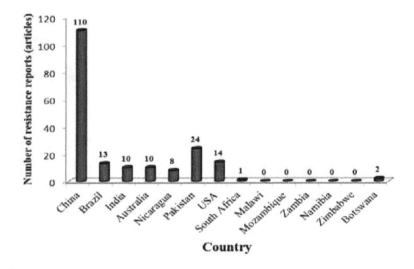

Figure 1. Selected country published reports on diamondback moth (DBM) insecticide resistance [48].

Consequently, farmers lack information on DBM resistance status in their respective localities to aid their pest control planning. This forces them to make their own, often ill-informed, decisions, mainly influenced by chemical manufacturers' advertisements, agro-dealers, and sometimes pesticide vendors ([40,43]; personal observation, 2014). Due to a lack of active pesticide control policies, farmers practice independent chemical choices and application (personal observation, 2014) without adequate consultation, resulting in 'dangerous' experimentations and haphazard chemical use with no regional or area-wide territorial regulations to aid Integrated Resistance Management [42,60]. In some cases, farmers smuggle 'effective' chemicals with noble modes of action from other countries into their home countries, where the chemicals have not yet been registered. Uncontrolled and inefficient use of these new pesticides results in early resistance development [54], which renders the modern pesticides ineffective by the time they are officially registered in the farmers' countries e.g., Hunter 500EC (Chlorfenapyr (pyrrole) 240 g/L) in Botswana (personal observation, 2014).

3.4. Low Research Attention

DBM research in SA is dominated by the public sector [2,24,25] where the agricultural ministries are custodians of agriculture and related work. In SA, countries with active DBM research are limited (see Figure 2). Conventionally research findings are delivered to the farmers through extension departments. This system is increasingly becoming inefficient due to declining public sector resources, the lack of farmer empowerment, and a lack of specialist staff in the sector [2,4]. The majority of SA research and development grants are funded externally [4,13], often coming with specific thematic areas that restrict researchers' flexibility. This may be a setback, as it limits scientists on tackling locally critical issues affecting small-scale farmers. This, coupled with low per capita funding and low capacity–building, exacerbated by 'brain-drain' to developed countries, limits research achievements [4,13]. Only a few SA countries can afford to keep specialist staff in the public sector, resulting in the disproportional distribution of research among SA member countries (Figure 2). South Africa and Kenya seem better off than the other countries, probably because of the presence of the Agricultural Research Council (ARC) and International Centre for Insect Physiology and Ecology (ICIPE), respectively, where DBM genetic, ecological, and IPM studies have mostly been conducted [23,28,71–76]. In South Africa, the ARC, in collaboration with industry and academic institutions, has conducted numerous studies on the DBM (Figure 3) on aspects including population dynamics, ecology, parasitism and predation, tritrophic interactions, and resistance breeding to *Bt* brassicas [21,71,72,76,77].

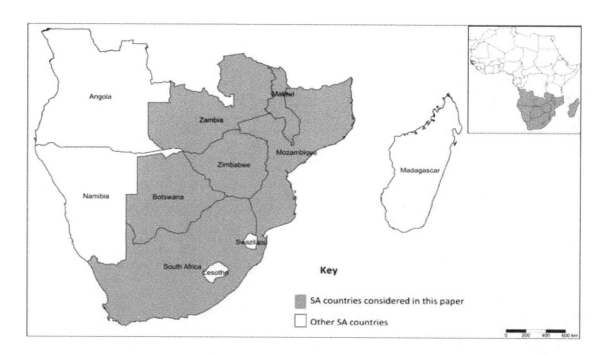

Figure 2. Countries in the Southern African region where DBM research has been conducted. This region appears in the high eco-climatic index of the world, where DBM persistence is year round and high [18]. * Namibia and Angola have very limited accessible research information on DBM, hence they were omitted from the map.

In contrast, the other SA countries have limited research on DBM, with Zimbabwe being the only country contributing just over 10% of DBM research. Most of the DBM research is conducted by incapacitated horticultural research institutions that are often poorly funded [24]. The bulk of the research was survey-based, covering general farmer practices, identification, and spatio-temporal distribution of the DBM natural enemies (Figure 4). These surveys brought about vast knowledge on DBM predation and parasitism rates in the region [5,19,22,77]. Crop systems approached through intercropping with mustard, *Brassica juncea* (L) (Czern); onion, *Allium cepa* (L.); and/or garlic, *Allium sativum* (L.) (also making 22% (Figure 4)), have been over studied and duplicated in many SA counties, due to a lack of research coordination and information sharing [12,78,79].

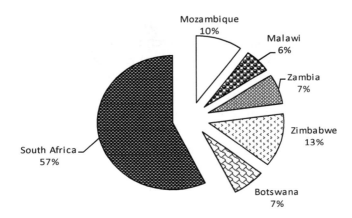

Figure 3. Proportion of publications on DBM in Southern African countries (1995–2015). (The data is based on physical counts of published papers and conference proceedings from respective countries).

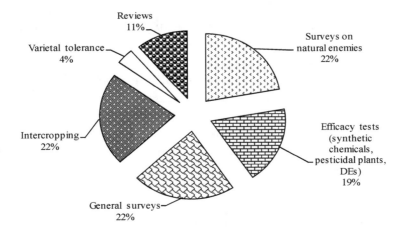

Figure 4. Proportion of published research articles on DBM in Southern Africa by theme (1995–2015). This is based on the physical checking of research themes for each of the publications in Figure 3).

Although SA has a long history of brassica production and an equally long acquaintance with DBM [25], research on its management seems to have started only about a decade ago with no data on the preceding years. Only recently, a synthetic pyrethroid (Cypermethrin), a *Bt* product (*B. thuringiensis* (*var. kurstaki*)), was tested for efficacy against DBM in Southern Africa, specifically Botswana [71,72]. Though this may be an important first step towards generating knowledge on DBM response to insecticides in the region, it needs to be expanded through testing area-specific populations for detailed territorial resistance profiling in all horticultural hotspots to aid planning on area-wide resistance management. Area-specific resistance assays may be critical in determining the susceptibility of DBM strains to current and future insecticides in different high production areas.

3.5. Lack of Regional Coordination in DBM Research

In 1984, the Southern African Development Community (SADC) (known as SADCC then) mooted and commissioned the Southern African Centre for Cooperation in Agricultural Research (SACCAR) for coordinated agricultural research in SA, which was partly funded by the Asian Vegetable Research and Development Centre (AVRDC) in the 1990s [80]. The AVRDC objectives in SACCAR were to coordinate vegetable research between and within SA member countries, develop novel vegetable postharvest preservation techniques, and, most importantly, develop an IPM program for the control of DBM in cruciferous vegetables for small-scale farmers [80]. Apart from coordinating regional research, SACCAR aimed to align agricultural research policies and priorities, identify constraints, promote cooperative research projects, strengthen national vegetable research centres, and encourage regional sharing and utilisation of scientific and technical information [80]. With headquarters in Botswana, SACCAR had sub-regional offices at reputable research institutions in Tanzania, Zimbabwe, Zambia, and Malawi in the early 1990s. However, as individual funding contributions from member states dwindled, independent donor organisations stepped in, diverting the organisation from its core mandates. To date, the organisation's activities are less visible on agriculture compared to the past, with high visibility on general economic constraints, labour-related issues, and the socio-economic welfares of selected member states. Thus, regional coordination and alignment of agricultural policies for concerted insect pest control efforts remain limited. However, there is hope in the recently formed Centre for Coordination of Agricultural Research and Development in Southern Africa (CCARDESA) (under SADC), which is targeting productivity and competitiveness of small-scale farmers across the region. The results of its activities are yet to be assessed.

The AVRDC, which is entirely committed to vegetable research, significantly sponsored regionally-coordinated research and capacitated national vegetable research centres in East Africa [2,4], but full expansion to SA was hampered by funding constrains [25,74]. Its major thrust was resistance breeding, farmer training, pest management, and general promotion of new technologies in SA that had proven successful in Asian nations [2,4]. To date, the results of its activities in the horticultural farming community in the region are certainly unclear, as is the case with SACCAR. The Asian Vegetable Network (AVNET), formed in 1989, successfully coordinated vegetable breeding and pest and disease control in Asia through the formation of strong dedicated sub-networks [81]. The advent of ICIPE in Kenya was an example of coordinated regional research in insect science, particularly in DBM crucifers [23]. Through this institution, Eastern Africa managed to conduct coordinated research aimed at DBM IPM [23,74,75]. ICIPE achieved DBM control in brassicas through the development and dissemination of biocontrol based IPM, using *Diadegma semiclausum* (Hellen) (Hymenoptera: Ichneumonidae) with complementary emphasis on a cropping systems approach [23]. Success was also achieved through a multidisciplinary approach, expert contributions, research funding, and supportive policies (national and regional) that enabled the granting of permissions to importations and releases of biocontrol agents [23]. Though this work did not effectively extend to SA, due to funding challenges (see discussion in [25]), the same model could be adopted in the DBM hard-hit SA region. Following the successful models of AVRDC and AVNET in Asia and ICIPE in East Africa, research networking may be a key mechanism for effective research aimed at achieving common goals for participating countries [23,81]. Southern Africa member states (Figure 2) can collaborate in the same manner for regionally consented efforts targeted at holistic DBM management. This networking is important to enable area-wide (regional) DBM management, as the pest's migration patterns and dispersal behaviour makes individual (farmer or country) methods ineffective [18,25]. A good example of this sub-regional collaboration is the recently-ended project aimed at combating the Asian fruit fly, *Bactrocera dorsalis* (Hendel), in Botswana, Namibia, Zambia, and Zimbabwe (BONAZAZI) under the technical assistance of FAO.

4. Possible Novel DBM Control Options

Climate-smart technologies aimed at maximizing production while promoting adaptation and mitigating the effects of changing environments are required [58]. IPM is a huge component of climate-smart agriculture, which, since the 1990s, has been generally agreed as the only sustainable and effective method of containing or managing economic pests, including the DBM [7,17,18,25,30,74]. Since synthetic chemicals offer short-time relief, several other management strategies have been investigated on a wide range of brassica agroecosystems, but IPM remains the most viable option [18,25,82]. IPM is that method of pest management that utilises all available and compatible techniques of pest management to reduce pest populations and maintain them below the crop economic injury levels [80,82]. The concept is aimed at eliminating the reliance on a single method of pest management in order to achieve better control and reduce or prevent development of pest resistance to a particular method [82]. This includes, but is not limited to, seasonal cropping (synchronised cropping calendar to minimize host plant availability), crop rotation, intercropping with non-host plants, enabling conducive environments for biological control agents, legislative plant host control (dead periods), the use of resistant varieties, and the judicious and minimal use (e.g., spot application) of environmentally benign insecticides (see [82]), which are applied only when absolutely necessary. In this system, insecticides from different chemical groups may also be rotated following legislation-enforced programs implemented and monitored by plant protection departments. Without legislative enforcement, synthetic insecticides continue to be used without due diligence despite widespread IPM awareness worldwide [43]. This is a practice that has caused deleterious consequences on DBM natural enemies including the reduction of their abundance and reduced efficacy in IPM systems [19,22,43,73].

Southern Africa is rich in natural enemies for DBM biological control [5,19,22,73]; therefore, the ecological consequence of widespread insecticide use, especially on these biological control agents, is a major concern [83,84]. Hence, a form of IPM aimed at reducing pesticide use and the promotion of selective soft insecticides (e.g., Pirimicarb, Pymetrozine and Spinosad (see Figure 5)) as its central tenets is the most crucial step in reducing the pesticide burden on the environment [84]. As explained earlier, biological control agents are currently dwindling due to intensive broad-spectrum chemical pesticide use and there is a danger that some of the natural enemy species may be completely lost unrecorded [83,84]. Therefore, unless the overreliance and unrestrained use of synthetic insecticides is significantly reduced, IPM and biological control measures in SA will continue to be hampered.

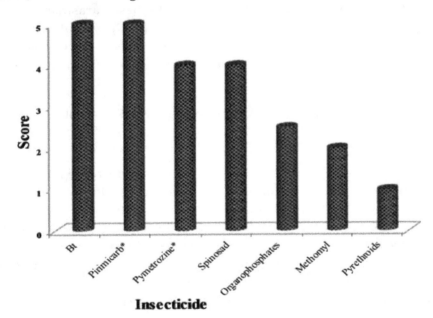

Figure 5. Common soft insecticides with high efficacy on the DBM and a low effect on its natural enemies (*Trichogramma* sp., *Cotesia* sp., spiders, lacewings and damsel bugs [84]). (Score: 5 = lowest effect on natural enemies, 1 = highest effect). *Insecticides not readily available on the market in Southern Africa (SA).

According to Walsh [84], *Bt*-based insecticides and Pirimicarb are the softest pesticides on DMB natural enemies, followed by Pymetrozine and Spinosad (Figure 5). Organophosphates, methomyl (a carbamate), and synthetic pyrethroids have high negative effects, particularly on *Trichograma* sp. and *Cotesia* sp., the most abundant and efficacious parasitoid species in SA [73,84,85]. Ironically survey results from SA, particularly Botswana [20], Zimbabwe [22], Malawi, Zambia [40], and Mozambique [5], show that synthetic pyrethroids, organophosphates, and carbamates are among the most commonly used insecticides in vegetable production. However, since genetically modified (*Bt*) brassicas were not accepted in SA due to social and environmental concerns [25], one of the sustainable management options is the rotation of *Bt*-based insecticides.

DBM is highly host-specific [86] (except in one observation of its survival on sugar-snap, *Pisum sativum* var. *macrocarpon* and snow peas, *P. sativum* var. *saccharatum* in Kenya [87]). Generally, moths do not oviposit on non-host plants; their host acceptance and oviposition is associated with a complicated integrated suite of chemical and physical cues [86]. Therefore, where soft insecticides are utilised, crop systems approached through the modification of agro-ecosystems and cropping practices can also be manipulated to confuse the adults' host finding techniques [88]. Research has shown partial DBM repellence success of cabbage intercrops with alliums through confusion in the chemical cues [12,78,79]. In such intercrops, natural enemies were shown to disperse and parasitise DBM at similar rates as in monocrops [88], evidence of compatibility between natural enemies and a cropping systems approach. There is potential in further improving this concept into a 'push-pull' cropping system technology by selecting appropriate repellent and attractant crops. Parasitoids are known to have originated from intricate mechanisms and are more efficient in heterogeneous than homogeneous landscapes. Push-pull intercropping that simultaneously improves habitat heterogeneity, conserves biodiversity by reducing hard chemical use, and improves refugia and nectar sources would improve parasitoid survival and efficiency [89]. This concept integrates climate-smart technologies as it utilizes ecosystem services for improved crop yields and quality.

Mass rearing and augmentative release of *Cotesia vestalis* (Haliday) (Hymenoptera: Braconidae) can be used to complement the conservation of existing faunal guilds [89] through the use of softer insecticides, as previously explained. Among the diverse range of DBM parasitoids, *C. vestalis* is the most widely distributed in SA [5,22,73,89], with the highest parasitism rates [5,21,22,73] and the only one tolerating the hot and arid tropical climates [5,17] typical of SA. The use of DBM entomopathogens naturally occurring in SA environments is also a novel possibility; for example, using *Metarhizium anisopliae* (Metchnikoff) [10] and a variety of other fungal microbes (as discussed in [25]).

5. Constraints to IPM Implementation and Adoption of Novel Sustainable Control Methods

5.1. Poor Understanding of the IPM Concept and Information Flow among DBM Management Actors

Currently, despite IPM being common, there is no evident decrease in pesticide usage even in areas where the concept is favourably viewed [83]. Farmers tend to adopt IPM based more on personal commitment level or influence by peers, rather than on recommendations from agricultural extension officers or researchers [52,82,83]. In Malaysia, [90] observed very little change in farming systems over a decade, particularly the use of synthetic insecticides despite widespread IPM campaigns. This can be attributed in part to lack of documented systematic IPM methodology or commercially prepared IPM packages with step-by-step instructions on how to use them [59,91]. Intensive research for a locally developed IPM system with simplified methodology, and inexpensive and accessible materials is therefore essential.

The major constraints to IPM adoption include a lack of awareness and knowledge [2,40], both of which are driven by the weak links and poor networking among the key players (Figure 6). Each player in the production and marketing chain has a crucial role to play; researchers develop the technology, extension officers transfer the technology to the end user (farmers), policy-makers create an enabling environment, and the agrochemical industry supplies the inputs (Figure 6). Vendors, supermarket chains, and horticultural export agents are key actors on the market and should be considered as part of the chain. Journalists, high profile multi-media agricultural reporters, and national broadcasters need to understand the principles of IPM for positive reporting to avoid misrepresentation of facts. The conceptual framework (Figure 6) shows that currently strong links (solid arrows) only exist between policy-makers and the agrochemical industry; researchers and funding agencies; and the agrochemical industry and media, all of which affect the farmers. Policy-makers have weak links with researchers, farmers, and the markets. The media also has weak links with researchers and extension agents, while having strong links with the agrochemical industry, explaining why horticultural programs on national broadcasters are currently dominated by product advertisements rather than IPM knowledge packages. It is therefore hypothesized that improving direct links between policy-makers and the markets, as well as the farmers, through pesticide residue limit assessment and enforcement, coupled with the development of knowledge packages that can also be passed through media and extension (Figure 6), would improve IPM adoption and reduce reliance on chemical pesticide usage. This would also improve consumer and worker safety against pesticide exposure. Knowledge packages may include case studies of successful local IPM programmes in vernacular languages to enable farmers to appreciate and fully understand the techniques, the principles, and the benefits of the IPM technology.

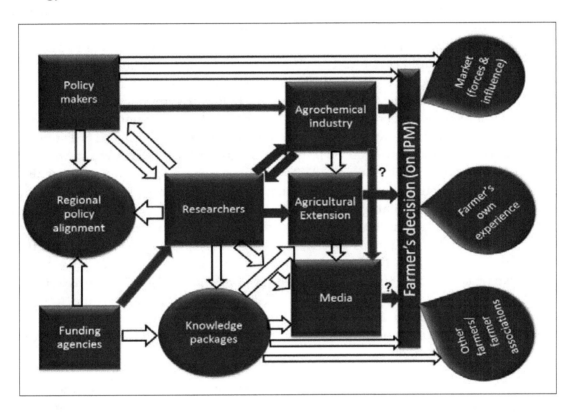

Figure 6. Perceived conceptual framework of links and information flow in DBM research in Southern Africa. The currently existing framework (solid arrows), the proposed framework (blank arrows), new suggested structures (circles), and links with both positive and negative influence on farmers' decision making (**?**) (Authors' own construction).

The introduction of IPM technology requires initial intensive training of the extension agents so that they cascade accurate and up-to-date information to farmers. IPM is complex process as it involves multiple components [82] and researchers often overestimate and equate their understanding of the concept with those of the extension agents, who also overestimate farmers' understanding [82]. In addition, donation of free agrochemicals by governments or donors and disproportional advertisements by the agrochemical industry, or a combination of such, does not only impede farmers' freedom of insect pest control options but also reduces their flexibility in decision-making [68,82]. Domination of synthetic pesticide research, manufacturing, and advertisements by the agrochemical industry, often in collaboration with academic researchers, coupled with lack of funding for research on non-chemical options, has further driven most agricultural extension agents and, subsequently, farmers to believe that the use of chemical pesticides is modern in agriculture [24,42,43,82]. This then overshadows the advances made in non-chemical pest control research, making farmers consider synthetic chemicals rather than non-chemical options as modern and first line of defence in DBM control [82].

Non-chemical control options, or a combination of such (in an IPM programme), are still largely considered as primitive due to a lack of understanding [40]. In most SA countries, extension work is dominated by the distribution of farming inputs (mainly fertilizers, seed, chemicals, etc.), with synthetic pesticides often being part of the package to the farmers [82]. This is also exacerbated by the farmers' high concern for access to inputs and the priority placed on these inputs [20] rather than the desire for knowledge or the use of non-chemical pest control methods [92]. This leaves little room for delivery of IPM knowledge packages through various training channels without input incentives. Requisite knowledge delivery to farmers is thus not valued as it should be, though it is key to understanding the concepts behind technologies enabling farmers to assess their risk and value for money invested, in order to make informed and independent adoption decisions [40,90,91]. Researchers and extension agents alike underestimate the amount of knowledge and information needed to convince small-scale farmers to adopt new technologies [91]. The latter's knowledge has not been able to keep pace with rapid agricultural technological changes, especially the dynamic DBM pest severity and management needs that continue to evolve in brassica production agro-ecosystems [42,91]. Increased knowledge has been proven to correlate with better pest management behaviour [43,75]. An understanding of the science behind building this knowledge in farmers is lacking among most extension agents in SA [86,91]. Knowledge is a dynamic system of cognition and is a sum of what has been learned, experienced, and perceived [91]. It involves observation, fact, and interpretative theory requiring intensive farmer participation [86,91,93]. As researchers and extension agents are more often providing information than knowledge, farmers' behaviour is unlikely to change under current scenarios [91,93]. Currently information is presented to farmers in a broad-spectrum format [91], but this has resulted in low uptake of technologies, as evidenced by low adoption. Information presented as such is often perceived by farmers as external rhetoric, associated with extension staff messages outside their farming systems [91,93]. This is so because most small-scale farmers in SA are risk-averse and unwilling to partake in voluntary schemes without immediate tangible incentives to which they are traditionally accustomed [82,91]. Until this mindset is changed through imparting knowledge and skills rather than information, for example through participatory IPM, the Farmer Field School (FFS) approach, or participatory action research/learning, co-learning and co-innovation approaches; adoption may remain a challenge.

There is a need to improve farmers' environmental knowledge base first, before the principles and practices of IPM can be emphasised [91,93]. For sustainability, the IPM packages need to be developed from local resources to avoid the constraints associated with external inputs and reduce strain on natural resources. For example, through development of participatory IPM in FFS, farmers may need to be trained in tritrophic interactions (plant-pest-natural enemy), pesticide toxicity, and its ecological consequences using farmer-tailored IPM curricula and approaches [90]. To foster positive attitudes towards IPM and improve its eventual implementation and adoption, there is need for awareness campaigns along the whole chain of stakeholders, alongside regular farmer trainings. As part of the reinforcement, it may also be beneficial if governments could feed eco-toxicological data into national pesticide registration policies to improve the adoption of IPM through the enforced use of softer and safer insecticides [43,62].

5.2. Weak Links between the Players in the Agroindustry

Parastatals, non-governmental organisations (NGOs), public national and international research institutions, independent researchers, private companies, and universities are not linked in a synergistic coordinated network, resulting in individual researchers and/or institutions independently presenting different technologies to farmers [2], sometimes with conflicting messages being conveyed. Sometimes host farmers may entertain a couple of researchers whose objectives are contradictory (personal observation, 2014), creating confusion and lack of trust among the farmers and extension agents alike [2]. The activities of the private sector, particularly the agrochemical industry, are scarce in literature. However, they are key to the procurement and distribution of chemicals and have a strong direct link with the farmers, which can be harnessed to propel other pest management options. Hence, there is need for collaboration of all stakeholders doing similar research and development work to fine-tune the broad-spectrum recommendations to specific relevant practices that enhance the fusion of the *emic* (inner perspective of the farmer) and the *etic* (outer perspective of the research/extension) [91,93]. Unfortunately, such platforms are rare.

5.3. Lack of Locally-Developed Well-Packaged IPM Practices and Procedures

The introduction of IPM should touch on various technical and social interventions [82]. The technical aspects mainly involve the techniques that farmers need to use to implement IPM in their brassica production systems. The development of step-by-step IPM methodology for cabbages in Asia through AVRDC and AVNET was key in the implementation and success of IPM in that region [82]. However, this has not been the case for SA. Direct adoption of Asian methodology may not necessarily apply in Africa due to different biophysical conditions, farmer practices, and socio-economic perceptions and circumstances [91]. Consequently, a SA IPM methodology tailored to specific local needs must be developed using participatory approaches to get farmers' buy-in. Furthermore, IPM monitoring tools to determine DBM economic threshold levels need to be scientifically investigated [29]. Local scientists and institutions have not developed IPM programs with regulatory and territorial chemical use boundaries for area-wide IRM, hence they still 'encourage' the use of any new chemicals [42,44]. These technical aspects also need to be locally refined and packaged within the small-scale farmer's contextual framework before a full IPM package can be presented and adapted for dissemination.

5.4. Lack of Policy Support

In Asia, the success of IPM programs was partly attributed to the crafting of enabling policies [80]. These included country agreements and harmonised policies to enable collaborative research, information-sharing, and the importation of natural enemies for key regional horticultural pests [79]. However, to the best of our knowledge, such enabling policies may still be lacking in SA. Global politics, as regards chemical use controls, is such that toxic pesticides are first banned in developed nations with effective regulatory and legislative policies. As regulations tighten in these countries, chemical manufacturing is reduced and the burden is passed on to developing countries by relocating factories and establishing subsidiaries in poor countries with governments that do not have effective regulatory controls [59,86,94]. In SA, this results in the uncontrolled use of extremely hazardous compounds, even years after they have been banned [5,20,61,94]. Some of the banned pesticides include DDT (only limited to mosquito control), Chlordane, Monochrotophos, Dieldrin, and Arsenic [94–96]. Therefore, this calls for strong technical and legislative capacitation of SA governments on issues of pesticide harmonized regulation and financial resources needed to develop and implement such legislations [42,81,82,94]. A classic example was Zimbabwe's successful development and implementation of a within-season pesticide rotational scheme and a closed season for the cotton bollworm *Helicoverpa armigera* (Hübner), achieved after a few years of strong legislative enforcement [24]. In SA, brassica farmers independently decide on the type of pesticide to buy, where to buy it, and when to apply it without any enforced regulation or legislation to consider. Though some general chemical regulatory frameworks may exist on paper, implementation is still a challenge in the region. Since brassicas are produced all-year-round, this promotes all-year-round unrestrained insecticide use on fresh vegetables that are supplied for public consumption, most of which are sometimes eaten raw. We therefore recommend a strong policy regulatory framework to control, minimize, and synchronize chemical use across all major horticultural production areas and markets. Mechanisms to implement and monitor the policy may also need to be clearly laid out right from the outset.

In some developing middle-income countries (e.g., Malaysia), threshold levels of pesticide residues permissible in crop products are well-laid out and monitored at different levels of the market value chain [88]. The lack of such policies in SA and the subsequent lack of regulatory frameworks account for the high pesticide residues in fresh products [2,42]. This is exacerbated by the cosmetic urban consumers' unconscious demand for damage-free brassicas [2,39,42]. The chronic nature of accumulated pesticide effect in humans makes the danger 'invisible' [63,64]. Though implementing residue-monitoring systems through the whole production and supply chain may prove logistically and financially infeasible for SA governments, the development of policy, legislation, and relevant monitoring tools may allow government officials to implement checkpoint systems across the vegetable production and supply chains.

Due to DBM notoriety and economic importance, we suggest that it may be necessary for SA governments to declare it a pest of regional economic importance, warranting policy recognition and consented governments' intervention, as is the case for tsetse flies, *Glossina morsitans* (Wiedemann); invasive fruit flies *B. dorsalis*; the larger grain borer, *Prostephanus truncatus* (Horn); migratory insect pests like the African armyworm, *Spodoptera exempta* (Walker); and African migratory locusts *Locusta migratoria migratorioides* (Fairmaire & Reiche). We recommend regional policy synchronisation, collaborative research, public awareness, farmer training, and IPM through Farmer Field School (IPM-FFS) initiatives in the management of *P. xylostella* synonymous with efforts applied to these other economic pests.

5.5. Taxonomic Confusion and Insufficient Adaptation of Biocontrol Agents to Release-Sites Climate and Bio-Ecological Conditions

The introduction of efficacious natural enemies has been marred by parasitoid taxonomic confusion and misidentification [18,97,98]. We have not found any reports of SA field-sourced parasitoid populations reared for mass release in DBM biological control programs in the region. *Diadegma semiclausum* and *C. vestalis* are currently the most common and efficacious DBM parasitoids in Africa [17,18,97,98]. *Diadegma semiclausum*, used for east African biocontrol programs, was once misidentified and exported as *D. mollipla* [99]. Due to misidentifications, release populations for the DBM control were imported from Taiwan, regardless of its local abundance in the horticultural hot-spots of Kenyan Eastern Highlands [74,98]. *Cotesia vestalis* is the most abundant and most efficient DBM parasitoid in SA [5,19,22,74,89], but some literature still refer to it as *Cotesia plutellae* (Kurdjumov) [18]. However, currently, *C. vestalis* populations from different climates are lumped together and considered as one species, despite observed biological differences [98]. Thus, molecular methods that can reliably separate biologically distinct but morphologically identical populations may be useful tools that can reliably confirm species' identity and hence improve the success of future biological control programmes [18,99].

Climate mismatching between parasitoid source areas and target release sites has led to the failure of most foreign reared but African released natural enemies [26,85]. This has now been exacerbated by unpredictably variable weather, increasing temperatures and fluctuating humidity caused by global climate change [9]. For DBM control, climate mismatching has previously been reported as a major setback for most biological control attempts [85]. Under the circumstances, climate matching between source area and target release site becomes an integral component of biological control programs based on parasitoid mass releases. This can only be achieved by a careful study of the thermal biology of the target parasitoid species. Mass introductions may then be targeted through acclimation, to suit areas of release [26,99,100]. Indeed, previous studies have recommended that thermal acclimation can significantly improve the fitness of laboratory reared insects upon introduction to wild conditions [100–103], and this approach has even been recommended for field releases using Sterile Insect Technique (SIT) [26]. It has been documented that biological control using predators and parasitoids should aim at developing resilient agro-ecosystems which maintain species' evolutionary potential to improve efficacy. This may be done through direct improvement in natural enemy genetic diversity and processes that encourage continuous in situ evolutionary adaptation [54].

5.6. Limited Alternative Control Options

In SA, the use of *Bt* transgenic brassicas have so far only been done in South Africa [96]. However, due to socio-political and controversial environment-related risks, it is yet to be commercialised in other SA countries [86]. Field and market observations from 2014 to 2016 showed that *Bt*-based insecticides are slowly filtering into the regional market. For example, pioneer *B. thuringiensis* (*var. kurstaki*) bioassays in Botswana showed 85.7%–94.6% reduction in DBM damage [71,72], but market availability still remains a challenge. However, this efficacy may also be short-lived, due to resistance development [69] as the insecticide is applied without a technical insecticide rotating scheme [68].

SIT has been successfully used for DBM management in Myanmmar [104], but its implementation in SA requires huge capital investment and substantial financial backup in addition to specialised human resources [105]. Furthermore, SIT is only effective in an area-wide approach, which may be challenging due to scattered distribution of small-scale subsistence farmers in SA.

Similarly, genetic engineering, through the release of insects carrying a male-selecting transgene, has equally managed to suppress DBM populations through the prevention of female progeny survival [106]. The same technique has been used successfully to control the fruit flies, *B. oleae* [107] and *Ceratitis capitata* (Wiedemann) [108], and the mosquito, *Aedes aegypti* (L.) [109]. However, this has not yet been considered in SA, probably due to the controversy surrounding genetically modified organisms. Nevertheless, it is an option worth considering in future DBM management programmes.

Strategies for developing varietal resistance in brassicas against DBM have not yet been fully exploited [82]. Modification of biochemical and morphological plant characteristics has also been unsuccessful [25,83]. Thus, despite its potential as an alternative non-chemical DBM control method, resistant variety development is still a huge challenge to biochemists and plant breeders [82] in SA. We have not found any research identifying chemical compounds or genes that are necessary to manipulate and cause brassicas to be completely non-preferred hosts for DBM [82].

6. Future Prospects and Research Needs

Future prospects in the sustainable management of DBM in SA lie in two principles, as outlined by [110].

(1)   'Do no harm'—the use of biologically- and environmentally-safe pest control methods with no or selective soft and safe insecticide use.
(2)   'Do good'—Improving farmer knowledge, consumer, agrochemical industry, and policy-maker awareness; policy reforms and regional policy harmonization; strengthening regulatory frameworks; and national and regional institutional capacitation.

Based on these principles, future research needs to identify and develop IPM-compatible components for the sustainable management of the DBM applicable to small-scale farmer circumstances in SA. Complementary to this, baseline information on the spatio-temporal population dynamics of DBM in relation to climatic parameters is needed. This may assist the area-specific determination of current population and pest management trends and how they correlate with environmental factors. This knowledge is important for the identification of gaps where development of new or improvement in existing IPM interventions is needed. Modelling the population and climate data will also assist in the development of predictive models that can be used in early warning systems to prepare farmers for possible outbreaks.

Farmer and extension staff capacity development can be achieved through participatory research using IPM in the FFS approach, on-farm farmer-managed, and researcher-managed trials. This will not only connect scientific findings with farmer's traditional knowledge and experience but will develop sustainable farmer-to-farmer knowledge-sharing platforms [58]. This also promotes co-learning, co-innovation and ownership of findings amongst all stakeholders which are essential ingredients for adoption. Success in technology adoption in various areas of agriculture was achieved in Asia through FFSs [44,91]. Farmer behavioural change may be possible through training, mass media awareness, legislative enforcement, and market condemnation of plant products exceeding set thresholds of pesticide residues [42,43,90]. Therefore, regular pesticide residue analysis may provide convincing evidence for governments to enforce regulations on chemical use on fresh vegetables. Where the regulations do not exist, they should be developed.

The future abundance and efficacy of *C. vestalis* under climate change remains uncertain. In addition, the synchrony and co-evolutionary adaptation between the host and the parasitoid also remains unpredictable [26]. Therefore, comparative abiotic stress tolerance studies of both the host and the parasitoid will provide insights into the needs for improvement of environmental fitness and efficiency of the potential parasitoids as an integral component of future IPM designs. Climate change was observed to impact insects negatively on the timing of life history traits, geographical shifts in species ranges, and the alteration of ecosystem interactions [55]. In addition, there are high-predicted rates of extinction in some species [111]. Therefore, comparative abiotic stress tolerance studies will not only enable the determination of whether it is the host or the parasitoid that is at high risk of extinction due to the impact of climate change, but will also be necessary to improve parasitoid field fitness for future release programs. Insect thermal biology and the ability to predict the impact of climate change on insect species are some of the most noble research findings of our time, yet adequate utilisation of this knowledge to improve pest management is still lacking, especially in Africa [35,55]. In the case of DBM, IPM systems need to have climate-resilient parasitoids capable of absorbing the 'shock' associated with the environmental changes due to global warming as a critical component of a broader climate-resilient IPM-FFS pest management systems approach [26,55].

Non-chemical control of *P. xylostella* may also be achieved through the manipulation of insect–host interactions. This may be achieved through brassica varietal resistance breeding and/or modification of the habitat by careful intercropping with attractant and repellent crops. As DBM larvae is generally monophagous, a varietal resistance option is promising if given full attention [6,82]. Research has also shown that *P. xylostella* moths do not oviposit on non-hosts. This means habitat management through agro-ecosystem manipulation may be an effective strategy to incorporate in IPM systems [81]. In light of the current knowledge, mere intercropping without careful selection of the repellent and attractant crops to enhance a 'push-pull' effect has not been very effective [12,78,79]. The 'push-pull' technology has been used for the successful management of cereal stem borers in eastern Africa [112,113]. This technology may be expanded, improved, and applied to economic pests such as the DBM in SA. For sustainability and cost-effectiveness, this technology may need to be geographically flexible in repellent crop selection to enable farmers to choose repellent crops naturally occurring and readily available in their localities. However, initial investment may be needed to conduct farmer participatory field research in the initial selection of potential candidate repellent and attractant crops.

7. Conclusions

SA is facing a serious DBM challenge and efforts towards its management are characterised by a variety of constraints. These vary from farmers' behaviour regarding insecticide choice and its use and/or misuse, a lack of health and environmental consciousness, a lack of locally-developed alternative control methods, a lack of regulatory enforcements, weak policy frameworks, and low research attention that is neither regionally-coordinated nor aligned for the achievement of common goals, all exacerbated by climate change and variability. The future of sustainable DBM control lies in IPM-FFS holistic approaches that include territorial IRM, cropping systems approaches (push-pull intercrops), soft and selective insecticides, area-wide pest management, biological control, the use of entomopathogens, and varietal resistance breeding developed in an IPM package. This should be supported by farmer and extension staff training as the founding principle of the approach to enhance in-depth knowledge and understanding of the IPM concepts, principles and procedures in a changing climate. There is also a need for exploring institutional or structural transformations to facilitate effective information flow and collaboration, sustainable uptake of IPM packages for improved crop protection systems, especially with respect to DBM and overall sustainable development.

Acknowledgments: The authors would like to acknowledge financial and technical support from Botswana International University of Science and Technology (BIUST) and technical support from the University of Zimbabwe (UZ). Many thanks to several anonymous referees, and the Eco-Physiological Entomology Research Team (BIUST) for the comments on an earlier version of the manuscript.

Author Contributions: All authors contributed equally to the manuscript. All authors have read and approved the final manuscript.

References

1. Sibanda, T.; Dobson, H.M.; Cooper, J.F.; Manyangarirwa, W.; Chiimba, W. Pest management challenges for small-holder vegetable farmers in Zimbabwe. *Crop Prot.* **2000**, *19*, 807–815. [CrossRef]

2. Saka, A.R.; Mtukuso, A.P.; Mbale, B.J.; Phiri, I.M.G. The role of research-extension-farmer linkages in vegetable production and development in Malawi. In *Vegetable Research and Development in Malawi. Review and Planning Workshop Proceedings, Lilongwe, Malawi, 23–24 September 2003*; Chadha, M.L., Oluoch, M.O., Saka, A.R., Mtukuso, A.P., Daudi, A., Eds.; World Vegetable Center (AVRDC): Shanhua, Taiwan, 2003.

3. Munthali, D.C. Evaluation of cabbage varieties to cabbage aphid. *Afr. Entomol.* **2009**, *17*, 1–7. [CrossRef]

4. Food and Agriculture Organization (FAO). *Evolving a Plant Breeding and Seed System in Sub-Saharan Africa in an Era of Donor Dependence*; FAO Plant Production and Protection Paper 210; FAO: Rome, Italy, 2011.

5. Canico, A.; Santos, L.; Massing, R. Development and adult longevity of diamondback moth and its parasitoids *Cotesia plutellae* and *Diadegma semiclausum* in uncontrolled conditions. *Afr. Crop Sci. Conf. Proc.* **2013**, *11*, 257–262.

6. Khonje, A.A. Research trends in horticultural crops in Malawi. *J. Crop Weed* **2013**, *9*, 13–25.

7. Ekesi, S.; Chabi-Olaye, A.; Subramanian, S.; Borgeimeister, C. Horticultural pest management and African Economy: Successes, Challenges and Opportunities in a changing global environment. *Acta Hortic.* **2009**, *911*, 165–183. [CrossRef]

8. Stathers, T.; Lamboll, R.; Mvumi, B.M. Postharvest agriculture in changing climate: Its importance to African smallholder farmers. *Food Secur.* **2013**, *5*, 361–392. [CrossRef]

9. Intergovernmental Panel on Climate Change (IPCC). *Climate Change 2014: Synthesis Report*; Fifth Assessment Report (AR5); Contribution of Working Groups I, II and III to the Fifth Assessment Report of the Intergovernmental Panel on Climate Change; IPCC: Geneva, Switzerland, 2014.

10. Maniania, N.K.; Takasu, K. Development of microbial control agents at the International Centre of Insect Physiology and Ecology. *Bull. Inst. Trop. Agric.* **2006**, *29*, 1–9.

11. Food and Agriculture Organization (FAO). *Growing Greener Cities in Africa*; First Status Report on Urban and Peri-Urban Horticulture in Africa; Food and Agriculture Organisation of the United Nations: Rome, Italy, 2009.

12. Katsaruware, R.D.; Dubiwa, M. Onion (*Allium cepa*) and garlic (*Allium sativum*) as pest control intercrops in cabbage based intercrop system in Zimbabwe. *J. Agric. Vet. Sci.* **2014**, *7*, 13–17.

13. Abate, T.; van Huis, A.; Ampofo, J.K.O. Pest Management Strategies in traditional agriculture: An African Perspective. *Ann. Rev. Entomol.* **2000**, *45*, 631–659. [CrossRef] [PubMed]

14. Momanyi, D.; Lagat, K.J.; Ayuya, O.I. Determinants of smallholder African indigenous leafy vegetables farmers' market participation behaviour in Nyamira County, Kenya. *J. Econ. Sustain.* **2015**, *16*, 212–217.

15. Department of Agriculture, Forestry and Fisheries. *A Profile of the South African Cabbage Market Value Chain*; Department of Agriculture, Forestry and Fisheries: Arcadia, South Africa, 2014.

16. Lubinga, M.; Ogundeji, A.; Jordaan, H. East African community trade potential and performance with European Union: A perspective of selected fruit and vegetable commodities. *ESJ* **2014**, *1*, 430–443.

17. Talekar, N.S.; Shelton, A.M. Biology, ecology and management of the diamondback moth. *Ann. Rev. Entomol.* **1993**, *38*, 275–301. [CrossRef]

18. Furlong, M.J.; Wright, D.J.; Dosdall, L.M. Diamondback moth ecology and management: Problems, progress and prospects. *Ann. Rev. Entomol.* **2013**, *58*, 517–554. [CrossRef] [PubMed]

19. Sithole, R. Life History Parameters of *Diadegma mollipla* (Holmgren), Competition with *Diadegma semiclausum* Hellen (Hymenoptera: Ichneumonidae) and Spatial and Temporal Distribution of the Host, *Plutella xylostella* (L.) and Its Indigenous Parasitoids in Zimbabwe. Ph.D. Thesis, University of Zimbabwe, Harare, Zimbabwe, 2005.

20. Obopile, M.; Munthali, D.C.; Matilo, B. Farmers' knowledge, perceptions and management of vegetable pests and diseases in Botswana. *Crop Prot.* **2008**, *27*, 1220–1224. [CrossRef]

21. Nofemela, R.; Kfir, R. The pest status of Diamondback moth (Lepidoptera: Plutellidae) in South Africa: The role of parasitoids in suppressing the pest populations. In Proceedings of the Fifth International Workshops on the Management of Diamondback Moth and other Crucifer Pests, Beijing, China, 21–24 October 2008.

22. Manyangarirwa, W.; Zehnder, G.W.; McCutcheon, G.S.; Smith, J.P.; Adler, P.H.; Mphuru, A.N. Parasitoids of the diamondback moth on brassicas in Zimbabwe. *Afr. Crop Sci. Conf. Proc.* **2009**, *9*, 565–570.

23. Nyambo, B.; Sevgan, S.; Chabi-Olaye, A.; Ekesi, S. Management of alien invasive insect pest species and diseases of fruits of vegetables: Experiences from East Africa. *Acta Hort.* **2009**, *911*. [CrossRef]

24. Tibugari, H.; Mandumbu, R.; Jowah, P.; Karavina, C. Farmer knowledge, attitude and practice on cotton (*Gossypium hirsutum* L.) pest resistance and management practices in Zimbabwe. *Arch. Phytopathol. Pflanzenschutz.* **2012**, *45*, 2395–2405. [CrossRef]

25. Gryzwacz, D.; Rosbach, A.; Rauf, A.; Russel, D.A.; Srivansan, R.; Shelton, A.M. Current control methods for diamondback moth and other brassica insect pests and the prospexcts for improved management with lepidopteran resistant *Bt* vegetable brassicas in Asia and Africa. *J. Crop Prot.* **2010**, *29*, 68–79. [CrossRef]

26. Chidawanyika, F.; Mudavanhu, P.; Nyamukondiwa, C. Biologically based methods for pest management under changing climates: Challenges and future directions. *Insects* **2012**, *3*, 1171–1189. [CrossRef] [PubMed]

27. Zalucki, M.P.; Shabbir, A.; Silva, R.; Adamson, D.; Shu-shen, L.; Furlong, M.J. Estimating the Economic Cost of One of the World's Major Insect Pests, *Plutella xylostella* (Lepidoptera: Plutellidae): Just How Long Is a Piece of String? *J. Econ. Entomol.* **2012**, *105*, 1115–1129. [CrossRef] [PubMed]

28. Macharia, I.; Mithofer, D.; Waibel, H. Health effects of pesticide use among vegetable farmers in Kenya. In Proceedings of the 4th International Conference of the African Association of Agricultural Economists, Hammamet, Tunisia, 22–25 September 2013.

29. Ayalew, G. Comparison of yield losses on cabbage from diamondback moth, *Plutella xylostella* L. (Lepidoptera: Plutellidae) using two insecticides. *Crop Prot.* **2006**, *25*, 915–919. [CrossRef]

30. Safraz, M.; Keddie, A.B.; Dosdall, L.M. Biological control of the diamondback moth, *Plutella xylostella*. A Review. *Biocontrol Sci. Technol.* **2005**, *15*, 763–789. [CrossRef]

31. Marchioro, C.A.; Foerster, L.A. Development and survival of diamondback moth, *Plutella xylostella* L. (Lepidoptera: Plutellidae) as a function of temperature: Effect on the number of generations in tropical and sub-tropical regions. *J. Neotrop. Entomol.* **2011**, *40*, 533–541.

32. Chown, S.L.; Nicolson, S. *Insect Physiological Ecology: Mechanisms and Patterns*; Oxford University Press: Oxford, UK, 2004.

33. Denlinger, D.L.; Lee, R.E., Jr. *Low Temperature Biology of Insects*; Cambridge University Press: Cambridge, UK, 2010.

34. Bahar, M.H.; Hegedus, D.; Soroka, J.; Coutu, C.; Bekkaoui, D. Survival and *Hsp70* gene expression in *Plutella xylostella* and its larval parasitoid *Diadegma insulare* varied between slowly ramping and abrupt extreme temperature regimes. *PLoS ONE* **2013**, *8*, e73901. [CrossRef] [PubMed]

35. Hill, M.P.; Bertelsmeier, C.; Clusella-Trullas, S.; Garnas, J.R.; Robertson, M.P.; Terblanche, J.S. Predicted decrease in global climate suitability masks regional complexity of invasive fruit fly species response to climate change. *Biol. Invasions* **2016**, *18*, 1105–1119. [CrossRef]

36. Andrew, N.R.; Hill, S.J.; Binns, M.; Bahar, M.H.; Ridley, E.V.; Jung, M.P.; Fyfe, C.; Yates, M.; Khusro, M. Assessing insect responses to climate change: What are we testing for? Where should we be heading? *Peer J.* **2013**, *1*, e11. [CrossRef] [PubMed]

37. Shirai, Y. Temperature tolerance of diamondback moth, *Plutella xylostella* (Lepidoptera: Yponomeutidae) in tropical and temperate regions of Asia. *Bull. Entomol. Res.* **2000**, *90*, 357–364. [CrossRef] [PubMed]

38. Nguyen, C.; Bahar, M.H.; Baker, G.; Andrew, N.R. Thermal tolerance limits of DBM in ramping and plunging assays. *PLoS ONE* **2014**, *9*, e87535. [CrossRef] [PubMed]

39. Kuntashula, E.; Silesh, G.; Mafongoya, P.L.; Bond, J. Farmer participatory evaluation of the potential for organic vegetable production in the wetlands of Zambia. *Outlook Agric.* **2006**, *35*, 299–305. [CrossRef]

40. Nyirenda, S.P.; Sileshi, G.W.; Belmain, S.R.; Kamanula, J.F.; Mvumi, B.M.; Sola, P.; Nyirenda, G.K.C.; Stevenson, P.C. Farmers' ethno-ecological knowledge of vegetable pests and pesticidal plant use in Northern Malawi and Eastern Zambia. *Afr. J. Agric. Res.* **2011**, *6*, 1525–1537.

41. Madisa, M.E.; Obopile, M.; Assefa, Y. Analysis of horticultural production trends in Botswana. *J. Plant Stud.* **2012**, *1*. [CrossRef]

42. Wandaat, E.Y.; Kugbe, J.X. Pesticide misuse in rural-urban agriculture: A case study of vegetable production in Tano South of Ghana. *AJAFS* **2015**, *3*, 343–360.

43. Williamson, S.; Ball, A.; Pretty, J. Trends in pesticide use and drivers for safer pest management in four African countries. *Crop Prot.* **2008**, *27*, 1327–1334. [CrossRef]

44. Timprasert, S.; Datta, A.; Ranamukhaarachchi, S.L. Factors determining adoption of integrated pest management by vegetable growers in Nakhon Ratchasima Province, Thailand. *Crop Prot.* **2014**, *62*, 32–39. [CrossRef]

45. Lobell, D.B.; Burke, M.B.; Tebaldi, C.; Mastrandrea, M.D.; Falcon, W.P.; Naylor, R.L. Prioritizing climate change adaptation needs for food security in 2030. *Science* **2008**, *319*, 607–610. [CrossRef] [PubMed]

46. Jack, C. *Climate projections for United Republic of Tanzania*; Climate Systems Analysis Group (CSAG)—University of Cape Town: Cape Town, South Africa, 2010.

47. Steynour, A.; Jack, C.; Taylor, A. *Information on Zimbabwe's Climate and How It Is Changing*; Climate Systems Analysis Group—University of Cape Town: Cape Town, South Africa, 2012.

48. Arthropod Pesticide Resistance Database (IRAC). Michigan State University. 2015. Available online: http://www.pesticideresistance.com/display.php?page=speciesarId=571 (accessed on 22 August 2015).

49. Mosiane, S.M.; Kfir, R.; Villet, M.H. Seasonal phenology of the diamondback moth, *Plutella xylostella* L. (Lepidoptera: Plutellidae) and its parasitoids on canola *Brassica napus* (L.) in Gauteng Province, South Africa. *Afr. Entomol.* **2010**, *11*, 277–285.

50. Dosdall, L.M.; Zalucki, M.P.; Tansey, J.A.; Furlong, M.J. Developmental responses of the diamondback moth parasitoid *Diadegma semiclausum* (Hellén) (Hymenoptera: Ichneumonidae) to temperature and host plant species. *Bull. Entomol. Res.* **2012**, *102*, 373–384. [CrossRef] [PubMed]

51. Sgrò, C.M.; Lowe, A.J.; Hoffmann, A.A. Building evolutionary resilience for conserving biodiversity under climate change. *Evol. Appl.* **2010**, *4*, 326–337. [CrossRef] [PubMed]

52. Kopper, B.J.; Lindroth, R. Effects of elevated carbon dioxide and ozone on the phytochemistry of aspen performance of an herbivore. *Oecologia* **2003**, *134*, 95–103. [CrossRef] [PubMed]

53. Sanders, N.T.; Belote, R.T.; Weltzin, K.F. Multitrophic effects of elevated carbon dioxide on understory plant and arthropod communities. *Environ. Entomol.* **2004**, *33*, 1609–1616. [CrossRef]

54. Gill, H.K.; Garg, H. Pesticide: Environmental impacts and management strategies. In *Pesticides—Toxic Effects*; Solenski, S., Larramenday, M.L., Eds.; Intech: Rijeka, Croatia, 2014; pp. 187–230.

55. Polson, K.A.; Brogdon, W.G.; Rawlins, S.C.; Chadee, D.D. Impact of environmental temperatures on resistance to organophosphate insecticides in Aedes aegypti from Trinidad. *Rev. Panam. Salud Publ.* **2012**, *32*, 1–8. [CrossRef]

56. Liu, F.; Miyata, T.; Wu, Z.J.; Li, C.W.; Wu, G.; Zhao, S.X.; Xie, L.H. Effects of temperature and fitness costs, insecticide susceptibility and heat shock protein in insecticide resistant and susceptible *Plutella xylostella*. *Pestic. Biochem. Physiol.* **2008**, *91*, 45–52. [CrossRef]

57. Metzger, M.J.; Bunce, R.G.H.; Trabucco, A.; Sayre, R.; Jangman, R.H.G.; Zomer, R.J. A high resolution bioclimate map for the world: A unifying framework for global biodiversity research and monitoring. *Glob. Ecol. Biogeogr.* **2013**, *22*, 630–638. [CrossRef]

58. Food and Agriculture Organization (FAO). *Identifying Opportunities for Climate-Smart Agriculture Investment in Africa*; Economics & Policy Innovations for Climate-Smart Agriculture, FAO: Rome, Italy, 2012.

59. Harvey, C.D. Integrated Pest Management in temperate horticulture: Seeing the wood for trees. *CAB Rev.* **2015**, *10*, 028. [CrossRef]

60. Ngowi, A.V.; Maeda, D.W.; Partanen, T.J. Knowledge, Attitudes and Practices (KAP) among agricultural extension workers concerning the reduction of the adverse impact in agricultural areas in Tanzania. *Crop Prot.* **2007**, *26*, 1617–1624. [CrossRef] [PubMed]

61. Baliga, S.S.; Repetto, R. *Pesticides and the Immune System: The Public Health Risks*; World Resources Institute: Washington, DC, USA, 1996.

62. Tsimbiri, P.F.; Moturi, W.N.; Sawe, J.; Henley, P.; Bend, J.R. Health impact of pesticides on residents and horticultural workers in the Lake Naivasha Region, Kenya. *Occup. Dis. Environ. Med.* **2015**, *3*, 24–34. [CrossRef]

63. Magauzi, R.; Mabaera, B.; Rusakaniko, S.; Chimusoro, A.; Ndlovu, N.; Tshimanga, M.; Shambira, G.; Chadambuka, A.; Gombe, N. Health effects of agrochemicals among farm workers in commercial farms of Kwekwe district, Zimbabwe. *Pan Afr. Med. J.* **2011**, *9*, 26. [CrossRef] [PubMed]

64. Khoza, S.; Nhachi, C.F.B.; Chikumo, O.; Murambiwa, W.; Ndudzo, A.; Bwakura, E.; Mhonda, M. Organophosphate and organochlorine poisoning in selected horticultural farms in Zimbabwe. *JASSA* **2003**, *9*, 7–15.

65. Mvumi, B.M.; Giga, D.P.; Chiuswa, D.V. The maize (*Zea mays* L.) post-production practices of smallholder farmers in Zimbabwe: Findings from surveys. *JASSA* **1995**, *1*, 115–130. [CrossRef]

66. Horna, D.; Falk-Zepeda, J.; Timpo, S.E. *Insecticide Use on Vegetables in Ghana. Would GM Seeds Benefit Farmers*; International Food Policy Research Institute (IFPRI), Environment and Production Technology Division: Accra, Ghana, 2008.

67. Abang, A.; Kouame, C.M.; Abang, M.M.; Hanna, R.; Kuate, A.F. Vegetable growers perception of pesticide use practices, cost and health effects in the tropical region of Cameroon. *Int. J. Agron. Plant Prod.* **2013**, *4*, 873–883.

68. Mudimu, G.D.; Waibel, H.; Fleischer, S. *Pesticide Policies in Zimbabwe: Status and Implications for Change*; Pesticide Policy Project; Special Issue Publication Series 1; Institute of Horticultural Economics: Hannover, Germany, 1999.

69. Tabashnik, B.E.; Malvar, T.; Liu, Y.B.; Finson, N.; Borthakur, D.; Shin, B.S.; Parck, S.H.; Masson, L.; Maard, R.A.; Bosch, D. Cross resistance of the diamondback moth indicates altered interactions with domain II of *Bacillus thuringiensis* Toxins. *J. Appl. Environ. Microbiol.* **1996**, *62*, 2839–2844.

70. Xia, Y.; Lu, Y.; Shen, J.; Gao, X.; Qiu, H.; Li, J. Resistance monitoring for eight insecticides in *Plutella xylostella* in central China. *Crop Prot.* **2014**, *63*, 131–137. [CrossRef]

71. Legwaila, M.M.; Munthali, D.C.; Kwerepe, B.C.; Obopile, M. Effectiveness of cypermethrin against diamondback moth (*Plutella xylostella* L.) eggs and larvae in cabbage under Botswana conditions. *Afr. J. Agric. Res.* **2014**, *9*, 3704–3710.

72. Legwaila, M.M.; Munthali, D.C.; Kwerepe, B.C.; Obopile, M. Efficacy of *Bacillus thuringiensis* (var. kurstaki) against diamondback moth *Plutella xylostella* (L.) Eggs and larvae on cabbage under semi controlled greenhouse conditions. *Int. J. Trop. Insect Sci.* **2015**, *7*, 39–45.

73. Kfir, R. Effect of parasitoid elimination on populations of diamondback moth in cabbage. In *The Management of Diamondback Moth and Other Crucifer Pests, Proceedings of the 4th International Workshop, Melbourne, Australia, 26–29 November 2001*; Endersby, N.M., Ridland, P.M., Eds.; The Regional Institute Ltd.: Gosford, Australia, 2004.

74. Löhr, B. Toward biocontrol based IPM for the diamondback moth in Eastern and Southern Africa. In *The Management of Diamondback Moth and Other Crucifer Pests, Proceedings of the 4th International Workshop. Melbourne, Australia, 26–29 November 2001*; Endersby, N.M., Ridland, P.M., Eds.; The Regional Institute Ltd.: Gosford, Australia, 2004; pp. 197–206.

75. Nyambo, B.; Löhr, B. The role and significance of farmer participation in biological control based IPM for brassica crops in East Africa. In Proceedings of the Second International Symposium on Biological Control of Arthropods, Davos, Switzerland, 12–16 September 2005.

76. Tonnang, N.E.Z.; Nodorezov, L.V.; Owino, O.; Ochanda, H.; Löhr, B. Evaluation of Discrete host-parasitoid models for diamondback moth and *Diadegma semiclausum* field time population densities. *Ecol. Model.* **2009**, *220*, 1735–1744. [CrossRef]

77. Bopape, M.J. The Management of Diamondback Moth, *Plutella xylostella* (L.) (Lepidoptera: Plutellidae), Population Density on Cabbage Using Chemical and Biological Control Methods. Master's Thesis, University of South Africa, Pretoria, South Africa, 2013.

78. Luchen, S.W.S. Effects of Intercropping Cabbage with Alliums and Tomato on the Incidences of Diamondback Moth *Plutella xylostella* (L.). Master's Thesis, University of Zambia, Lusaka, Zambia, 2001.

79. Karavina, C.; Mandumbu, R.; Zivenge, E.; Munetsi, T. Use of garlic Allium sativum as a repellent crop to control diamondback moth (*Plutella xylostella*) in cabbage (*Brassica oleracia* var. *capitata*). *J. Agric. Res.* **2014**, *52*, 615–622.

80. Opena, R.T.; Kyomo, M.L. Vegetable Research and Development in SADCC Countries. Available online: http://trove.nla.gov.au/work/7530097?selectedversion=NBD7890552 (accessed on 5 January 2017).

81. Green, S.K.; Shanmugasundaram, S. *AVRDC's International Networks to Deal with the Tomato Leaf Curl Disease: The Needs for Developing Countries*; Springer: Dordrecht, The Netherlands, 2007; pp. 417–439.

82. Maredia, M.K.; Dakomo, D.; Mota-Sanchez, D. *Integrated Pest Management in the Global Arena*; Commonwealth Agricultural Bureau International (CABI): Wallingford, UK, 2003.

83. Devine, G.J.; Furlong, M.J. Insecticide use: Contexts and Ecological Consequences. *Agric. Hum. Values* **2007**, *24*, 281–306. [CrossRef]

84. Walsh, B. *Impact of Insecticides on Natural Enemies in Brassica Vegetables*; Horticulture Australia Ltd.: Sydney, Australia, 2005.

85. Kfir, R. Biological control of the diamondback moth *Plutella xylostella* in Africa. In *Biological Control in IPM Systems in Africa*; Neuenschwander, P., Borgemeister, J., Langewald, J., Eds.; CABI Publishing: Wallingford, UK; Cambridge, MA, USA, 2003; pp. 363–376.

86. Safraz, M.; Dosdall, L.M.; Keddie, B.A. Diamondback moth host plant interactions: Implications for pest management. *Crop Prot.* **2006**, *25*, 625–639. [CrossRef]

87. Rossbach, A.; Löhr, B.; Vidal, S. Host shift to peas in the diamondback moth *Plutella xylostella* (Lepidoptera: Plutellidae) and response of its parasitoid *Diadegma mollipla* (Hymenoptera: Ichneumonidae). *Bull. Entomol. Res.* **2006**, *96*, 413–419. [CrossRef] [PubMed]

88. Silva-Torres, C.S.A.; Torres, J.B.; Barros, R. Can cruciferous agro-ecosystems grown under variable conditions influence biological control of *Plutella xylostella* (L.) (Lepidoptera: Plutellidae). *Biocontrol Sci. Technol.* **2011**, *21*, 625–641. [CrossRef]

89. Sohati, H.P. Establishment of *Cotesia vestalis* (Haliday) and *Diadromus collaris* (Grav.) Parasitoids of the Diamondback Moth *Plutella xylostella* (L.) and Assessment of the Effectiveness of *Cotesia vestalis* as a Biological Control Agent in Zambia. Ph.D. Thesis, University of Zambia, Lusaka, Zambia, 2012.

90. Mazlan, N.; Mumford, J. Insecticide use in cabbage pest management in Cameron highlands, Malaysia. *Crop Prot.* **2004**, *24*, 31–39. [CrossRef]

91. Price, L.L. Demystifying farmers' entomological and pest management knowledge: A methodology for assessing the impacts on knowledge from IPM-FFS and NES Interventions. *Agric. Hum. Values* **2001**, *18*, 153–176. [CrossRef]

92. Stadlinger, N.; Mmochi, A.J.; Dobo, S.; Glyllback, E.; Kumblad, L. Pesticide use among smallholder rice farmers in Tanzania. *Environ. Dev. Sustain.* **2011**, *13*, 641–656. [CrossRef]

93. Williamson, S. Understanding natural enemies: A review of training and information in the practical use of biological control. *Biocontrol News Inf.* **1998**, *19*, 117–126.

94. Hough, P. *The Global Politics of Pesticides: Forging Consensus from Conflicting Interests*; Earthscan Publications Ltd.: London, UK, 1998.

95. Association of Veterinary and Crop Associations of South Africa (AVCASA). *Hands off Banned Pesticides*; Media Statement; AVCASA: Midrand, South Africa, 2008.

96. Quinn, L.P.; de Vos, B.J.; Fernandes-Whaley, M.; Roos, C.; Bouwman, H.; Kylin, H.; Pieters, R.; van den Berg, J. Pesticide Use in South Africa: One of the Largest Importers of Pesticides in Africa. 2012. Available online: http://www.intechopen.com/books/pesticides-in-the-modern-world-pesticides-use-and-management/pesticide-use-in-south-africa-one-of-the-largest-importers-ofpesticides-in-africa (accessed on 5 January 2017).

97. Löhr, B.; Kfir, R. Diamond back in Africa: A review with emphasis on Biological Control. In *Improving Biocontrol of Plutella xylostella*; Kirk, A.A., Bordat, D., Eds.; Agricultural Research for Development: Montpellier, France, 2004.

98. Momanyi, C.; Löhr, B.; Gitonga, L. Biological Impact of the Exotic Parasitoid *Diadegma semiclausum* (Hellen) of diamondback moth *Plutella xylostella* (L.) in Kenya. *Biol. Control* **2006**, *38*, 254–263. [CrossRef]

99. Löhr, B.; Kfir, R. Diamondback moth *Plutella xylostella* (L.) in Africa: A review with emphasis on biological control. In Proceedings of the International Symposium, Montpellier, France, 21–24 October 2002.

100. Chidawanyika, F.; Terblanche, J.S. Costs and benefits of thermal acclimation for codling moth, *Cydia pomonella* (Lepidoptera: Tortricidae): Implications for pest control and the sterile insect release programme. *Evol. Appl.* **2011**, *4*, 534–544. [CrossRef] [PubMed]

101. Sørensen, J.; Addison, M.; Terblanche, J.S. Mass rearing of insects for pest management: Challenges, synergies and advances from evolutionary physiology. *Crop Prot.* **2012**, *38*, 87–94. [CrossRef]

102. Overgaard, J.; Sørensen, J.G. Rapid thermal adaptation during field temperature variations in *Drosophila melanogaster*. *Cryobiology* **2008**, *56*, 159–162. [CrossRef] [PubMed]

103. Kristensen, T.N.; Hoffmann, A.A.; Overgaard, J.; Sørensen, J.G.; Hallas, R. Costs and benefits of cold acclimation in field released *Drosophila*. *Proc. Natl. Acad. Sci. USA* **2008**, *105*, 216–221. [CrossRef] [PubMed]

104. Htun, P.W.; Myint, M. Radiation induced sterility for biological control of diamondback moth *Plutella xylostella* (L.). *Int. J. Adv. Sci. Eng. Technol.* **2014**, *4*, 285–291.

105. Dyck, V.A.; Hendrichs, J.; Robinson, A.S. *Sterile Insect Technique: Principles and Practice in Area-Wide Integrated Pest Management*; Dyck, V.A., Hendrichs, J., Robinson, A.S., Eds.; Springer: Dordrecht, The Netherlands, 2005.

106. Harvey-Samuel, T.; Morrison, N.I.; Walker, A.I.; Marubbi, T.; Yao, J.; Collins, H.L.; Gorman, K.; Davies, T.G.E.; Alphey, N.; Warner, S.; et al. Pest control and resistance management through release of insects carrying a male sterile transgene. *BMC Biol.* **2015**, *13*, 49–64. [CrossRef] [PubMed]

107. Ant, T.; Koukidou, M.; Rempoulakis, P.; Gong, H.F.; Economopoulos, A.; Vontas, J.; Alphey, L. Control of the olive fruit fly using genetics-enhanced sterile insect technique. *BMC Biol.* **2012**, *10*, 51. [CrossRef] [PubMed]

108. Leftwich, P.T.; Koukidou, M.; Rempoulakis, P.; Gong, H.F.; Zacharopoulou, A.; Fu, G.; Chapman, T.; Economopoulos, A.; Vontas, J.; Alphey, L. Genetic elimination of field-cage populations of Mediterranean fruit flies. *Proc. R. Soc. Biol. Sci.* **2014**, *281*, 1792. [CrossRef] [PubMed]

109. De Valdez, M.R.W.; Nimmo, D.; Betz, J.; Gong, H.F.; James, A.A.; Alphey, L. Genetic elimination of dengue vector mosquitoes. *Proc. Natl. Acad. Sci. USA* **2011**, *108*, 4772–4775. [CrossRef] [PubMed]

110. Van Veen, S. The worldbank and pest management. In *Integrated Pest Management in the Global Arena*; Maredia, K.M., Dakouo, D., Mota-Sanchez, D., Eds.; Commonwealth Agricultural Bureau International (CABI): Wallingford, UK, 2003.

111. Gillson, L.; Dawson, T.P.; Jack, S.; McGeoch, M.A. Accommodating climate change contingencies in conservation strategy. *Trends Ecol. Evol.* **2013**, *28*, 135–142. [CrossRef] [PubMed]

112. Cook, S.M.; Khan, Z.R.; Pickett, J.A. The use of push-pull strategies in Integrated Pest Management. *Ann. Rev. Entomol.* **2007**, *52*, 375–400. [CrossRef] [PubMed]

113. Khan, R.Z.; Midega, C.A.O.; Bruce, T.J.A.; Hooper, A.M.; Pickett, J.A. Exploiting phyto-chemicals for developing a "push-pull" crop protection strategy for cereal farmers in Africa. *J. Exp. Bot.* **2010**, *10*, 1–12.

7

Managing New Risks of and Opportunities for the Agricultural Development of West-African Floodplains: Hydroclimatic Conditions and Implications for Rice Production

Aymar Yaovi Bossa [1,2,*], Jean Hounkpè [1,2], Yacouba Yira [1,3], Georges Serpantié [4], Bruno Lidon [5], Jean Louis Fusillier [5], Luc Olivier Sintondji [2], Jérôme Ebagnerin Tondoh [6] and Bernd Diekkrüger [7]

[1] West African Science Service Centre on Climate Change and Adapted Land Use (WASCAL), Ouagadougou, Burkina Faso; hounkpe.j@wascal.org (J.H.); yira.y@wascal.org (Y.Y.)

[2] National Water Institute, University of Abomey Calavi, Cotonou, Benin; o_sintondji@yahoo.fr

[3] Applied Science and Technology Research Institute–IRSAT/CNRST, P.O. Box 7047, Ouagadougou, Burkina Faso

[4] Institute for Research and Development—IRD-UMR GRED-UPV, 34090 Montpellier, France; georges.serpantie@ird.fr

[5] Centre for International Cooperation in Agronomic Research for Development—CIRAD-UMR G-eau, 34090 Montpellier, France; bruno.lidon@cirad.fr (B.L.); jean-louis.fusillier@cirad.fr (J.L.F.)

[6] UFR des Sciences de la Nature, Université Nangui Abrogoua, 02 BP 801 Abidjan 02, Cote D'Ivoire; jetondoh@gmail.com

[7] Department of Geography, University of Bonn, Meckenheimer Allee 166, 53115 Bonn, Germany; b.diekkrueger@uni-bonn.de

* Correspondence: bossa.a@wascal.org

Abstract: High rainfall events and flash flooding are becoming more frequent, leading to severe damage to crop production and water infrastructure in Burkina Faso, Western Africa. Special attention must therefore be given to the design of water control structures to ensure their flexibility and sustainability in discharging floods, while avoiding overdrainage during dry spells. This study assesses the hydroclimatic risks and implications of floodplain climate-smart rice production in southwestern Burkina Faso in order to make informed decisions regarding floodplain development. Statistical methods (Mann-Kendall test, Sen's slope estimator, and frequency analysis) combined with rainfall—runoff modeling (HBV model) were used to analyze the hydroclimatic conditions of the study area. Moreover, the spatial and temporal water availability for crop growth was assessed for an innovative and participatory water management technique. From 1970 to 2013, an increasing delay in the onset of the rainy season (with a decreasing pre-humid season duration) occurred, causing difficulties in predicting the onset due to the high temporal variability of rainfall in the studied region. As a result, a warming trend was observed for the past 40 years, raising questions about its negative impact on very intensive rice cultivation packages. Farmers have both positive and negative consensual perceptions of climatic hazards. The analysis of the hydrological condition of the basin through the successfully calibrated and validated hydrological HBV model indicated no significant increase in water discharge. The sowing of rice from the 10th to 30th June has been identified as optimal in order to benefit from higher surface water flows, which can be used to irrigate and meet crop water requirements during the critical flowering and grain filling phases of rice growth. Furthermore, the installation of cofferdams to increase water levels would be potentially beneficial, subject to them not hindering channel drainage during peak flow.

Keywords: inland valley development; hydroclimatic hazard; water control structure; sustainable rice production

1. Introduction

The future of West Africa, and its economic, political, and social balance, depend on the ability of the agricultural sector to adapt and ensure food security under multiple pressures, such as climate change and demographic growth. In Africa, only 12.5 million hectares are irrigated out of a total of 202 million hectares of cultivated land, or 6.2%. The proportion of irrigated land in the south of Saharan Africa is even smaller, with only 5.2 million hectares, or 3.3%, of cultivated land being irrigated [1]. The increase in population will have serious implications in terms of agricultural production and the availability of natural resources. Adaptive strategies to help cope with the potential decrease in crop yields include promoting the extensive development of inland valleys in West Africa.

This is because of their great potential as rice-based production systems due to the high and secure water availability and soil fertility [2]. As such, the West African floodplains are privileged places for agricultural intensification, but play a diminishing role in the face of droughts that affect rainfed crops. Key factors of concern for the agricultural development of floodplains include flood hazards, surface flow deficits due to dry spells, and early flood recession. The valorization of floodplains faces numerous technical, social, and economic constraints that involve an intensification of crops and hence new risks linked to climate change. These are characterized by increased irregularities in rainfall, onsets of extreme floods, and long-lasting droughts.

As a landlocked country, Burkina Faso is vulnerable to climate variability [3]. This variability not only occurs at a daily, seasonal, and interannual scale, but can also be multidecadal. A break in annual rainfall was observed during the 1970s, irrespective of latitude or longitude, in West Africa [4]. It was especially prominent in the savanna, which includes the study area of Dano. The causes of this prolonged drought, subcontinental in scale, remain controversial and undoubtedly multifactorial and multiscalar. Many authors have shown, in addition to the global natural variations (i.e., astronomical, oceanic, and volcanic), the anthropogenic effects at different scales. These are observed at a regional (increase in the albedo effect because of the rapid urbanization of the Sahel, and deforestation of the lower coast reducing real evapotranspiration (ETR) and increasing flow), intercontinental (European air pollution of the 1970s, favoring regional cooling, i.e., anticyclonic conditions in regulatory pathways), and global (greenhouse gas-related climate change and its effects on heat and excessive events) level [5]. A change in hydroclimatic conditions can substantially modify the hydrological regime of an inland valley and its drainage area [6].

This modification can result in flooding or drying conditions in the lowlands, implying a possible reduction in its productivity. Furthermore, land cover and land use change can alter the floodplain and impact its ecosystem [7], but investigating this is beyond the scope of this study. Given the uncertainties in climate model predictions (especially for precipitation), analyzing the current climate hazards using observed data and their possible implications for the future is required.

Burkina Faso's agricultural sector continues to generate approximately one-third of the country's GDP and employs 80% of the population [8], despite the harsh climatic condition. Notwithstanding the importance of agriculture in the economy of Burkina Faso, the sector is facing many challenges, including threats from many natural disasters, such as floods, droughts, and violent winds, which lead to low crop and livestock productivity [9]. Since 1970, investment has been made by the government of Burkina Faso to address the issue posed by hydroclimatic risks. This includes developing rice production intensification policies in inland valleys that encompass physical development, the social organization of production, material support, organization of the rice sector, subsidies, and legal connotations. Subsequently, 10% of the inland valleys suitable for agriculture have been developed in southwestern Burkina Faso. However, as reported by the regional agriculture extension service, up to 30% of the developed inland valleys have been abandoned due to increasing hydroclimatic hazards. There is therefore a need to describe the seasonal, average, and frequency characteristics of the climate that can impact rice production in the region.

The objective of this work is to analyze the hydroclimatic hazards by considering the period of 1922–2017 and their implications for rice production in southwest Burkina Faso to support agricultural policies for adequate water infrastructural development. Two research questions are considered: (i) What are the current trends in climatic and hydrological hazards and what are their implications for food production in inland valleys? (ii) What are farmers' perceptions of the hydroclimatic risks in the region, and what strategies have consequently been developed to face the challenges encountered?

2. Materials and Methods

This section is divided into five sub-sections, which are the study area and data used, the various modalities of lowland development (traditional vs. modern development), climate-related local knowledge and hydroclimatic variables' analysis, rainfall–runoff modeling and frequency analysis, and water availability evaluation during the critical phase of rice development.

2.1. Study Area and Data Used

The case study areas are the Lofin catchment and Lofin inland valley, located in the municipality of Dano in the southwest region (région du Sud-Ouest) of Burkina Faso, West Africa (Figure 1). Dano is situated in a tropical climate region with a unimodal rainfall regime (Figure 2). The mean annual rainfall is approximately 921 mm, with a standard deviation of 106 mm for the period 1980–2018. The annual rainfall regime is characterized by the alternation of two contrasting seasons: a dry season from November to March, in which rainfall is almost absent (58 mm in October, the driest month), and a rainy season from April to October (average 238 mm in August, the wettest month) [10].

The climatic water demand (ET0) is more stable during the year, but it varies, on average, from 123 mm in August to 175 mm in April [10]. Long-cycle crops (120 days), such as rice sown after the 10th of July, are at risk of water stress in the middle and end of the growing cycle. The average annual temperatures between 1970 and 2013 ranged from 25 to 33 °C and from 25 to 31 °C at the Boromo and Gaoua stations, respectively (Figure 2). The annual insolation varies between 6 and 8 h/day, and the air humidity ranges from 35% to 80%. The dominant vegetation comprises shrubs and/or the tree savanna type, and resulting successional vegetation from the degradation of cleared forests [11]. This is due to both human activities and the dry period since 1970.

In this region, wetlands have historically been used as pasture in the dry season. Rice was one of the first crops cultivated in these areas (Figure 3). From the 20th century onwards, the agricultural use of the inland valleys, referred to locally as 'bas-fonds', was fostered because of population growth and migratory flows. Currently, in addition to rice, wetland use has been diversified with other crop types, such as vegetables, fruits, and cereals [12]. Rice products are mainly intended for consumption, with an increasing share of rice in the local food.

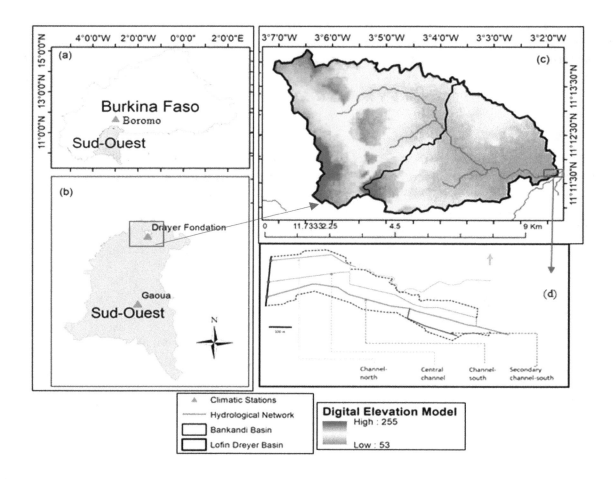

Figure 1. (**a**) Location of the study area in Burkina Faso. (**b**) The southwest region. (**c**) The Lofin catchment. (**d**) The Lofin inland-valley with irrigation/drainage channels.

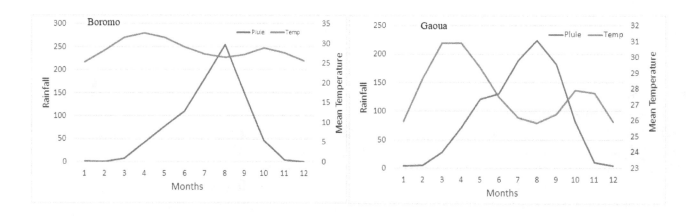

Figure 2. Rainfall and temperature at Boromo and Gaoua climatic stations.

The data used in this study are rainfall data of the Boromo and Gaoua stations from 1922 to 2016, and rainfall data of the Dreyer Foundation from 2017 to 2018. Discharge data of the Lofing-Radier station from 2017 to 2018 were measured by WASCAL (www.wascal.org) during project implementation. Other climate data, such as the minimum and maximum temperature, sunshine duration, wind speed, and minimum and maximum relative humidity from 1922 to 2016 of the Boromo and Gaoua stations were used and obtained from the Burkina Faso national meteorological directorate (including long-term rainfall data).

Figure 3. Different types of lowland development, including a rice field model with sprinkler drains (**a,b**), cyclopean concrete pouring dikes with a central cofferdam (**c**), and compacted clay bunds protected by geotextiles and rocks following the level curves with openings (**d**).

2.2. Various Modalities of Lowland Development: Traditional vs. Modern Development

Four types of lowland development have been observed and can be classified into two main groups: traditional and modern lowland designs. Traditional lowland development is a combination of the techniques developed by farmers for managing the agrarian space, the water, and the various types of production. The objective pursued by the farmers is to grow crops while minimizing the risks associated with drought and flooding. Schemes designed by farmers for the control and management of water in the lowlands include, firstly, large ridges arranged perpendicularly to the water flow direction, and secondly, large ridges in the shape of a contour dike with gaps.

Baffles are formed that not only slow the flow of water favoring infiltration, but that also enable the management of a water level in the grooves between the ridges. Upland crops (maize or tobacco) are placed on the ridges. Sorghum and taro are placed on the flank of the ridges. Rice, a water-demanding crop, is sown and transplanted into the furrows. This polyculture system is adapted to local conditions, has a low associated cost, and is more resilient to the risk of climatic disasters.

In contrast, modern lowland developments are those designed and implanted by external organisations, such as funded projects and programs, and NGOs. The principle objectives of such developments are multi-functional, aiming to

- Partially control water through the installation of hydraulic structures;
- Distribute the water at the landscaped site;
- Optimize the drainage of flood waters;
- Avoid and minimize the adverse effects of water shortages due to dry spells during the crop season; and
- Support non-seasonal crops, if possible.

To achieve these objectives, several types of development were designed, of which three (3) types of models are described. First is the rice field model with sprinkler drains (Figure 3). This model consists of channeling runoff by following preferential paths marked by the differentiation of surface elevation.

The canals are used for irrigation and drainage. The individual plots are partitioned by small bunds which the producers can open to irrigate their plants. This model is promoted in the area by the Dreyer Foundation. Second are compacted clay bunds protected by geotextiles and rocks that follow the level curves with openings. This model is a flood spreading arrangement, with the possibility of drainage being provided by the openings. The third model is the cyclopean concrete pouring dikes with a central cofferdam. This model is based on the threshold for slowing the flow of water on the course bed, which leads to a substantial change in the height of the water level, and is then managed using the cofferdam.

With these three models of landscape control, adding garden plants arranged with wells is necessary. The modern development models of lowlands strongly alter the hydrology of the valley bottom. Although this may be advantageous, it also adds new constraints that can become risks, depending on the physical and social environment.

2.3. Climate-Related Local Knowledge and Hydroclimatic Variables' Analysis

A survey of farmers' perceptions on climate and climatic changes was conducted in the Lofing lowland in 2017 using a questionnaire. A total of 17 farmers were randomly selected, and a questionnaire was administered individually. The hydroclimatic variables were statistically analyzed using the quantile method, Mann-Kendall test, and Sen's slope estimator. Different time steps were considered to aggregate the time series over 10-day, monthly, and annual time scales. Reference evapotranspiration was computed using the Food and Agriculture Organisation (FAO) ETo calculator software based on the Penman–Monteith formula [13]. The rainfall onset and cessation dates were defined using the Franquin (1969) [14] method. At the 10-day scale, the rainfall onset corresponds to the date from which rainfall is greater than half of the potential evapotranspiration (R > ET0/2). The methodological framework of the study is presented in Figure 4. It shows the different steps fulfilled to perform this study. The Mann-Kendall test [15,16] is a nonparametric trend detection method widely applied to hydroclimatic variables [17–19]. The null hypothesis H_0 for the test is that there is no trend in the time series, while for the alternative hypothesis H_1, there is a significant trend in the time series at the 0.05 significance level. This is a robust test in the sense that it does not make any assumptions about the distribution of variables. In addition, the Sen slope method [20] is considered for estimating the magnitude of the slope if a trend is detected in the time series.

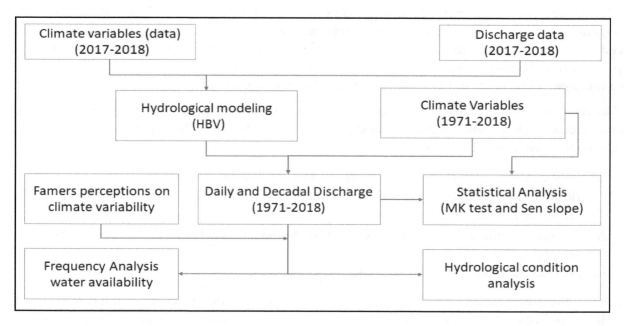

Figure 4. Methodological framework of the study.

2.4. Rainfall–Runoff Modeling and Frequency Analysis

To further access the hydrological aspect of the study area, an HBV model [21] was calibrated and validated for the Lofing basin at the outlet of Lofing Radier. HBV is a conceptual, lumped, and time-continuous hydrological model that simulates discharge using rainfall and potential evaporation as inputs. The HBV model has four main component routines: (1) snow (not used); (2) soil moisture (computes actual evapotranspiration, soil moisture, and groundwater recharge); (3) response function (calculates runoff and groundwater levels); and (4) routing (calculates the distribution of runoff for a given time series). Model calibration used 2018 data, and validation was performed with data from 2017. Performance criteria included the Nash–Sutcliffe efficiency (NSE) normalized statistic [22] combined with the coefficient of determination (R^2). The past hydrological condition of the basin was then simulated using climatic data from the Boromo and Dano stations (Figure 1) for the period 1971–2018 (Figure 4).

2.5. Water Availability Evaluation during the Critical Phase of Rice Development

The flowering and grain filling growth stages of rice are the most critical phases of its development. Water stress during these phases can be seriously detrimental to the rice yield [10]. In the Dano region, rice varieties with a growing period of 120 days are the most cultivated. By considering the following 10-day periods of rice sowing seed (21–30 June, 1–10 July, and 11–20 July), the critical periods for rice development ranged from 11–20 September to the 11–20 October. The total discharge of water during these 10-day periods was obtained. Trends in these data collection periods were evaluated using the Mann-Kendall and Sen slope tests. The availability of water in terms of discharge, level, and volume needed for effective rice development in the Lofing inland valley was then assessed.

3. Results and Discussion

3.1. Farmers' Perceptions of Changes in Climatic Hazards in the Lofing Inland Valley

Farmers have both positive and negative consensual perceptions of climatic hazards. According to them, no severe drought has been observed since 1974, except in 1984, when a famine was experienced. Since 1996, extreme rains able to destroy houses have not been observed, except in 2015, when a long-lasting and heavy rainfall event was recorded. Events perceived negatively were deemed to be of the greatest importance. Farmers consider that the heat waves initially experienced mainly in April have shifted to March and that they last almost the entire year. The weather is therefore perceived as becoming warmer. Furthermore, the farmers perceive there to be changes in annual rainfall patterns because, according to them, rainfall was previously well-distributed throughout the rainy season. Now, they believe that the dry season is longer, and the rainy season is shorter and irregular. Dry spells have become more frequent during crop growth and mainly occur during the rice grain formation stage. According to the interviewed farmers, there is currently a decrease in rainfall amount per event, with more thunderstorm winds, but limited rainfall, in comparison to past observations.

The events best-described by the farmers are the catastrophic years, the increasing heat, and the effect of wetland development over the last two years. The consensual perceptions of the degrading rainy season should be considered with caution, given the vagueness of the compared periods and the intensity of the variations. For example, it is virtually impossible to attribute a precise date to references such as "previously", "formerly", or "around 2000", when farmers stated that there have been more severe droughts in the past than at present. Was there a period of a small series of very regular years that was idealized and would now be "referenced"? Did the farmers identify recent problematic years to more comprehensively judge the past, thereby forgetting about the reality of the variability and the change in the process? An in-depth analysis of several timescales of the long climatological series, including a frequency analysis, is required to answer these questions.

Some perceptions are not consensual, namely, the rainfall onset date in 2017 and the effect of the dike and channel rehabilitation in 2017. In fact, 10 of the 17 people interviewed reported an early rainfall onset, 3/17 interviewees reported a normal onset, and 4/17 interviewees indicated a late rainfall onset in 2017. The least consensual perceptions are paradoxically related to the climate or the water regime of the year. These perceptions address, on the one hand, the location of the respondent's plot in relation to the newly constructed dike, and on the other hand, the expectations that are relative to a farmer's specific needs and workplan. The perceptions of climate risk also do not have the same levels of concern among individual respondents. Few have seen "no change" to the climate. In terms of the motivation for sowing rice rather than transplanting in 2017, respondents cited the climatic risk. However, the perception of climate risk varies, according to the respondents' gender and the level of development of the lowlands (Table 1). Indeed, excluding the developed lowlands, the climate risk was mentioned by 100% of women as the reason for rice transplanting. This result might be due a lack of knowledge about agricultural rainfall onset identification. Early sowing is adopted by farmers to free themselves from rainfall and hydrological hazards (i.e., uncertainty about the moisture conditions of the lowland region). Despite existing water control structures in the developed lowlands (which are designed to enable transplanting), up to 20% of farmers do not wait for good sowing conditions. Men are more restricted by their other farming operations. Rice is of a lower priority, and where there is a competition of labor against cotton, the lowlands are often abandoned. Indeed, the climate risk is less important to them, as it is shifted toward alternative activities. This local information is extremely valuable as it both identifies the concerns of local farmers and provides new information regarding their perceptions and likely responses as a result. It is nevertheless necessary to compare the local farmers' perceptions with measured data, which is independent of the farmers' gender, knowledge, and situation.

Table 1. Reasons for the choice of sowing (rather than transplanting) in 2017, according to gender and situation.

Reasons	Operational Constraints	Social Organization	Climate Risk	Lack of Know-How
Undeveloped lowland by men (%)	43	0	43	14
Undeveloped lowland by women (%)	0	0	100	0
Developed lowland by men (%)	63	13	25	0
Developed lowland by women (%)	80	0	20	0

3.2. Analysis of Rainfall over the Last 40 Years at the Regional Scale

The mean rainfall recorded at Dano during 2013–2017 by the Dreyer Foundation (951 mm) is similar to that of the period 1970–2013 at Boromo-Gaoua (962 mm). For that reason, the Boromo-Gaoua rainfall stations, which are the closest to Dano, have been used to provide a detailed understanding of the local climatic pattern. Increased water exceedance events (water available to recharge the reserve and water flows) have been observed in Boromo since 1984 (Figure 5a). There has been no increase in rainfall; however, an increase in water excess implies a change in rainfall regime and/or land use. More rainfall in a shorter time period results in an increase in flood risk and less actual evapotranspiration. In Gaoua, the change was small, but similar, to the change depicted in Boromo, except that the water excess did not increase. Figure 5b shows the climatic balance of the last five years in Dano. There was a high seasonal and interannual variability of rainfall during this period. In 2013, there was no pre-humid period, the water excess was limited, and an early cessation of the season occurred, while in 2014, the pre-humid period occurred in mid-July. In 2015 and 2016, there was no pre-humid period, and the risk of inundation was high. A pre-humid period was detected in early September 2017, with limited water excess.

The period of rainfall uncertainty is longest during the pre-humid season. The hazard zone corresponds to rainfall being less than half of the potential evapotranspiration between the 25 and 75 quantiles of ten days of rainfall. The season profile is asymmetric, meaning that the cessation of the season is more predictable than its onset and that the early rainy season provides more information on the rainy season duration. An ideal opportunity for an informed choice of season length, mainly if there is the potential to irrigate, is provided based on this data.

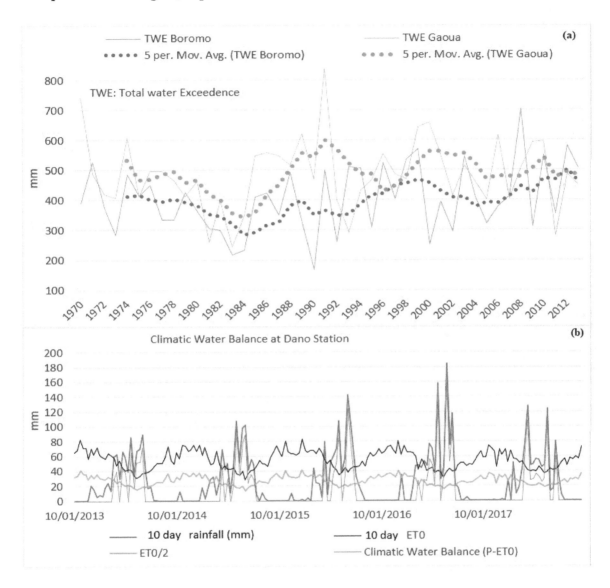

Figure 5. Water excess at Boromo and Gaoua rainfall stations from 1970 to 2012 (**a**) and the climatic water balance at Dano station (2013:2017) (**b**).

3.3. Temperature Analysis at the Regional Scale

The Sen slopes estimated for each month by considering the minimum and the maximum temperatures at Boromo and Gaoua from 1970 to 2013 indicate a tendency toward a warmer atmosphere. During the rainy season (August to October), the average minimum temperature and the average maximum temperature increased at a rate of 0.031–0.035 and 0.005–0.016 °C yr^{-1}, respectively, which is in line with the global observation [23,24]. The minimum temperature increased faster (around two times) than the maximum temperature during the rainy season. Climate change causes increasing air temperatures and evapotranspiration, increases the risk of intense rainstorms, and increases the risk of heat waves associated with drought [25]. An increase in temperature was also observed in the Beninese part of the Niger basin by Badou et al. (2017). An increase in temperature implies a higher level of evapotranspiration, a higher water demand for crops, and a lower level of water exceedance. An increasing temperature will exhibit a larger impact on the grain yield than on vegetative growth and will reduce the ability of the crop to efficiently fill the grain or fruit [26]. This output has the potential to inform famers and stakeholders in framing appropriate policies for rice intensification in the region. The minimum relative humidity (RH) increases from July to October at Boromo and from August to December at Gaoua. The maximum RH decreases from May to September at Boromo and from July to September at Gaoua.

3.4. Potential of Watering/Drainage of the Channels' System in the Lofing Inland Valley

High seasonal and interannual variabilities of discharge were observed at the Lofing-Radier outlet (Figure 6). Throughout the growing season, the river provides enough water to satisfy the requirements of rice (100 L s^{-1} is needed to irrigate 30 ha) (Figure A1). Irrigation should be possible whenever necessary, mainly during the dry spell period. Irrespective of the rice sowing date, the critical period (end of the rice cycle) requiring irrigation varies between the 10th of September and 20th of October. Channel dimensioning is challenging in rice cultivated in inland valleys. There must be a trade-off between the necessities of the discharge peak flow, while maintaining a water level in the channels required for direct irrigation (through, for instance, the use of cofferdams), but also maintaining the wetness of cultivated parcels. The drainage capacity of the channels is large enough for discharging the peak flows arriving from the Dano basin at the outlet of Lofing, while its irrigation potential is problematic. Although the inflow into the channel system was adequate, the water level in the channels is problematic. The minimum water level required in one of the main channels for irrigation is 35 cm. In 2017, the water level in the channels rarely exceeded 35 cm (Figure A2), suggesting that the installation of cofferdams to increase water levels would be beneficial. The implementation of cofferdams, however, may result in additional problems, such as hindering the channel drainage function during peak flow. Frequency analysis has shown that the likelihood of obtaining a high flow during the critical period is low, and mainly occurs after the 21st of September. To ensure that irrigation at the end of the rice cycle is sufficient, different mobile cofferdam types require testing to ensure their efficiency and acceptability in terms of cost and the capacity of farmers to implement the technology.

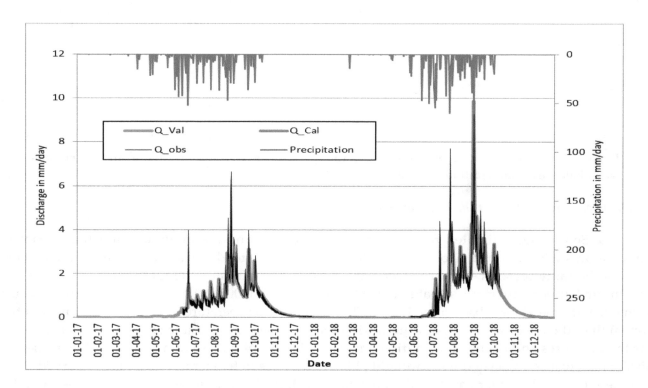

Figure 6. Observed and simulated discharges produced by the HBV model (Q_Val, Q_Cal, and Q_obs corresponding to the validation discharge, the calibration discharge, and the observed discharge, respectively).

3.5. Model Calibration and Validation

The HBV model was calibrated for the year 2018, and the simulated discharge was compared to the observed discharge using the numerical and visual criteria. The observed and simulated discharges are similar (Figure 6). NSE values for calibrated/validated data were 0.75/0.70, with a coefficient of determination of 0.75/0.73 and a logarithm of NSE of 0.74/0.85. The high values imply a strong performance of the HBV model for the two years of observed discharge. Both high discharge and recession discharge (lower discharge) series were accurately simulated, although it is acknowledged that a greater number of discharge observations are required to increase the accuracy in the future.

3.6. Hydrological Condition of the Lofing Upstream River

The calibrated and validated HBV model simulated discharge of the Lofing-Radier River for the period 1971–2018. The water balance components, precipitation, discharge, and actual evapotranspiration are shown in Figure 7. The results of the Mann-Kendall trend test applied to the discharge statistic are shown in Table 2. No statistically significant trend at the 5% level was found in the total annual discharge for the period 1971–2018, indicating that the hydrological regime of the catchment did not vary at the annual scale. Nevertheless, a small annual increasing rate of the discharge of 0.0074 mm per day (2.7 mm yr^{-1}) was evident, implying a constant availability of water at the annual scale. The flowering and grain filling of rice seeded between the 20th and 30th of June occurred between the 11th and 20th of September. This stage of rice growing is critical since water stress experienced in this period may drastically reduce the rice yield. No significant increase in the water stress level is found during this period. Between the 21st and 30th of September, the third quartile displays an increase in the risk of excess water levels. In October, there is an increase in water resource availability, which is mainly beneficial for rice production during this critical stage.

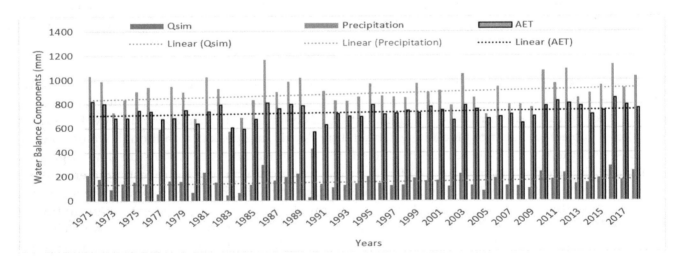

Figure 7. Water balance components: precipitation, simulated discharge (Qsim), and actual evapotranspiration (AET) for the period 1971–2018.

Table 2. Sen slope (SS) and total 10-day water discharges (see Section 3.3) in different conditions for 1971–2018. * indicates a significant trend obtained through the Mann-Kendall test.

	11–20 September		21–30 September		1–10 October		11–20 October	
Test implemented (1971–2018)	SS/MK *	10 day water (mm)	SS/MK *	10 day water (mm)	SS/MK *	10 day water (mm)	SS/MK *	10 day water (mm)
1st Quartile (Dry Condition)	0.006	<12.7	0.009	<11.7	0.009 *	<9.2	0.006 *	<6.4
Median (Normal Condition	0.009	19.5 [12.7, 24.1]	0.010	16.8 [11.7, 20.5]	0.010 *	12.6 [9.2, 5.7]	0.007 *	9.2 [6.4, 10.4]
3rd Quartile (Wet Condition)	0.014	>24.1	0.016 *	>20.5	0.013 *	>15.7	0.008 *	>10.4
Sum	0.120	-	0.125	-	0.127 *	-	0.070 *	-

The water level in the river at the gauging station decreased from the 10th of September to 10th of October (Table 2). Therefore, if there is enough rainfall, sowing the rice during 10–30 June will be optimal to take advantage of the higher surface water flows that can be mobilized to irrigate and meet the crops' water requirements during the critical phases of flowering and formation-filling of the grains. Lower flow rates can be utilized to irrigate the crop during the critical phases if sown between the1st and 20th of July.

4. Conclusions

Rainfall events exceeding 100 mm and flash flooding are becoming more frequent, leading to severe damage to crop production and water infrastructure. Special attention must therefore be given to the design of water control structures to ensure their flexibility and sustainability in discharging floods while avoiding overdrainage during dry spells. In this study, we analyzed the hydroclimatic conditions of the study area Dano, Burkina Faso, and the implication for rice production in the region. There was no significant increase in annual rainfall for the period of 1970–2013; however, an increasing delay in the onset of the rainy season (with a decreasing pre-humid season duration) was observed. This causes difficulties in predicting the onset due to the high temporal variability of rainfall in the studied region. As a result, a warming trend was observed for the past 40 years, raising questions about its negative impact on very intensive rice cultivation packages. During the rainy season (August to October), the average minimum and maximum temperatures increased by 0.031 and 0.016 °C yr^{-1}, respectively, comparable to global observations.

The maximum relative humidity decreased due to this increase in temperature, while the sunshine duration also decreased. Farmers have both positive and negative consensual perceptions of climatic hazards. The HBV hydrological model indicated no significant increase in water discharge; however, the total 10-day water level observed between the 11th of September and 20th of October, corresponding to the critical flowering and grain filling phases of rice growth, showed an increasing trend for the period 1971–2018.

The sowing of rice during the 10–30 June has been identified as optimal in order to benefit from the higher surface water flows, which can be used to irrigate and meet the crop water requirements during the critical phase outlined. The installation of cofferdams to increase water levels would be beneficial, subject to them not hindering channel drainage during peak flow, although water flow after the 21st of September was generally insufficient to be deemed an issue. To ensure that irrigation at the end of the rice cycle is sufficient, different mobile cofferdam types require testing to ensure their efficiency and acceptability in terms of cost and the capacity of farmers to implement the technology. The results of this study will be useful to rural communities, as well as decision makers, in framing agricultural risk management in the study region and devising policy for rice intensification in lowland areas. Further data collection is required to improve the HBV model output and to account for climate and land change effects on rice production in Dano, Burkina Faso.

Author Contributions: Conceptualization, A.Y.B. and G.S.; methodology, A.Y.B., G.S., J.H., and B.D.; formal analysis, A.Y.B., J.H., Y.Y., G.S., and B.D.; writing—original draft preparation, review, and editing, A.Y.B., J.H., G.S., Y.Y., B.L., J.L.F., L.O.S., J.E.T., and B.D. All authors have read and agreed to the published version of the manuscript.

Acknowledgments: The authors are grateful for the financial support provided by the French Agency for Development (AFD) under the auspices of the AGRICORA initiative and GENERIA project. They thank the German Federal Ministry of Education and Research (BMBF) for supporting the WASCAL program.

Appendix A

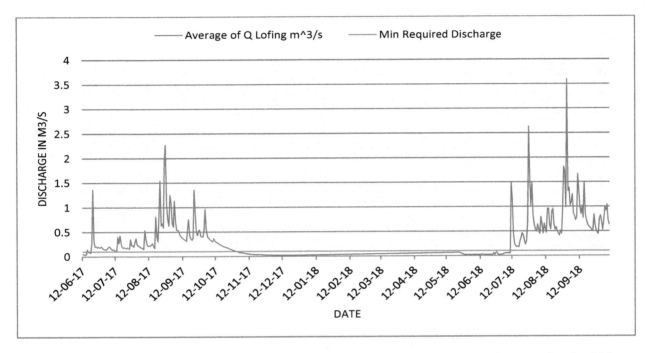

Figure A1. Comparison of the available discharge and irrigation water requirement for the 30 ha Lofing inland-valley.

Appendix B

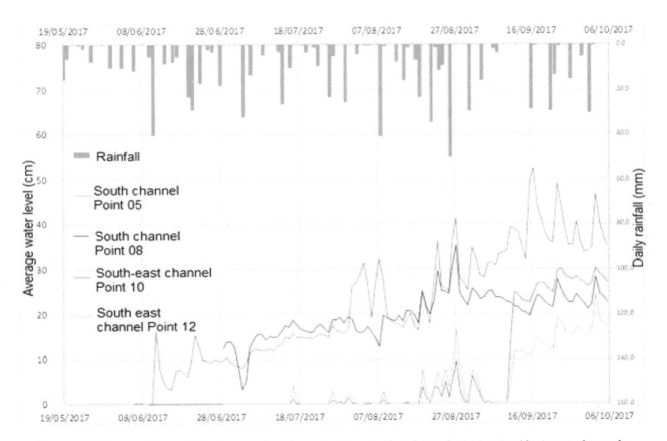

Figure A2. Comparison of the observed and required water height in the irrigation/drainage channels of Lofing inland-valley. Points represent the water level measurement locations.

References

1. Faurès, J.; Sonou, M. *Les aménagements hydro-agricoles en Afrique Situation actuelle et perspectives*; FAO: Rome, Italy, 2005.

2. Danvi, A.; Giertz, S.; Zwart, S.J. Rice Intensification in a Changing Environment: Impact on Water Availability in Inland Valley Landscapes in Benin. *Water* **2018**, *10*, 74. [CrossRef]

3. Ouédraogo, M.; Dembele, Y.; Somé, L. Perceptions et stratégies d'adaptation aux changements des précipitations: Cas des paysans du Burkina Faso. *Cah. Agric.* **2010**, *21*, 87–96. [CrossRef]

4. Badou, D.F.; Kapangaziwiri, E.; Diekkrüger, B.; Hounkpè, J.; Afouda, A. Evaluation of recent hydro-climatic changes in four tributaries of the Niger River Basin (West Africa). *Hydrol. Sci. J.* **2017**, *62*, 715–728. [CrossRef]

5. Jacob, D.; Kotova, L.; Teichmann, C.; Sobolowski, S.P.; Vautard, R.; Donnelly, C.; Koutroulis, A.G.; Grillakis, M.G.; Tsanis, I.K.; Damm, A.; et al. Van Earth's Future Climate Impacts in Europe Under + 1.5 °C Global Warming. *Earth's Future* **2018**, *6*, 264–285. [CrossRef]

6. Yacouba, Y.; Aymar, B.Y.; Fusillier, J.; Thomas, Y.B. Failure of inland valleys development: A hydrological diagnosis of the Bankandi valley in Burkina Faso. *Model. Earth Syst. Environ.* **2019**, *5*, 1733–1741. [CrossRef]

7. Leemhuis, C.; Thonfeld, F.; Näschen, K.; Steinbach, S.; Muro, J.; Strauch, A.; Ander, L.; Daconto, G.; Games, I. Sustainability in the Food-Water-Ecosystem Nexus: The Role of Land Use and Land Cover Change for Water Resources and Ecosystems in the Kilombero Wetland, Tanzania. *Sustainability* **2017**, *9*, 1513. [CrossRef]

8. USAID Agriculture and Food Security. Available online: https://www.usaid.gov/burkina-faso/agriculture-and-food-security (accessed on 29 April 2019).

9. Yameogo, T.B.; Bossa, A.Y.; Torou, B.M.; Fusillier, J.; Da, D.E.C.; Yira, Y.; Serpanti, G.; Some, F.; Dama-balima, M.M. Socio-Economic Factors Influencing Small-Scale Farmers' Market Participation: Case of Rice Producers in Dano. *Sustainability* **2018**, *10*, 4354. [CrossRef]

10. Raherizatovo, T. Conception d'aménagements à maîtrise partielle de l'eau pour l'irrigation du riz dans les bas-fond du Sud-Ouest du Burkina Faso. Master's Thesis, CIRAD Montpellier—UMR G-EAU, Montpellier, France, 2018.

11. Bellefontaine, R.; Gaston, A.; Petrucci, Y. *Aménagement des forêts naturelles des zones tropicales sèches*; FAO: Rome, Italy, 1997.

12. Da, S.J. *Etude des usages et de la regénération d'une plante alimentaire au Sud-Ouest du Burkina Faso*; Polytechnic University of Bobo-Dioulasso: Bobo-Dioulasso, Burkina Faso, 2009.

13. Allen, R.; Pereira, L.; Raes, D.; Smith, M. *Crop Evapotranspiration—Guidelines for Computing Crop Water Requirements—FAO Irrigation and Drainage Paper 56*; FAO: Rome, Italy, 1998.

14. Franquin, P. Analyse agroclimatique en régions tropicales, saison pluvieuse et saison humide, applications. *Cah. ORSTOM* **1969**, *9*, 65–95.

15. Mann, H. Nonparametric tests against trend. *Econometrica* **1945**, *13*, 245–259. [CrossRef]

16. Kendall, M. *Rank Correlation Methods*; Griffin: London, UK, 1975.

17. Ahmad, I.; Tang, D.; Wang, T.; Wang, M.; Wagan, B. Precipitation trends over time using Mann-Kendall and spearman's Rho tests in swat river basin, Pakistan. *Adv. Meteorol.* **2015**, *2015*, 431860. [CrossRef]

18. Shadmani, M.; Marofi, S.; Roknian, M. Trend Analysis in Reference Evapotranspiration Using Mann-Kendall and Spearman's Rho Tests in Arid Regions of Iran. *Water Resour. Manag.* **2011**, *26*, 211–224. [CrossRef]

19. Zhang, W.; Yan, Y.; Zheng, J.; Li, L.; Dong, X.; Cai, H. Temporal and spatial variability of annual extreme water level in the Pearl River Delta region, China. *Glob. Planet. Chang.* **2009**, *69*, 35–47. [CrossRef]

20. Nash, J.E.; Sutcliffe, J.V. River flow forecasting through conceptual models: Part I—A discussion of principles. *J. Hydrol.* **1970**, *10*, 282–290. [CrossRef]

21. Sen, P.K. Estimates of the Regression Coefficient Based on Kendall's Tau. *J. Am. Stat. Assoc.* **1968**, *63*, 1379–1389. [CrossRef]

22. Bergström, S. *The HBV Model—Its Structure and Applications*; RH No 4; SMHI: Norrköping, Sweden, 1992.

23. IPCC Climate Change. 2001: The scientific basis. In *Contribution of Working Group I to the Third Assessment Report of the Intergovernmental Panel on Climate Change*; Houghton, J.T., Ding, Y., Eds.; Cambridge University Press: Cambridge, UK, 2001; p. 881.

24. Arnell, N.W. Relative effects of multi-decadal climatic variability and changes in the mean and variability of climate due to global warming: Future streamflows in Britain. *J. Hydrol.* **2003**, *270*, 195–213. [CrossRef]

25. Touré, H.A.; Kalifa, T.; Kyei-Baffour, N. Assessment of changing trends of daily precipitation and temperature extremes in Bamako and Ségou in Mali from 1961–2014. *Weather Clim. Extrem.* **2017**, *18*, 8–16. [CrossRef]

26. Hat, J.L.; Prueger, J.H. Temperature extremes: Effect on plant growth and development. *Weather Clim. Extrem.* **2015**, *10*, 4–10.

8

Financing High Performance Climate Adaptation in Agriculture: Climate Bonds for Multi-Functional Water Harvesting Infrastructure on the Canadian Prairies

Anita Lazurko [1] and Henry David Venema [2,*]

[1]   Department of Environmental Sciences and Policy, Central European University, Budapest 1051, Hungary; anita.lazurko@mespom.eu

[2]   Prairie Climate Centre, International Institute for Sustainable Development, Winnipeg, MB R3B 0T4, Canada

*   Correspondence: hvenema@iisd.ca

Abstract: International capital markets are responding to the global challenge of climate change, including through the use of labeled green and climate bonds earmarked for infrastructure projects associated with de-carbonization and to a lesser extent, projects that increase resilience to the impacts of climate change. The potential to apply emerging climate bond certification standards to agricultural water management projects in major food production regions is examined with respect to a specific example of multi-functional distributed water harvesting on the Canadian Prairies, where climate impacts are projected to be high. The diverse range of co-benefits is examined using an ecosystem service lens, and they contribute to the overall value proposition of the infrastructure bond. Certification of a distributed water harvesting infrastructure bond under the Climate Bond Standard water criteria is feasible given climate bond issue precedents. The use of ecosystem service co-benefits as additional investment criteria are recommended as relevant bond certification standards continue to evolve.

Keywords: climate change; agriculture; climate bonds; investment; distributed infrastructure; water harvesting; Canada

---

1. Introduction

The political success achieved by the 2015 Paris Climate Accord with respect to a broad political consensus to reduce greenhouse gas emissions and accelerate adaptation to climate change, was followed by further political commitments in 2016 to increase climate financing. The 2016 G20 Hangzhou Leader's summit communique stated, "We believe efforts could be made to ... provide clear strategic policy signals and frameworks, promote voluntary principles for green finance, support the development of local green bond markets and promote international collaboration to facilitate cross-border investment in green bonds" [1].

The G20 leaders expressed support for a well-established trend—the rise of a new class of labeled infrastructure investment bond aligned with de-carbonization and climate de-risking objectives. Between 2011 and the 2015, the volume of "green" or "climate" labelled bonds issued increased from $3 billion to $95 billion, a large increase but still a small fraction of the estimated $93 trillion infrastructure investment requirements frequently cited as necessary to meet Paris accord objectives of limiting global warming to under 2 °C [2].

The large majority of labeled green and climate bonds have been designated for renewable energy, energy efficiency and low-carbon transport. In 2015 these sectors comprised 79% of the value of bond issues [3], whereas bonds specifically designated for climate adaptation had only a 4% market share—despite compelling evidence that investments in adaptation can provide very high rates of return [4]. The underlying issue is that although climate change is a global issue and its mitigation requires collective global action, climate change impacts are inherently localized and adaptation is necessarily a granular design process requiring highly localized climatic, socio-economic and ecosystem information—a challenge for harnessing the larger scale investment flows commensurate with the scale of the opportunity. In addition, bond financing requires that a large number of relatively small individual projects be aggregated to reach a sufficient scale. The scale at which local adaptation projects require financing is typically two to four orders of magnitude lower than the scale at which bonds are issued [3].

The Canadian Prairies are an interesting geographic context to analyse the logic for increasing market share for climate adaptation bonds and the associated challenges, by referencing the specific case of multi-functional water retention structures for agriculture. The Canadian Prairies comprise about 90% of Canada's agricultural land base, produce approximately 20% of internationally traded grains and oilseeds and thus are an important component of world food security. The Canadian Prairies also have a history of high vulnerability to climate shock for anthropogenic and climatological reasons, and a history of innovative ecosystem and water resources management based on distributed water harvesting (DWH) that could be revived in the context of climate adaptation [5]. Berry et al. [6] review a multi-purpose surface water retention system at Pelly's Lake, in the Canadian Prairie province of Manitoba that illustrates the economic case for water harvesting. Berry et al. conclude that when all economic benefits are evaluated; flood and drought risk reduction, irrigation and other ecosystem service benefits, the net value of retention storage (more than CAD $25,000/hectare) far exceeded its land value as conventional agriculture. Nonetheless, the total investment requirement for this high performance, but highly local, climate adaptation project at under CAD $1 million falls below the threshold for prioritization as conventional infrastructure spending. The urgency and logic for aggregating large numbers of such "precision infrastructure" projects for innovative climate financing through bond issues on the Canadian Prairies is, therefore, the focus of this paper.

This paper aims to explain and analyse the opportunity to finance high performance climate adaptation projects like multi-functional DWH infrastructure with certified climate bonds under the Water Criteria of the Climate Bond Standard, and to explore the concept of informing the project or bond value proposition with the economic value of ecosystem services and co-benefits. In addition, this paper aims to demonstrate the logic for aggregating a large number of relatively small projects to a scale appropriate for bond financing. This paper uniquely combines concepts and provides a new iteration upon leading solutions from seemingly disparate entities: engineers and scientists turning to distributed, localized, green infrastructure solutions, climate modelers increasingly understanding the importance of temporal variability and downscaling data to regional impacts, financers seeking to open new markets for green infrastructure and to find ways to aggregate localized projects into large-scale financing structures, and new entities like the Climate Bonds Initiative providing a new platform to set standards and increase visibility. The methodology of this paper includes articulating the direct benefits and enhanced ecosystem services of DWH solutions, presenting a general framework for a project and bond value proposition that aggregates those benefits using downscaled climate change data for assessing the value generated over future scenarios, and providing recommendations for the institutional, regulatory, and technical elements needed to finance this solution with government-issued bonds certified under the Water Criteria Climate Bond Standard Phase 1: Engineered Infrastructure [7]. This paper concludes with recommendations for implementation of DWH systems on the Canadian prairies and future development of CBS criteria for natural and semi-natural water infrastructure. The broad conclusions drawn in this report can be used to disseminate the DWH solution to other regions with similar climatic stressors and agricultural conditions.

2. Distributed Water Harvesting on the Canadian Prairies

2.1. Introduction to the Canadian Prairies

Climate change on the Canadian prairies manifests as temperature increases and changes to precipitation patterns that demand greater climate resilience in the agricultural sector. The size and shape of the continent of North America, its proximity to the Arctic Ocean, and other factors accelerate the climatic warming felt on the prairies. The Prairie Climate Centre has shown that Winnipeg may experience summer temperatures similar to the panhandle of Texas by the year 2080 [8]. The prairies are also vulnerable to precipitation changes, including an increase of spring precipitation and decrease of rainfall during the summer. Farmers will be forced to adapt their farming practices to stretch a variable hydrologic budget across a long, dry growing season. These rainfall challenges will be further exacerbated by the heightened temperatures through increased evapotranspiration rates [9]. In Saskatchewan and Manitoba, a large majority of agriculture is rain fed [10], and the patchwork of 150-acre quarter-sections of land separated by drainage ditches and culverts is designed to allow for limited groundwater percolation and rapid runoff into large reservoirs or natural water bodies. The use of fertilizer inputs in the region also results in accumulation of nutrients in runoff water and water bodies resulting in frequent eutrophication problems [11]. New precipitation patterns have already begun to strain the agriculture sector and government risk management practices, as seen during the Manitoba floods of 2011 [12]. Evidently, the current 'drainage culture' is in tension with the rainfall variability that will be introduced with the climatic pressures of the future, presenting the 21st century challenge of adaptation for farmers and governments.

2.2. The Engineered Solution

Multi-functional DWH infrastructure is a semi-natural climate adaptation solution that aims to overcome the climatic stresses that challenge the excessive drainage culture of agriculture in the region. It is a system of many small, controllable earthen dams that have been located and sequenced to enable control over current and future hydrologic cycles based on aggregated hydrologic and climate data. DWH mitigates floods in a similar manner to wetlands, but with a higher degree of control to overcome the risk of saturation and snow melt patterns that inhibit the ability of wetlands to buffer peak flows. By encouraging more groundwater percolation, maintaining a potentially higher groundwater table, and retaining standing water throughout the landscape, farmers will have the ability to access water during drought conditions. DWH is expected to have significantly less environmental disruption than hard infrastructure like dams and reservoirs, as well as a much lower infrastructure cost. Farmers upstream of the water harvesting system could have the option to drain their land more quickly to take advantage of early seeding dates, while farmers downstream of the system will be protected from seasonal flooding via controlled, intentional drainage patterns. Though innovative for the Canadian prairies, this solution is not new. India has met demand for seasonal water storage and lack of food security with similar technologies for millennia, though these systems were left abandoned or unmaintained in favor of groundwater irrigation in recent decades [13]. Sustainable development principles, cost-effectiveness, and environmental considerations are incenting a shift back toward such common-sense, localized solutions. Fortunately, the 21st century context of modern DWH systems presents new opportunities with this historic solution. For example, farmers may harvest biomass from "low spots" for energy generation, nutrient recovery, and profit, expanding the "bioeconomy" demonstrated in the Lake Winnipeg delta [14]. The multi-functional distributed water harvesting infrastructure as a climate adaptation solution inherently generates co-benefits and a business case at the intersection of the water–food–energy nexus.

2.3. Climate Change Adaptation and Enhanced Ecosystem Services

Climate change introduces new risks for governments, demanding innovative techniques for assessing and mitigating risk through adaptation. A higher frequency and severity of floods and droughts introduces significant challenges for governments, including infrastructure damage and loss of productivity in the agriculture sector. The 2011 floods in Manitoba caused CAD $1.2 billion of distributed infrastructure damage [12], triggering financial and stakeholder management challenges for the Province of Manitoba and the Government of Canada. Droughts may not directly cause property damage, but they have the potential to severely strain the agricultural sector and rural economies [15]. Assessing the impact of these climate change effects in terms of property and crop damage merely scratches the surface of the potential value of a well-managed flood mitigation and drought resilience program; assessing multiple dimensions of ecosystem services can highlight the full value of climate adaptation solutions. In addition, the economic valuation of such ecosystem services can inform a water pricing scheme that incorporates externalities and reflects full cost recovery [16], further incenting change toward water conservation and more appropriate water management. Robust assessments of risk and proposed value enable innovative solutions to emerge. These solutions demand resources, presenting the challenge of financing climate adaptation projects—a challenge insurance companies and the broader financial sector continue to grapple with. Balancing traditional institutional financing structures with the need to encourage granularity of high-performance adaptation projects informed by robust data and climate projections presents a unique design challenge for engineers, governments, and financers.

The main functional purpose of a multi-functional water harvesting system is to increase control over the hydrologic cycle to overcome climate change challenges to the agricultural sector. Climate change adaptation, a benefit derived from direct use of the infrastructure, is only part of the equation. An ecosystem services lens generates a more well-rounded picture of benefits and supporting services derived from DWH, generating a much stronger value proposition and informing better water management. Figure 1 depicts the network of potentially quantifiable climate change adaptation benefits and enhanced ecosystem services generated by a DWH system. The benefits in this figure could manifest similarly in different watersheds across the Canadian prairies, and so should be interpreted as a broad estimate of direct and co-benefits generated. In addition, this list of direct use and co-benefits could vary depending on the presence of agricultural irrigation or other climate adaptation measures in the region. The co-benefits in black typeface are significant and potentially quantifiable, while the co-benefits in grey typeface exist but are more difficult to quantify in economic terms in the value propositions described later in this paper. The following sections describe Figure 1 in more detail, which includes brief descriptions of the ecosystem services classified under the Millennium Ecosystem Assessment [17].

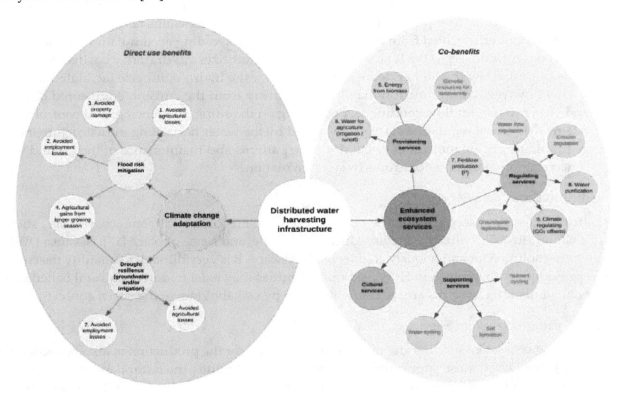

Figure 1. Distributed water harvesting infrastructure system as a network of direct use benefits and enhanced ecosystem services described in the following sections.

2.3.1. Flood Risk Mitigation and Drought Resilience

Climate change adaptation for flood risk mitigation and drought resilience can be easily connected to risk identification and management for governments and insurance entities. The need to consider climate change impacts, particularly property damages and crop loss but also ecosystem service benefits, will be increasingly important as governments begin to feel the monetary impacts. The flood risk mitigation benefit of the water harvesting system manages or avoids multiple hazards described in Figure 1, including agricultural losses due to loss of cultivable land or crop yield damages, property damages due to severe flood events or longer-term changes to the regional hydrology, and employment losses due to a decline in or local industry. The drought resiliency function of water harvesting systems manages similar hazards, including agricultural losses from lack of precipitation events that diminish crop yield and employment losses from reduced agricultural activity. DWH introduces the ability to control the hydrologic cycle with greater precision, presenting a valuable opportunity to increase crop yields with earlier seeding times and a longer growing season.

2.3.2. Provisioning Ecosystem Services

Provisioning ecosystem services are defined as 'the products obtained from ecosystems' [17]. These are the most relevant services provided in agriculture-based regions because of the direct economic benefit. Beyond agricultural crop yields, DWH may allow for provision of water for other uses such as irrigation or controlled runoff. The accumulation of biomass in low spots where water is retained by small earthen dams is an opportunity for farmers or private entities to harvest biomass seasonally for energy generation, similar to the bioeconomy of Lake Winnipeg [14]. This can lead to the secondary provisioning of phosphorus nutrients from the ash. Lastly, avoiding the environmental disruption of large dams and reservoirs may have a positive impact on the natural provision of biodiversity and genetic resources in the region, though this is difficult to quantify.

2.3.3. Regulating Ecosystem Services

Regulating ecosystem services are 'the benefits obtained from regulation of ecosystem processes' [17]. Water harvesting systems behave as a wetland during high water flow conditions, which can facilitate the natural purification of water and buffer peak water flows. Additional water purification functions are derived from biomass harvesting, by avoiding accumulation of phosphorus nutrients that are introduced to the landscape as chemical fertilizers in drainage basins. Water flow regulation is optimized by the higher degree of control over the hydrologic cycle facilitated by DWH systems. This flow regulation function may be a step away from the current, engineered drainage culture and closer to natural flow conditions, depending on the siting, sequencing, and control design of the system. Additional regulating services enhanced by the water harvesting infrastructure include erosion regulation from the more intentional drainage patterns and maintenance of the ground water table by encouraging more time for groundwater percolation.

2.3.4. Cultural Ecosystem Services

Cultural ecosystem services are 'the non-material benefits obtained from ecosystems', such as existence value, altruism, cultural benefits, educational value, and sense of place [17]. Because DWH is an engineering solution for a previously engineered landscape, it is very difficult to quantify the cultural services provided by this solution. However, opportunities may exist to derive cultural benefits, like educational value, if the systems are used intentionally by stakeholders in the social context.

2.3.5. Supporting Ecosystem Services

Supporting ecosystem services are 'the services necessary for the production of all other ecosystem services' [17]. For DWH, these supporting ecosystem services include the natural cycles enhanced by partially reversing or altering the current engineered drainage culture of the agricultural landscape on the Canadian prairies. This should improve the function of several supporting ecosystem services, including water cycling, nutrient cycling, and soil formation.

It is important to note that in addition to established monetary valuation techniques of many direct use and co-benefits, cultural ecosystem services are difficult to value in monetary terms. 'Willingness-to-pay' and related techniques have been used to justify monetary value of intangible assets. However, it cannot be assumed that an unwillingness to pay for an ecosystem service means that the service does not have value [18]. Several non-monetary valuation techniques exist, including Social Network Analysis, preference ranking, or the Q-methodology [18]. There is significant need for plural valuation that considers non-monetary value from such techniques alongside monetary values. However, until financing institutions are restructured to absorb such value into their more rigid frameworks, other important stakeholders may need to compromise and continue to use more easily quantified, less nuanced, monetary valuation techniques. The full list of ecosystem services depicted in Figures 1 and 2 is shown in Table 1.

Table 1. Key ecosystem services and monetization options from Figures 1 and 2.

Theme	Service	Examples of Service Monetization
Climate adaptation	Flood mitigation & drought resilience	Avoided agricultural losses *(estimated area loss x $ yield per unit area)* Avoided employment losses *(estimated job loss x employment insurance)* Avoided property damage *(estimated property damage as function of flood risk)* Crop yield increase from longer growing season *(Estimated yield increase x total affected area)*

Table 1. *Cont.*

Theme	Service	Examples of Service Monetization
Provisioning services	Irrigation water	Cost of equivalent agricultural irrigation *(Estimated irrigation costs for affected crop area)*
	Biomass harvesting	Cost of equivalent energy production *(Estimated energy from biomass x cost of alternative production)*
Regulating and supporting services	Nutrient cycling	Cost of purchasing chemical phosphorus fertilizers *(Estimated kg equivalent nutrient harvest from biomass x market price per kg)*
	Water purification	Cost of equivalent water treatment *(Estimated water quality improvement x cost of conventional water treatment methods)*
	CO_2 offsets	Cost of equivalent CO_2 offsets *(Estimated CO_2 offsets x price of carbon)*
Cultural services	Educational value, intrinsic natural value	Monetary valuation of cultural services *Willingness-to-pay* Non-monetary valuation of cultural services *Q-methodology, social network analysis, mental models, etc.*

2.4. The Design and Value Proposition of Climate Adaptation

Government risk management and strategic planning requires a balance of priorities. Robust quantification of the value proposition of climate change adaptation projects in economic terms, considering the direct benefits of flood risk mitigation and drought resilience, and the co-benefits of enhanced ecosystem services, can drive planning that reflects the multidimensional interests of society. This planning can feed into the project value proposition for DWH and better inform integrated water resource management via water pricing and other market-based mechanisms. The value proposition for DWH requires breaking down complexity and uncertainty with models informed by decades of detailed climate data that has been aggregated, downscaled to the appropriate region, and analyzed. The results of these models should quantify the difference between the impacts of future climate change scenarios with and without climate adaptation measures, such as a proposed distributed water harvesting system. The difference, in monetary terms, generates the measurable climate adaptation benefit over the long term with a relatively high degree of certainty.

Figure 2 below provides a broad framework to quantify the broad benefits derived from a DWH system. Internal rate of return (IRR) is the primary measure of the value or worth of an investment based on yield over the long term. Rather than quantifying the present worth or annual worth as separate entities, IRR calculates the break-even interest rate for which the project benefits are equal to the project costs [19]. In other words, IRR sets the sum of the Net Present Value (NPV) of all cash flows of a particular project equal to zero. The characterization of the NPV functions that make up the larger IRR function inherently takes into account the time-value of money, as the present value of each discrete Present Value function requires discounting the future value. This type of calculation is critical for DWH harvesting; without considering the up-front capital cost alongside the gradual increase of benefits over time, the true value of the project will not be revealed. The suggested formula for internal rate of return (IRR) as a function of {infrastructure cost, flood damage reduction, reservoir cost, drought resiliency benefit, employment benefits, crop yield benefits, ecosystem benefits from biomass, P, CO_2} offset on the diagram is thus an expansion of the more traditional IRR of flood mitigation infrastructure, with IRR as a function of {infrastructure cost, flood damage reduction}. In addition, the ability of governments to establish an institutional environment that supports innovation for biomass harvesting, energy production, and nutrient recovery significantly increases this project value proposition. There is uncertainty inherent in any IRR calculation given the use of NPV, which uses assumed interest rates. A robust assessment of uncertainty requires assessment of fluctuations of various categories of localized data, which can be assessed according to various interest rates. For example, Holopainen et al. [20] perform an uncertainty assessment for NPV calculations of forests. The study relates uncertainty to inventory data, growth models, and timber price fluctuation under assumed of 3, 4, and 5% interest rates. Similar studies must be performed to understand fluctuations of NPV, and ultimately IRR calculations, based on project valuation of DWH systems. For example, variability in hydrologic data or climate change projections will present uncertainty that must be addressed and understood to present a well-rounded assessment of present value and rate of return.

The mathematical expression for internal rate of return Figure 2 above is intentionally general, but further characterization of the mathematical expression may reflect the following, where r is the rate of return of the project, C_t is the net cash inflow during the period t, and C_0 is the net cash outflow during the same time period. As previously mentioned, the calculated IRR will be subject to uncertainty, which must be assessed on a case-by-case basis.

$$IRR = r \text{ when } \left[\sum_{t=1}^{T}\frac{C_t}{(1+r)^t} - C_0\right]_{\text{employment losses}} + \left[\sum_{t=1}^{T}\frac{C_t}{(1+r)^t} - C_0\right]_{\text{purchasing fertilizer}}$$
$$+ \left[\sum_{t=1}^{T}\frac{C_t}{(1+r)^t} - C_0\right]_{CO_2 \text{ offsets}} + \ldots = 0 \tag{1}$$

The overall value derived from the methods described above inherently require a long-term view. This is particularly important considering the need for comparability between more conventional solutions for flood mitigation as governments choose between alternatives. High performance climate adaptation solutions require that the boundaries around the cost benefit analysis expand to include the co-benefits previously described, with an understanding of the full value proposition over several decades, hence the logic of a long-term view and bond finance. The threats of climate change manifest as significant costs for governments and individuals, but only if quantified over long time horizons informed by accurate data [8]. The IRR calculation described above helps capture this characteristic in monetary terms. Figure 3 below attempts to visualize the net increasing benefits over time, by separating the short term, medium term, and long term costs and benefits. The figure clearly shows that the peak monetary costs would likely occur within the first five years of the DWH project, while the maximum benefit may be realized on a much longer time horizon. The Red River Floodway in Manitoba, Canada, is a proven historical example of such benefits. The original floodway was built to protect the City of Winnipeg between 1962 and 1968 at a cost of CAD $63 million (in 2011 Canadian dollars) [21]. Premier Duff Roblin spearheaded project development, which required significant political persistence due to the massive project scale. Since 1969, "Duff's Ditch" has prevented over CAD $40 billion of flood damage in the City of Winnipeg [21]. The Red River Floodway is an excellent example of high up-front capital costs reaping long-term benefits, grounding the concept of Figure 3 in historical context.

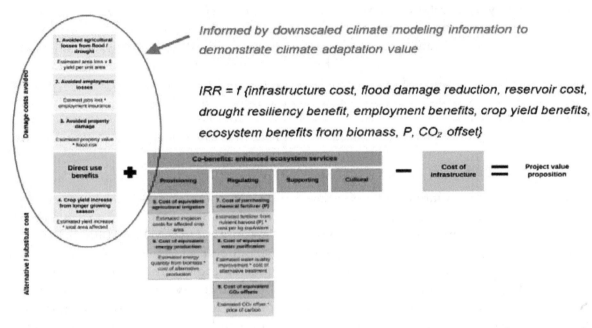

Figure 2. Framework for value proposition of distributed water harvesting infrastructure for consideration when quantifying project value in comparison to more traditional flood risk mitigation methods.

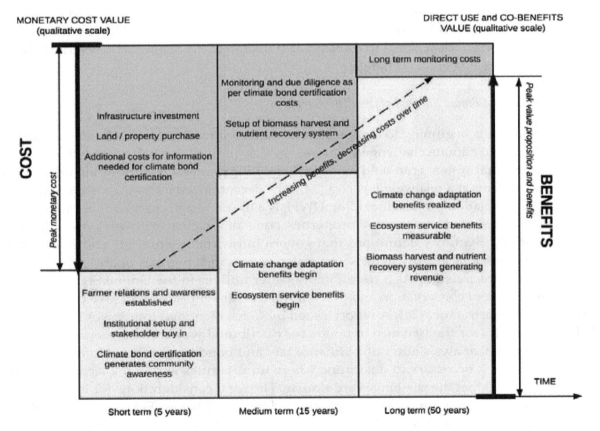

Figure 3. Temporal diagram depicting increasing benefits and decreasing costs over time, emphasizing the need to integrate the long term to understand the changing cost:benefit ratio of climate adaptation.

3. Climate Bonds for Financing Distributed Water Infrastructure

Multi-functional distributed water harvesting lies at the intersection of many challenges that are difficult for traditional debt instruments and government institutions to finance. Better climate adaptation solutions demand the sustainable development principle of subsidiarity, which in turn demands granularity in adaptation projects. Taking advantage of access to robust climate data and projections enables better engineering solutions, but it also places high demands on most aspects of financing including internal rate of return calculations, comparability to conventional projects, and the nuances of risk assessment. An emerging financing solution for climate-resilient and low carbon solutions is to use "climate-aligned bonds"—a twist on the traditional bond, a debt instrument when an investor loans money to a corporation or government for a predefined period of time on a fixed or variable interest rate [22]. These climate-aligned bonds are often unlabeled, but increasingly these bonds are certified as either "green" or "climate" bonds to provide a clear, reliable signal to investors.

3.1. Water Climate Bonds

The Climate Bonds Standard from the Climate Bonds Initiative ear-marks bonds that fund projects with very specific climate change adaptation and mitigation qualities [23]. The Canadian green bond market is growing, with Canadian labeled green bonds amounting to CAD $2.9 billion and Canadian unlabeled climate-aligned bonds amounting to CAD $30 billion [22]. The green bond label has been called into question recently, with some stakeholders questioning whether its criteria are restrictive enough to avoid "greenwashing" [22]. The Climate Bonds Initiative (CBI) uses its Climate Bond Standard (CBS), a rigorous certification and reporting process for climate adaptation and mitigation projects, to demonstrate the value of certification, incent a shift in public and investor perception, and provide a platform to highlight innovative climate-related projects. The Water Criteria under the Climate Bonds Standard were released in 2016, providing investors with "verifiable, sector-specific eligibility criteria to evaluated water-related bonds for low-carbon, climate resilient criteria" [23], with the first phase targeted toward engineered infrastructure. Adherence to the standard is determined after bond originators submit water-related issuances for certification of third party auditors [23]. Successful certification is a clear signal to investors that the project has rigorously considered its role in adapting to and mitigating climate change.

3.2. Government-Issued Bonds for Distributed Infrastructure

The water sector is beginning to embrace decentralized infrastructure as an emerging solution for modern water and climate challenges. For example, water utilities have found that distributed natural or semi-natural systems can help manage fluctuating demand and the strain on storm water and wastewater systems at a relatively low cost [24]. The decentralized nature of these systems, shared by many climate adaptation projects including DWH, is a major design challenge for financers. DWH systems are also distributed across many properties, some of which are privately owned, adding to the legal complexity. Statutory definitions that govern infrastructure projects and management of water systems have a long history, with some water governance regimes unable to accommodate for these project characteristics. As a result, many water utilities in the United States are forced to rely on cash financing of conservation and green infrastructure efforts and to save debt instruments for conventional infrastructure [24]. A report issued by Ceres identified four major themes that may enable legal authority for the issuance of bonds for distributed water infrastructure in the United States [24]. More legal analysis into public finance law and bond issuance requirements in various provinces in Canada is necessary to determine where uncertainties within the legal framework lie, but it can be assumed that the challenges are similar. The legal considerations for issuing bonds for distributed water harvesting infrastructure are outlined in Table 2 below. Financing distributed water infrastructure with bonds issued by public authorities presents some challenges, but to move forward with high performance climate adaptation and mitigation projects it is important to tap into these liquid markets.

Table 2. Legal considerations for issuance of bonds for distributed infrastructure [24].

Legal Consideration	Applicability to Distributed Water Harvesting
Bond issuer must have the legal authority to issue bonds for distributed infrastructure on private property.	Water harvesting requires financing to construct earthen dams on private property or to directly acquire the land.
The bond issuer or water utility must not be legally restrained from using enterprise revenue bonds to finance distributed infrastructure on private property, if applicable.	The provincial and federal government financing structure in Canada may limit acquisition of certain types of debt until existing debts are repaid.
	Constitutional clauses may prohibit the use of public credit for private benefit, though justifying based on the public benefit is possible (see Case Study Section 3.3.2)

Table 2. *Cont.*

Legal Consideration	Applicability to Distributed Water Harvesting
Bond issuer must structure the bond to maintain federal income tax exemptions.	Care must be taken to understand the role of farmers as private business, and to intentionally highlight and quantify public benefit.
Bond issuer must establish 'control' of the financed asset to conform to Generally Accepted Accounting Principles.	Conservation easements may act as intangible assets to ensure intended function of property and infrastructure.
	(Rebates have also been constituted as contracts with final customers in water efficiency programs.)

3.3. Case Studies

The available literature does not contain a precedent for funding DWH systems with bonds, in Canada or elsewhere. However, case studies from a variety of angles may inform the feasibility and methods for approaching the structure of a bond for this application.

3.3.1. Water Climate Bond Certified—San Francisco Public Utilities Commission [25]

The San Francisco Public Utilities Commission issued the first bond certified under the Water Criteria for the Climate Bond Standard in May of 2016. The USD $240 million will help fund projects under the Sewer System Improvement Program. The sewer and storm water systems in San Francisco are currently nearly 100 years old, and the aging infrastructure is expected to present increasingly significant risks to the region. In addition, San Francisco is located in a seismic zone and the aging structures are seismically vulnerable. By investing in large scale capital improvements now, the utilities commission hopes to avoid emergency repairs and regulatory fines, while creating broader public benefit from the improved system design. From a climate change perspective, San Francisco will experience increasing temperatures and greater intensity of downpours and storm systems that directly threaten the storm and waste water systems [26]. Certification of this project under the CBS Water Criteria is a positive signal for the possible certification of a bond financing DWH systems. Storm water and wastewater systems are distributed and decentralized by nature, involve many stakeholders, require long time horizons, and are informed by significant hydrologic complexity. These factors all exist as key institutional and technical considerations with DWH systems.

3.3.2. Bond Distributed on Private Property: Southern Nevada Water Authority [24]

The Southern Nevada Water Authority has financed its Water Smart Landscapes Program with government issued bonds. The water authority rebates customers USD $2 per square foot of grass removed and replaced with desert landscaping up to the first 5000 square feet converted per property per year. To satisfy the legal requirement to maintain control of the 'financed asset', a conservation easement is recorded against the property if the converted landscape is funded by bond funds. Again, this unique bond structure is a positive signal for the possibility to finance DWH with government bonds. The Southern Nevada Water Authority has justified the individual private benefit with the claim that public funds generate much greater public benefit. In addition, the use of conservation easements is a pertinent example of a legal structure that can overcome the legal requirement to maintain control of the asset being financed, which is also a pertinent consideration for DWH systems.

3.3.3. Canadian Green Bond: Province of Ontario [27]

The Province of Ontario Green Bond Program is leading the green bond market in Canada. The first bond issued as part of this program was a CAD $500 million bond to fund the Eglington Crosstown Light Rail Transit (LRT) project, which aims to generate public benefit and mitigate climate change impacts from multiple angles [28]. The new transit corridor will move people up to 60 percent faster than the current bus system. The LRT vehicles are electric and produce zero emissions, reducing the greenhouse gas footprint compared to the bus system. In addition, the shift of transport mode from auto to LRT is expected to further reduce the carbon footprint of the transport system. This project, and the successful issuance of a second CAD $750 million bond through the Province of Ontario Green Bond Program, demonstrates the potential liquidity of the market for financing rural projects certified under the international Certified Climate Bond Standard.

3.3.4. Asian and the Pacific Climate Bond: Asian Development Bank [29]

In early 2017 the Asian Development Bank (ADB) backed a climate bond for AP Renewables, Inc. of the Philippines. The local currency bond, equivalent to USD $225 million, is the first bond certified by the Climate Bonds Initiative to any country in Asia and the Pacific, and it is also the first ever single-project Climate Bond issued in an emerging market. The bond will finance AP Renewables' Tiwi-MakBan geothermal power generation facilities in the form of a guarantee of 75% of the principle and interest on the bond, in addition to a direct local currency ADB loan of USD $37.7 million equivalent. This landmark project demonstrates innovation in the financing realm from multiple dimensions—the opportunity for development institutions to assist developing and emerging economies in accessing new capital, the use of credit enhancement risk from the Credit Guarantee Investment Facility that has been established by ASEAN+3 governments and ADB to develop bond markets, and the proven importance of 'green' financing in emerging economies. The applicability of this financing mechanism, in addition to the DWH concept, is clearly transferable to economies all of the world with similar climatic and agricultural challenges, despite their different institutional structures and capacities.

4. Designing the System to Support Multi-Functional Distributed Water Harvesting Infrastructure and Climate Bond Certification

Implementation of a distributed water harvesting system is a complex design challenge with consideration of the engineering, property rights, environmental, institutional, and regulatory contexts. The following sections outline the starting point for implementing a DWH system on the Canadian prairies and ensuring that this setup increases the likelihood of successful bond certification under the Water Criteria of the Climate Bond Standard.

4.1. Engineering Considerations, Land and Property Ownership, and the Environment

There are several practical considerations when moving to implement water harvesting infrastructure. The list in Table 3 is not exhaustive but begins to frame the types of considerations to be made to successfully design and implement the technology solution, while incorporating the needs of various stakeholders and the technical requirements listed under the Climate Bond Standard.

Table 3. Considerations for technical/practical factors in implementing water harvesting infrastructure.

Theme	Relevant Factors to Consider
Engineering considerations	Hydrological modeling project boundaries must operate within provincial boundaries while considering river basin boundaries.
	Hydrological modeling must consider present and multiple climate change impact scenarios.

Table 3. *Cont.*

Theme	Relevant Factors to Consider
Engineering considerations	Hydrological modeling and engineering must take into account changes to water quality and water supply to all downstream.
	Siting and sequencing of location and scale of water harvesting dams and flow patterns should be optimized for physical context.
	Siting and sequencing of water harvesting dams and flow patterns should be adjusted based on external social or environmental factors if optimized physical considerations does not fit.
	Siting and sequencing of projects must meet regulated hydrological budgets based on current and future projections of water allocations.
Land and property ownership	Farmers or other property owners must be willing to sell land to municipal or provincial government.
	Farmers must be consulted on willingness to lease land back during periods when land is suitable for cultivation.
	Governments must be willing to consider easements or other mechanisms to incent farmers to allow for modifications to land and the landscape.
Environmental considerations	Siting and sequencing of projects must meet regulations on minimum environmental flows, water quality, etc.
	Water quality and flow monitoring must be in place to enable due diligence in project design and implementation.
Profit generating activities	System for harvest of biomass for local heating and/or sale for energy production must be set up for farmers to take advantage of the possible business case.

4.2. Institutional and Legal Structure

A multi-functional distributed water harvesting system requires the coordination of various stakeholders. The proper institutional and legal structure can ease project implementation and increase the likelihood of sustainable project outcomes. In addition to the institutional environment within Canada, it will be critical to consider the transboundary effects, given the shared water basins along the Canada–US border and the potential for changes to transboundary water allocation and environmental impacts. In addition to designing and implementing the technology solution, issuing bonds for the distributed, rural infrastructure and receiving certification for the bonds under the Climate Bond Standard requires an additional layer of stakeholder coordination. Table 4 identifies and explains key stakeholders involved and includes suggestions for possible stakeholders who may be well positioned to take on these roles and functions.

In addition to the key stakeholders in Table 3, the institutional environment for financing infrastructure includes several limitations and challenges. Provincial and federal governments may have limits to their debt, and bonds are only one of many avenues from which to obtain funding. If local governments are included in financing considerations, many municipal governments also face a patchwork of funding sources including provincial and federal grants. Perhaps most importantly, governments generally expect a 'net drain' on investments from infrastructure, unlike investments in other sectors such as electricity. This 'net drain' highlights the importance of implementing the biomass harvest and nutrient recovery system as soon as possible once the DWH system is operational [30]. A fundamental consideration for project design is the uncertainty of future system performance given future climate uncertainty, therefore, IRR estimates will necessarily have estimates of uncertainty that associate with the range of future climate projection, which investors should recognize and understand. The current state-of-the-art in hydraulic design is to use ensemble climate projections to analyze expected performance and variability [31,32]. A key hypothesis with respect to DWH design, and its bond value and risk management proposition is that the higher the degree of climate impact, the greater the system benefit as this class of infrastructure is designed specifically to modulate climate impacts.

Table 4. Stakeholders involved with institutional and legal structure of water harvesting infrastructure.

Role	Function	Possible Stakeholders
Project initiator	A government entity to initiate project under mandate to protect public and manage hydrology of a region.	Relevant municipal and provincial branches of governance, such as the Province of Manitoba, Province of Saskatchewan, or relevant municipalities.
Financing authority	A public lending institution that issues bonds on behalf of government entities.	Provincial lending institutions like Alberta Capital Financing Authority (ACFA), Ontario Financing Authority (OFA), or Infrastructure Ontario (IO).
Watershed management and environmental agencies and advisory committees	A broad role, this covers all agencies involved in watershed management, hydrological planning and monitoring of the region.	Canadian watershed-level entities such as Alberta Watershed Planning and Advisory Committees, Saskatchewan Watershed Advisory Committees, Manitoba Conservation Districts, Manitoba Water Council; Inter-province entities such as the Prairie Provinces Water Board; United States watershed-level entities such as North Dakota Water Resource Boards.
Regulator	Regulatory agencies that operate within and between jurisdictions with regulatory power.	A federal government agency such as Environment Canada; provincial government agencies such as Alberta Environment and Sustainable Resource Development and Department of Conservation and Water Stewardship in Manitoba; United States agency such as United States Environmental Protection Agency; transboundary agency such as International Joint Commission.
Property owners	Any individual or agency with private or public property involved with water harvesting project.	Individual property owners such as farmers; other property owners such as Ducks Unlimited.
Monitoring and verification	An agency that provides ongoing oversight into the operations, maintenance, and upgrades involved with water harvesting project.	An entity that already has monitoring responsibilities such as the Prairie Provinces Water Board, provincial water and environmental government bodies.

4.3. Climate Bond Standard Certification

Upon examination of the Climate Bond Standard Phase 1 Water Criteria [7] and the San Francisco Public Utilities Commission case, the DWH concept has the potential to be an eligible candidate for certification. Certainty requires a more in-depth analysis of the river basin in question and full scoring by the independent third party auditors commissioned by the CBI. In the case of the Canadian prairies, key stakeholders for certification include governmental stakeholders including the Government of Canada, the environmental departments of the provincial governments of Alberta, Saskatchewan, and Manitoba, inter-provincial or international (US-Canada) agencies of interest and all others listed in Table 3 above. If these stakeholders approach the project with the intention of bond financing and climate bond certification, several unique considerations emerge. For example, the CBS requires that the project boundaries for assessment only include the direct effect of the proceeds of the bond [33]. It is likely that the most suitable project boundary for a DWH system is a river basin, with additional consideration of provincial boundaries prompted by the CBS criteria. The project must also qualify under criteria for all certified bonds, criteria for sector-specific bonds, and broader human rights and environmental considerations for water management before being considered for CBS certification [33]. This requirement may also prompt more intentional engagement with community members and civil society.

The CBS Water Criteria are separated into two streams: projects primarily for climate adaptation and projects primarily for climate mitigation. Water harvesting clearly falls under the climate adaptation criteria. Evaluation for CBS certification is based on a Scorecard system, in which a range of criteria are evaluated for no points, half points, or full points. The evaluation starts with a Vulnerability Assessment, followed by an Adaptation Plan if deemed necessary by the Vulnerability Assessment. Rough consideration of the criteria and the integrated nature of DWH systems indicate that they would likely require the Adaptation Plan. The Vulnerability Assessment is split into three major categories described in Table 5 below.

In some cases, the water harvesting concept may exceed the criteria in the way they are currently written, while in other cases the criteria are limiting. In addition, DWH projects are inherently climate adaptation projects, and thus the requirement for an Adaptation Plan presents an opportunity to highlight this functional purpose. The following sections are based on the CBS Water Criteria for Phase 1: Engineered Infrastructure [7], and may inform upcoming iterations of the criteria for natural or semi-natural systems. The applicability of CBS water criteria to the water harvesting system is broken out in more detail in the following sections. This evaluation is partially informed by the 2015 Organization for Economic Cooperation and Development (OECD) report, *Water Resources Allocation: Sharing Risks and Opportunities* [34], which evaluates institutional gaps in water allocation policy in Alberta and Manitoba. The sections below focus on these two provinces.

Table 5. Vulnerability Assessment section themes (as per Climate Bonds Initiative (CBI) requirements).

Theme	Description
Allocation	Assesses how water is shared by users within a given basin or aquifer, concentrating on the potential impacts of bond proceeds on water allocation.
Governance	Assesses how or whether the proceeds of the bond take into account the ways in which water will be formally shared, negotiate, and governed.
	Assesses compliance with allocation mechanisms that protect water resources.
Diagnostic	Assesses how or whether the use of the proceeds takes into account changes to the hydrologic system over time.
Adaptation Plan	If Vulnerability Assessment reveals significant climate change impacts on the project, the Adaptation Plan must be created as a management response plan to the conclusions and findings of the Vulnerability Assessment, noting how identified climate risks will be addressed.

4.3.1. Meeting the Criteria

A strong institutional environment on the Canadian prairies already exists, increasing the likelihood for a DWH system on the Canadian prairies to be certified under the CBS criteria. Accountability mechanisms for management of water allocation at different institutional, spatial, and temporal scales are established by water management plans, water code statutes, and compliance mechanisms that are in place in the regions in question. For example, water monitoring is performed by the Prairie Provinces Water Board, Alberta Environment and Sustainable Resource Development (ESRD), and the Department of Water Conservation and Stewardship in Manitoba. Scientific hydrological services that inform monitoring of adherence to codes already exist in current institutions like Manitoba's Water Stewardship Division. Furthermore, some elements of water allocation policies are already designed as required by the CBS criteria. For example, Alberta and Manitoba have differentiated entitlements based on the level of security of supply or risk of water shortage [34]. Both provinces have sanctions for withdrawal over limits. New entitlements or the increase of existing entitlements requires assessment of third party impacts, an environmental impact assessment, and that existing users forgo use [34]. In Alberta, minimum environmental flows are considered, and monitoring and enforcement mechanisms are in place in both Manitoba and Alberta [34]. Manitoba's Water Use Licensing Section monitors compliance for agriculture, domestic, and industrial water use by metering [34]. Allocation is enforced through sanctions with fines, and conflicts are resolved through the normal application of principles of good governance [34]. Alberta ESRD monitors and enforces water allocation for agriculture, domestic use, energy production, and the environment through metering and drawing penalties for contravening the enforcement order. Part of the sanction actions may also include fines or imprisonment, and formal conflict resolution is included under Section 93 of the Alberta Water Act. These existing institutional frameworks are key components of climate bond certification.

4.3.2. Exceeding the Criteria

The nature of the multi-functional water harvesting solution for the Canadian prairies exceeds the CBS criteria in several ways, though these are not necessarily captured in the formal CBS Scorecard.

For example, the CBS criteria requires a connection between water resource management at the project and hydrologic scale. Because a DWH system is based entirely upon the hydrologic scale, the boundaries of the bond proceeds and the hydrologic scale are one and the same. The criteria also include requirements for specific data, flow criteria, modeling scenarios, and water users to be included in hydrologic modeling. The hydrologic models used to design the DWH systems on the Canadian prairies would easily integrate these requirements in a manner that complies with the CBS criteria. For example, a dynamic simulation model of a DWH climate adaptation system was recently conducted for a portion of a watershed downstream of Pelly's Lake, Manitoba, Canada. This simulation model integrates physical variables related to the landscape, energy balance, moisture fluxes, hydrologic cycle with operational climate forecasting tools to understand the multi-purpose benefits of the system and to estimate their economic value [6]. Furthermore, the use of downscaled climate data and quantification of future climate impact scenarios with and without the system increases certainty about the future success of the system, beyond the requirements of the CBS criteria. The quality and breadth of information put into these hydrological models, environmental impact assessments, and other assessment mechanisms that are part of the planning and design process benefit the climate bond certification process by informing a rigorous Adaptation Plan. More importantly, the use of downscaled climate change data with rigorous hydrologic modeling to design DWH systems demonstrates a fundamental shift towards greater certainty for context-specific system functionality as a climate adaptation solution under long range climate impacts.

4.3.3. Challenges with the Criteria

Some institutional gaps in water management on the Canadian prairies and the current structure of the CBS Water Criteria present some challenges for certification. Water allocation agreements must be dynamic to accommodate changes to flow scenarios with new water harvesting infrastructure, so adherence to the criteria may not be clear until the planning process is mature. Additionally, inconsistent provincial water allocation policies reveal weaknesses in water governance in some provinces. Manitoba does not define its environmental flows, and while freshwater biodiversity is considered on a project-by-project basis, terrestrial biodiversity is not considered. Return flow obligations are not specified, and the nature of water entitlements is based on the purpose of water allocation, maximum area irrigated, and the maximum volume removed, rather than as a proportion of total flow conditions. Alberta has a more rigorous policy framework, but its water allocation is currently classified as 'over-allocated'. These institutional gaps should not only be addressed to allow for climate bond certification, but also as part of an effort to establish best practices for water management.

5. Distributed Water Harvesting and Climate Bonds in the International Context

The Canadian Prairies are not the first or only agricultural region to be confronted with increasing pressure driven by climate change impacts—globally 80 percent of agricultural land is rainfed making up 65 to 70 percent of staple food crops [35]. Model output of mean climatic changes are far more robust than changes to climate variability, meaning that the full impacts of climate change are likely seriously underestimated [35]. However, just as with the Canadian example, the interactions of different climatic stresses on biological and food systems over time in different regions all over the world require investigation of localized changes over time. Variability in rainfall is demonstrated as the principle cause of inter-annual variability in crop yields at both aggregate and plot level [35]. Semi-arid and arid environments around the world are projected to face similar challenges that may be solved by DWH solutions or other distributed agricultural adaptation solutions financed by ear-marked climate bonds. For example, rainfall variability in the Middle East and the Mediterranean region is projected to result in an overall drier climate, with an impact on major river systems and food productivity [36]. Specific impacts are disparate across this region—rainfall is expected to decrease in southern Europe, Turkey, and the Levant, while rainfall in the Arabian Gulf may increase [36]. Still, in the former example, rainfall is expected to increase in the winter and decrease in the summer [36], affecting crop productivity differently in each growing season. In another locale, studies have also shown one of the highest agricultural productivity losses due to climate change scenarios is predicted in India [37]. Though temperatures are expected to rise and annual precipitation rates to remain stable, regional variability is expected to result in extreme changes to both surface and groundwater due to the changes in temporal rainfall variability [37]. Several countries in sub-Saharan Africa also rely heavily on rainfed agriculture, and expect a higher frequency of droughts and rainfall variability in the future [38]. DWH solutions, or some derivate of the technology, is likely to be necessary in regions with high dependence on rainfed agriculture and projected rainfall variations.

The existence of rainfed agriculture and current or projected climate change impacts is not enough to determine the suitability of DWH solutions or financing via the use of labeled or unlabeled climate bonds. An institutional environment conducive to such multi-stakeholder, rural-based solutions must exist or be managed to achieve the maximum return on project investment and ensure the system is used appropriately. Any institution or entity that is set up to issue a bond has the ability to issue a green bond, and if institutional capacity meets the requirements, may be certified under the Climate Bond Standard. Southern Europe and the Middle East may be well-served by such distributed engineering solutions, and may also be set up to access the pool of capital offered by green or climate bonds. In addition, developing countries face low visibility on low carbon projects because of the high cost of capital and higher interest rates, despite a significant need for climate-friendly infrastructure investment [39]. Development institutions, as demonstrated by the Asian Development Bank, are well positioned to facilitate and support such enabling environments. This paper has demonstrated the application of DWH harvesting and climate bond certification and financing in one locale, but several other contexts requires a similar approach, adapted to the local agricultural and climate system, institutional circumstance, and financing environment.

6. Conclusions and Recommendations

A multi-functional distributed water harvesting system on the Canadian prairies financed with government-issued bonds that are certified under the Water Criteria for the Climate Bond Standard presents a feasible, innovative climate adaptation solution for the increased temperatures and variable precipitation expected to strain agriculture in the region in the coming decades. Successfully implementing this solution requires stakeholder coordination, an institutional lens, and innovative engineering methods. In addition, lessons learned from the analysis contained in this paper can inform the establishment of CBS criteria for natural and semi-natural water infrastructure.

It is recommended that institutions involved with water management and public infrastructure on the Canadian prairies think creatively about their role in driving and supporting innovative climate adaptation projects. Taking advantage of the growing green bond market potential and learning from the success of the green bond initiatives in the Province of Ontario requires that more financial institutions recognize their value and build programs to support them. For example, the Liberal government's proposed Canadian Infrastructure Bank and other existing financers can consider green bonds as an opportunity to aggregate projects for risk reduction and public benefit and to access an otherwise exclusive pool of private capital. Assessing the true value of innovative solutions, particularly distributed climate adaptation projects, requires that governments consistently establish a long-term view that quantifies direct monetary ecosystem service benefits and co-benefits. This lens should not only be adopted to inform the full economic value for projects with direct environmental or climate adaptation benefits. A report from the Ministry of Environment in Sweden recommends the inverse view; that "...government should investigate different strategies to improve transparency regarding the dependence and impact of bond investments on the ecosystem services, including investments by the national pension funds" [40]. Taking care to involve existing stakeholders through all phases of visioning and implementation of a DWH system will take advantage of existing institutional capacity and help anticipate demands to fill institutional gaps. Stakeholder involvement should also include a comprehensive community benefits framework and active community engagement, as was established alongside the Eglington Crosstown LRT project under the Province of Ontario Green Bond program. Prairie Provinces may need to also consider tightening up water allocation policies to fill the identified gaps. Engineers, hydrologists, and environmental scientists must also consider their role in designing an effective system and using the requirements of the CBS Water Criteria to inform robust hydrological modeling and engineering practices. These stakeholders must also take care to build the business case and supply chain connections for farmers to harvest biomass, generate bioenergy, and recover nutrients, in order to capitalize on long-term project value and protect downstream water bodies from excess nutrient accumulation. All stakeholders that have a potential role in the design and implementation of a water harvesting system, financing the project under certified climate bonds, or creating an appropriate policy environment, must be aware of the complexity of the space and importance of demonstrating effective climate adaptation solutions.

A multi-functional distributed water harvesting system can enable agricultural productivity on the Canadian prairies in the face of climate change. Successfully implementing and financing a DWH project requires that stakeholders understand the value of the direct climate adaptation benefits and enhanced ecosystem services, actively pursue the business case generated alongside the public benefit, and generate buy-in and momentum through active institutional and community engagement. Financing a DWH project, and other distributed water infrastructure, with government bonds is possible if the bond is structured with consideration of the legal authority of the bond issuer. Seeking Climate Bond Standard certification creates an additional incentive for robust project design, takes advantage of an untapped pool of private capital, and demonstrates the full value that decades of climate data and refined hydrologic knowledge can bring to infrastructure solutions. Lastly, the Phase 1 Water Criteria for the CBS rewards water and wastewater projects that have shown adequate proof that climate adaptation and mitigation have been considered as design constraints. It is recommended that as the Climate Bonds Initiative develops water criteria for natural or semi-natural infrastructure, it might consider finding ways to explicitly reward projects that have a functional purpose of climate adaptation or mitigation rather than simply as a design consideration of a project with a different functional purpose. The analyses and recommendations contained in this paper are directed toward implementation of a DWH systems on a hypothetical river basin on the Canadian prairies, but it is evident that this solution is transferable to many regions with similar climate change effects and agricultural systems that will cause climate adaptation challenges in the future.

Acknowledgments: The manuscript was conceived through a partnership between the International Institute for Sustainable Development–Prairie Climate Centre and the Central European University Budapest under the direction of Laszlo Pinter at the Central European University. No grant funds were allocated to this project.

Author Contributions: Anita Lazurko and Henry David Venema both contributed to the analysis and wrote the manuscript together. All authors approved the final manuscript.

References

1.   G20 Leaders' Communique. Hangzhou Summit. 4–5 September 2016. Available online: http://www. consilium.europa.eu/press-releases-pdf/2016/9/47244646950_en.pdf (accessed on 9 April 2017).
2.   Granoff, I.; Hogarth, J.R.; Miller, A. Nested barriers to low-carbon infrastructure investment. *Nat. Clim. Chang.* **2016**, *6*, 1065–1071. [CrossRef]
3.   OECD. *Mobilising Bond Markets for a Low-Carbon Transition*; OECD Publishing: Paris, France, 2017.
4.   Neumann, J.E.; Strzepek, K. State of the literature on the economic impacts of climate change in the United States. *J. Benefit Cost Anal.* **2014**, *5*, 411–443.
5.   Gray, J.H. *Men against the Desert*; Western Producer Prairie Book: Saskatoon, SK, Canada, 1967.
6.   Berry, P.; Yassin, F.; Belcher, K.; Lindenschmidt, K. An economic assessment of local farm multi-purpose surface water retention systems under future climate uncertainty. *Sustainability* **2017**, *9*, 456. [CrossRef]

7.   Climate Bonds Initiative (CBI). The Water Criteria of the Climate Bonds Standard: Phase 1: Engineered Infrastructure. Available online: https://www.climatebonds.net/files/files/Water_Criteria_of_the_ClimateBondsStandard_October2016.pdf (accessed on 15 January 2017).

8.   Blair, D.; Mauro, I.; Smith, R.; Venema, H. Visualising Climate Change Projections for the Canadian Prairie Provinces. Available online: http://dannyblair.uwinnipeg.ca/presentations/blair-ottawa.pdf (accessed on 5 February 2017).

9.   Betts, A.; Desjardins, R.; Worth, D.; Cerknowiak, D. Impact of land use change on diurnal cycle climate on the Canadian prairies. *J. Geophys. Res. Atmos.* **2013**, *118*, 11996–12011. [CrossRef]

10. Statistics Canada. Environment Accounts and Statistics Division. Agricultural Water Survey (Survey Number 5145). 2011. Available online: http://www.statcan.gc.ca/pub/16-402-x/2011001/ct002-eng.htm (accessed on 5 February 2017).

11. Environment and Climate Change Canada. Canadian Environmental Sustainability Indicators: Nutrients in Lake Winnipeg. Environment and Climate Change Canada, 2016. Available online: https://www.ec.gc.ca/indicateurs-indicators/55379785-2CDC-4D18-A3EC-98204C4C10C4/LakeWinnipeg_EN.pdf (accessed on 5 February 2017).

12. Manitoba 2011 Flood Review Task Force Report: Report to the Minister of Infrastructure and Transportation. Available online: https://www.gov.mb.ca/asset_library/en/2011flood/flood_review_task_force_report.pdf (accessed on 10 March 2017).

13. Van Meter, K.; Steiff, M.; McLaughlin, D.; Basu, N. The sociohydrology of rainwater harvesting in India: Understanding water storage and release dynamics across spatial scales. *Hydrol. Earth Syst. Sci.* **2016**, *20*, 2629–2647. [CrossRef]

14. Grosshans, R.; Grieger, L.; Ackerman, J.; Gauthier, S.; Swystun, K.; Gass, P.; Roy, D. Cattail Biomass in a Watershed-Based Bioeconomy: Commercial-Scale Harvesting and Processing for Nutrient Capture, Biocarbon, and High-Value Bioproducts. Available online: https://www.iisd.org/sites/default/files/publications/cattail-biomass-watershed-based-bioeconomy-commerical-scale-harvesting.pdf (accessed on 10 January 2017).

15. Quiring, S.; Papakryiakou, T. An evaluation of agricultural drought indices for the Canadian prairies. *Agric. For. Methodol.* **2003**, *118*, 49–62. [CrossRef]

16. Aylward, B.; Bandyopadhyay, J.; Belausteguigotia, J.; Borkey, P.; Cassar, A.; Meadors, L.; Saade, L.; Siebentritt, M.; Stein, R.; Sylvia, T.; et al. Chapter 7 Freshwater ecosystem services. In *Millennium Ecosystem Assessment*; Island Press: Washington, DC, USA, 2005.

17. Millenium Ecosystem Assessment (MEA). *Ecosystems and Human Well-Being Synthesis*; Island Press: Washington, DC, USA, 2005.

18. Gomez-Baggethun, E.; Martin-Lopez, B.; Barton, D.; Braat, L.; Kelemen, E.; Lorene, M.; Saarikoski, H.; van den Bergh, J. State-of-the-art report on integrated valuation of ecosystem services Deliverable D.4.1/WP4. 2014. Available online: http://www.openness-project.eu/sites/default/files/Deliverable%204%201_Integrated-Valuation-Of-Ecosystem-Services.pdf (accessed on 15 March 2017).

19. Newnan, D.; Eschenbach, T.; Lavelle, J. *Engineering Economic Analysis*, 9th ed.; Oxford University Press: New York, NY, USA, 2004.

20. Holopainen, M.; Mäkinen, A.; Rasinmäki, J.; Hyytiäinen, K.; Bayazidi, S.; Vastaranta, M.; Pietilä, I. Uncertainty in forest net present value estimations. *Forests* **2010**, *1*, 177–193. [CrossRef]

21. Province of Manitoba. Flood Information: Red River Floodway. 2011. Available online: http://www.gov.mb.ca/flooding/fighting/floodway.html (accessed on 10 February 2017).

22. CBI. Bonds and Climate Change: The State of the Market. Climate Bonds Initiative, 2016. Available online: https://www.climatebonds.net/files/files/reports/cbi-hsbc-state-of-the-market-2016.pdf (accessed on 10 January 2017).

23. CBI. Climate Bonds Standard, Version 2.1. 2016. Available online: https://www.climatebonds.net/standards/standard_download (accessed on 10 January 2017).

24. Leurig, S.; Brown, J. Bond Financing Distributed Water Systems: How to Make Better Use of Our Most Liquid Market for Financing Water Infrastructure. Report for Ceres. 2014. Available online: https://www.ceres.org/resources/reports/bond-financing-distributed-water-systems-how-to-make-better-use-of-our-most-liquid-market-for-financing-water-infrastructure (accessed on 10 January 2017).

25. San Fransisco Public Utilities Commission, Climate Bond Certified. Announcement from CBI. Available online: https://www.climatebonds.net/standards/certification/SFPUC (accessed on 10 January 2017).

26. Ekstrom, J.A.; Susanne, C.M. Climate Change Impacts, Vulnerabilities, and Adaptation in the San Francisco Bay Area: A Synthesis of PIER Program Reports and Other Relevant Research. California Energy Commission; Publication Number: CEC-500-2012-071; 2012. Available online: http://www.energy.ca.gov/2012publications/CEC-500-2012-071/CEC-500-2012-071.pdf (accessed on 15 March 2017).

27. Ontario Financing Authority (OFA). Green Bond Presentation: Province of Ontario. Available online: https://www.ofina.on.ca/pdf/ontario_greenbonds_presentation_jan2016_en.pdf (accessed on 5 February 2017).

28. Stear Davies Gleave. *Eglington Crosstown Rapid Transit Benefits Case Final Report*; Prepared for Metrolinx; Stear Davies Gleave: Toronto, ON, Canada, 2009; Available online: http://www.metrolinx.com/en/regionalplanning/projectevaluation/benefitscases/Benefits_Case-Eglinton_Crosstown_2009.pdf (accessed on 15 March 2017).

29. Asian Development Bank (ADB). ADB Backs First Climate Bond in Asia in Landmark $225 Million Philippines Deal. 2016. Available online: https://www.adb.org/news/adb-backs-first-climate-bond-asia-landmark-225-million-philippines-deal (accessed on 15 March 2017).

30. Siemiatycki, M. Creating an Effective Canadian Infrastructure Bank. Independent Research Study prepared for Residential and Civil Construction Alliance of Ontario. 2016. Available online: http://www.rccao.com/research/files/02_17_RCCAO_Federal-Infrastructure-Bank2016WEB.pdf (accessed on 20 February 2017).

31. Tabari, H.; Troch, R.D.; Giot, O.; Hamdi, R.; Termonia, P.; Saeed, S.; Brisson, E.; Lipzig, N.V.; Willems, P. Local impact analysis of climate change on precipitation extremes: Are high-resolution climate models needed for realistic simulations? *Hydrol. Earth Syst. Sci.* **2016**, *20*, 3843–3857. [CrossRef]

32. Kavvas, M.L.; Ishida, K.; Trinh, T.; Ercan, A.; Darama, Y.; Carr, K.J. Current issues in and an emerging method for flood frequency analysis under changing climate. *Hydrol. Res. Lett.* **2017**, *11*, 1–5. [CrossRef]

33. Matthews, J.; Timboe, I. Guidance Note for Issuers and Verifiers: Phase 1: Engineered Infrastructure. Supplementary note to the Water Criteria. Climate Bonds Initiative, 2016. Available online: https://www.climatebonds.net/files/files/Water_Criteria_Guidance_Note_to_Issuers%26Verifiers_October_2016(1).pdf (accessed on 10 January 2017).

34. OECD. *Water Resources Allocation: Sharing Risks and Opportunities*; OECD Studies on Water; OECD Publishing: Paris, France, 2015; Available online: http://www.oecd.org/fr/publications/water-resources-allocation-9789264229631-en.htm (accessed on 5 February 2017).

35. Thornton, P.; Ericksen, P.; Herrero, M.; Challinor, A. Climate variability and vulnerability to climate change: A review. *Glob. Chang. Biol.* **2014**, *20*, 3313–3328. [CrossRef] [PubMed]

36. Lelieveld, J.; Hadjinicolaou, P.; Kostopoulou, E.; Chenoweth, J.; El Maayar, M.; Giannakopoulos, C.; Hannides, C.; Lange, M.A.; Tanarhte, M.; Tyrlis, E.; et al. Climate change and impacts in the Eastern Mediterranean and the Middle East. *Clim. Chang.* **2012**, *114*, 667–687. [CrossRef] [PubMed]

37. Asha latha, K.V.; Gopinath, M.; Bhat, A.R.S. Impact of climate change on rainfed agriculture in India: A case study of Dharwad. *Int. J. Environ. Sci. Dev.* **2012**, *3*, 368–371.

38. Cooper, P.J.M.; Dimes, J.; Rao, K.; Shapiro, B.; Shiferaw, B.; Twomlow, S. Coping better with current climatic variability in the rain-fed farming systems of sub-Saharn Africa: An essential first step in adapting to future climate change? *Agric. Ecosyst. Environ.* **2008**, *126*, 24–35. [CrossRef]

39. Nelson, D.; Shrimali, G. Finance Mechanisms for Lowering the Cost of Renewable Energy in Rapidly Development Countries. Report from Climate Policy Initiative Series 2014. Available online: https://climatepolicyinitiative.org/wp-content/uploads/2014/01/Finance-Mechanisms-for-Lowering-the-Cost-of-Clean-Energy-in-Rapidly-Developing-Countries.pdf (accessed on 10 January 2017).

40. Schultz, M. *Making the Value of Ecosystem Services Visible*; Summary of the Report of the Inquiry M 2013:01 Ministry of the Environment; Swedish Government Inquiries: Stockholm, Sweden, 2013. Available online: https://www.cbd.int/financial/hlp/doc/literature/sammanfattning_engelska_1301105.pdf (accessed on 5 February 2017).

9

Warming Winters Reduce Chill Accumulation for Peach Production in the Southeastern United States

Lauren E. Parker [1,2,*] and John T. Abatzoglou [3]

[1]   USDA California Climate Hub, Davis, CA 95616, USA
[2]   John Muir Institute of the Environment, University of California, Davis, CA 95616, USA
[3]   Department of Geography, University of Idaho, Moscow, ID 83844, USA
*   Correspondence: leparker@ucdavis.edu

Abstract: Insufficient winter chill accumulation can detrimentally impact agriculture. Understanding the changing risk of insufficient chill accumulation can guide orchard management and cultivar selection for long-lived perennial crops including peaches. This study quantifies the influence of modeled anthropogenic climate change on observed chill accumulation since 1981 and projected chill accumulation through the mid-21st century, with a focus on principal peach-growing regions in the southeastern United States, and commonly grown peach cultivars with low, moderate, and high chill accumulation requirements. Anthropogenic climate change has reduced winter chill accumulation, increased the probability of winters with low chill accumulation, and increased the likelihood of winters with insufficient chill for commonly grown peach cultivars in the southeastern United States. Climate projections show a continuation of reduced chill accumulation and increased probability of winters with insufficient chill accumulation for cultivars with high chill requirements, with approximately 40% of years by mid-century having insufficient chill in Georgia. The results highlight the importance of inter-annual variability in agro-climate risk assessments and suggest that adaptive measures may be necessary in order to maintain current peach production practices in the region in the coming decades.

Keywords: chill accumulation; climate change; peaches; perennial crops; Georgia; South Carolina

1. Introduction

The peach industry in the southeastern United States (SEUS) has been a part of the regional iconography since at least the mid-1920s, and was historically an important part of the agricultural economy [1]. While California's current peach production dwarfs that of Georgia and South Carolina [2], the industry in the SEUS continues to contribute millions to regional, state, and local economies [3], and peaches remain important to regional identity [1]. In 2017, approximately 80% of Georgia's peach crop and 90% of South Carolina's peach crop were damaged due to warm winter temperatures. The warm conditions resulted in insufficient winter chill accumulation in some areas, while other parts of the SEUS were impacted when an early bloom, due to unseasonably warm temperatures, was followed by a mid-March freeze. In Georgia, an estimated 70% of the total 2017 peach losses were attributed to inadequate chill and 10–15% of the losses, the result of a spring freeze [4]. The combined impacts of anomalously low chill accumulation and spring freeze yielded substantial economic damage across the region [5]. Given the role of the peach industry in both the economy and culture of the SEUS, the 2017 crop failure garnered much public interest including whether such warm winters and impacts to perennial agriculture may become more commonplace in the coming decades.

Like other fruit trees, peaches undergo a series of physiological changes during the fall that allow for the onset of dormancy, when growth and development are slowed or stopped and the plant is better able to tolerate cold temperatures. Many perennial crops must be exposed to a certain amount of cold temperatures, or chill, during this period of dormancy to continue their development in the spring [6]. Peach cultivation is governed by a number of climatic factors such as cold hardiness, frost tolerance, and sufficient heat accumulation. Peach cultivars are frequently selected based on climatological chill accumulation [7] as insufficient chill accumulation can reduce flower quality, inhibit pollination and fruit development, and lower fruit quality and yield [6,8,9], with subsequent economic impacts to both growers and consumers [10].

Observational studies have shown warming in both the mean and extreme cold winter temperatures over the past half century across the US [11–13], much of which is consistent with anthropogenic forcing [14] and is expected to continue under climate change [15,16]. The exceptions of observed warming trends are primarily found in the warming hole across parts of the SEUS where winter temperatures cooled and spring onset trended later over the latter half of the 20th century [17,18]. The warming hole is likely a consequence of internal variability of the climate system that has buffered the influence of anthropogenic forcing to date, but is not expected to persist into the coming decades [17]. While it is acknowledged that chill accumulation is only one of many thermal-metrics that might directly impact crop suitability in a changing climate [19], declines in chill accumulation have been observed in some regions [20] and are projected to decline further [21]. Likewise, increases in winter temperatures are projected to reduce chill accumulation below the thresholds needed for peach cultivars in many peach-growing portions of the US [20,22].

In view of recent crop impacts due to warm winters, we examine chill accumulation across the SEUS in the context of ongoing climate change with a focus on implications for peach cultivation. First, a first-order estimate is provided of the contribution of anthropogenic climate change to observed low chill accumulation winters in the SEUS and years with insufficient chill in prime peach-growing areas in Georgia and South Carolina during 1981-2017. Secondly, using a suite of downscaled climate projections, changes in chill accumulation, the frequency of low-chill winters, and changes in the risk of winters with insufficient chill for common peach cultivars in the coming decades were investigated. Comprehensively, this study presents methodologies that may be applied to agro-climate metrics for conducting climate change risk and impact analyses for perennial crop systems globally, and provides a risk assessment of insufficient chill for peaches—and general chill accumulation for other perennials—in the SEUS, presenting information useful for climate-informed decision making.

2. Materials and Methods

Two primary datasets were used in this study (available at https://data.nkn.uidaho.edu/). First, the observed daily maximum and minimum temperature (T_{max}, T_{min}) at a ~4-km spatial resolution for the period 1981–2017 for the SEUS [25°–35.2° N, 78.5°–88.5° W (see Figure 1a)] were acquired from the gridded surface meteorological dataset (gridMET) of [23]. Previous validation of gridMET showed high correlation and low bias of temperature when compared to meteorological station observations across the US [23], and comparisons in chill accumulation between gridMET and data from 50 SEUS meteorological stations from 1980–2017 showed strong spatial correlation ($r = 0.99$), with a mean absolute error of 50 chill hours and a median bias of -17 chill hours (analysis not shown). Second, the projections of daily T_{max} and T_{min} from 20 global climate models (GCMs) that participated in the fifth phase of the Climate Model Intercomparison Project (CMIP5) were statistically downscaled using the multivariate adaptive constructed analogs (MACA) method [24]. The MACA used gridMET as training data, thereby ensuring compatibility in contemporary climate statistics between the downscaled GCM experiments and gridded observations. The analysis of climate projections was constrained to simulations for the early (2010–2039) and mid- (2040–2069) 21st century periods given the limited ability for developing meaningful management strategies relevant to the end-of-century projections. Further, we focused on future experiments run under the Representative Concentration Pathway 4.5 (RCP 4.5) to provide a conservative estimate of projected changes in chill accumulation. The projections using RCP 8.5 would likely show similar qualitative changes, but with larger magnitudes, particularly for the mid-21st century where multi-model mean changes in winter mean temperatures show an additional 0.6 °C warming above RCP 4.5, although the variability among models exceeds the difference between RCP 4.5 and RCP 8.5 for the time horizons highlighted herein.

A first-order estimate is provided on the influence of anthropogenic climate change on observed 1981–2017 chill accumulation using a large ensemble of CMIP5 simulations and a pattern scaling approach that allows for comparisons between rates of local and global change [25]. The differences in monthly T_{max} and T_{min} as simulated by 23 different GCMs at their native spatial resolution were taken between two 30-year periods, 1850–1879 and 2070–2099. The pattern scaling approach allows the expression of modeled rates of regional change for an individual variable and month to modeled rates of change in the global mean annual temperature. This approach assumes a linear relationship between the variables, which is reasonable for climate change timescales [25]. The pattern scaling was calculated separately for each model, as well as for the 23-model median. For each model, the anthropogenic climate change signal was defined for monthly T_{max} and T_{min} by multiplying the monthly varying pattern scaling function by an 11-year moving average of the change in the modeled global mean annual temperature relative to each model's 1850–1879 baseline. It is acknowledged that this is one of several first-order approaches for approximating the modeled influence of anthropogenic climate change over the historical record [26,27].

Following [26], a time series of daily T_{max} and T_{min} for 1981–2017 for the SEUS was created that preserves the observed interannual climate variability, but removes the influence of modeled anthropogenic climate change by subtracting the estimated difference in modeled monthly temperature anomalies (relative to the 1850–1879 baseline) using pattern scaling from the observed temperatures. These counterfactual scenarios do not make an effort to discern the sources of change in the observed data. Rather, they provide an approach for estimating the proximal effects of modeled anthropogenic climate change in the context of real-world observations.

The peach location data were obtained from the 2016 United States Department of Agriculture—National Agricultural Statistics Service Cropland Data Layer (CDL, available at https://www.nass.usda.gov/Research_and_Science/Cropland/Release/index.php) for the SEUS states of Alabama, Georgia, South Carolina, and Florida [28]. Approximately 94-km^2 were classified as peach in the 2016 Southern CDL with nearly all of the orchards located in Georgia (~34.5-km^2) and South Carolina (~58-km^2) (Figure 1a). The 30-m resolution CDL data was aggregated to the common 4-km resolution of the climate data for analyzing chill accumulation over peach-growing locations, summing the number of 30-m peach cells within each 4-km grid cell. The peach-growing locations were classified as those 4-km grid cells with >0.01% peach density. Finally, in order to provide locally-relevant results in addition to the regional analysis, our peach cultivar-specific analysis focused on peach locations within a 4-county area of central Georgia and a 3-county area in the Piedmont region of South Carolina that are responsible for ~75% and ~50% of each state's peach production, respectively.

Estimates of chill accumulation derived from chilling models are used for selecting appropriate crop species and cultivars, and to track plant phenology for farm management practices [29,30]. While there are multiple modeling approaches for calculating chill, the Weinberger Chilling Hours Model [31] was utilized as chill requirements for SEUS peaches are most commonly reported in chilling hours. The chill thresholds for peach cultivars examined in this study were quantified using the Weinberger model in central Georgia. Further, this model is commonly used to track winter chill accumulation across the SEUS as part of the online tools available through regional university consortiums and university extension programs (e.g., http://agroclimate.org/; http://weather.uga.edu/), and as such, using this chill model allows for the most direct translation of this work to end users.

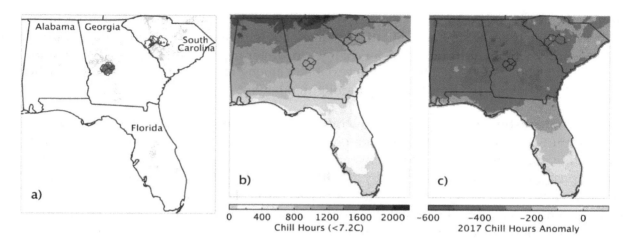

Figure 1. (^a) The southeastern US study area. 4-km cells with $^>$0.01% peach density are highlighted in grey. Georgia and South Carolina peach-growing counties examined explicitly in this study are outlined in grey and those cells with >0.01% peach density within these counties are highlighted in red (Georgia) and purple (South Carolina). (**b**) The average annual number of chill hours for the 1981–2017 observed period. Areas with <100 chill hours are masked in grey. (**c**) The winter chill accumulation anomaly in 2017 compared to the 1981–2017 average. Areas masked in grey as in (**b**).

The Chilling Hours Model sums the number of hours per day with temperatures <7.2 °C; hourly data were temporally disaggregated from daily T_{max} and T_{min} using a modified sine curve model [32]. Annual chill accumulation was considered from 1 October to 15 February, as is standard in the SEUS peach industry [33]. Peach chill requirements were obtained from the University of Georgia [34] for three cultivars grown in the SEUS. Gulfprince and Juneprince peaches require 400 and 650 chill hours, respectively, and are hereafter referred to as low- and moderate-chill cultivars. The Elberta peach cultivar (hereafter referred to as high-chill) requires 850 chill hours and is a cultivar standard to which the phenology of other peach cultivars is compared [34]. It is noted that not all of these cultivars are grown across all peach-growing locations of Georgia and South Carolina. Gulfprince is a cultivar grown primarily in southern Georgia, while central Georgia principally grows peaches with chill requirements ≥600 chill hours (Dario Chavez, University of Georgia Extension Specialist, personal communication). However, these three cultivars have been included as exemplary of the range of chill requirements across SEUS-grown peaches. By including the low-chill cultivar in our analyses of selected South Carolina and Georgia peach-growing counties, we show the capacity for these counties to continue to produce peaches under future climate conditions should future chill accumulation limit the productivity of the currently-grown moderate- and high-chill cultivars.

Chill accumulation was calculated over the 1981–2017 period with the observational data, and for the counterfactual scenarios using the observed data from 1981–2017 after removing the influence of anthropogenic climate change. The 1981–2017 data were further used to quantify changes in the frequency of low-chill winters, defined as the bottom decile (10th percentile). This provides both additional context for the peach-focused analysis herein and may be of broader interest to the SEUS fruit and nut industry reliant on understanding exposure of low-chill winters as it pertains to the economics of orchard operations [29]. The observed and counterfactual scenarios for 1981–2017 were used to quantify the degree to which modeled climate change influenced the average chill accumulation, the probability of experiencing a low-chill winter, and the risk of insufficient chill accumulation for the three peach cultivars across the key Georgia and South Carolina peach-growing regions. Chill accumulation was also calculated for the 2010–2069 period for each of the 20 downscaled climate datasets. A similar set of tests were applied to projections including changes in average chill accumulation and the probability of experiencing a low-chill winter across the SEUS. Finally, the probability of insufficient chill was estimated for the early and mid-21st century conditions for the key peach cultivars and regions in order to highlight the potential risk to peach cultivation. Given our focus on the changes to chill accumulation with respect to perennial fruit cultivation, areas with <100 chill hours over the 1981–2017 observed period were masked out.

3. Results

The average chill accumulation for the observed period 1981–2017 across the SEUS ranged from less than 100 h in southern Florida, to more than 2000 h in the Blue Ridge mountains of northeastern Georgia (Figure 1b). The majority (>65%) of the region—from northern Florida to northern Alabama, Georgia, and South Carolina—averaged 500–1500 chill hours, including approximately 1100 h in the central Georgia peach-growing region and 1350 h in the Piedmont peach-growing region of South Carolina. With the exception of southern Florida, the 2017 chill accumulation was substantially lower than the 1981–2010 normal. The accumulated chill in 2017 showed an SEUS average anomaly of approximately 330 h below normal. The Georgia peach regions showed an anomaly of ~430 h below normal, and South Carolina peach regions showed an anomaly of ~360 h below normal (Figure 1c).

The observed average chill accumulation over 1981–2017 was less than that modeled in the absence of anthropogenic climate change, consistent with the expectations from modeled warming (Figure 2a). A distinct geographic pattern of the reduced chill hours due to climate change was evident across the SEUS, with nominal differences in southern Florida and reductions of more than 120 h in northern Alabama, Georgia and South Carolina. The peach-growing regions showed average reductions of ~115–120 chill hours in Georgia and South Carolina. Notably, these reductions are averages over the 37-year period as the modeled estimate was larger in more recent years. Complementary to average reductions in chill hours, the percent of years experiencing low winter chill was substantially higher across the SEUS over the 1981–2017 period than it would have been in the absence of climate change (Figure 2b). These trends were found across models. The 23-model range for declines in chill was ~68–140 h, while the range for the probability of low-chill winters was 1.6–4.6% of years (from a reference of 10% of years).

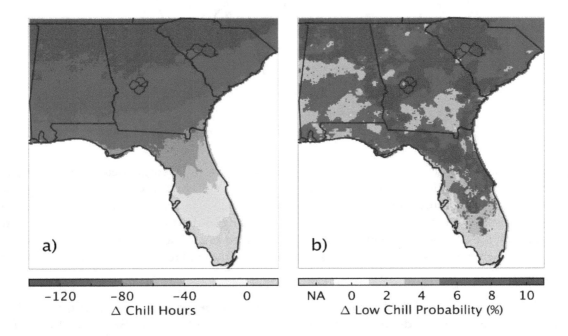

Figure 2. (**a**) The average change in 1981–2017 observed winter chill hours due to the influence of anthropogenic climate change (23-model median). (**b**) The change in the probability of a low-chill winter as a result of climate change, shown as 1981–2017 observed minus 1981–2017 counterfactual (23-model median). For both panels, the areas with <100 chill hours for the 1981–2017 observed climatology are masked in grey.

The reductions in chill accumulation and increases in the occurrence of low-chill winters may be inconsequential for agriculture unless there are direct impacts to plant physiology or indirect crop impacts (e.g., pathogens, pests). For the three peach cultivars, we show that the Georgia peach-growing region had five winters from 1981–2017 that did not accumulate sufficient chill for the high-chill cultivar (Figure 3a). No winters in the South Carolina peach-growing regions had insufficient chill for the cultivars considered from 1981–2017 (Figure 3b). By contrast, the counterfactual scenarios all showed greater chill accumulation and reduced occurrence of winters with insufficient chill for high-chill cultivars in Georgia. Notably, we show that the chill accumulation in 2017 would have been the lowest in the 37 year period in Georgia without climate change, suggesting that it was primarily driven by natural variability. However, the estimated 2017 chill accumulation excluding the modeled first-order influence of climate change for the peach-growing area of Georgia ranged from ~760 to ~920 chill hours across 23 models, with a median of ~825 h, well above the threshold of 650 chill hours required for moderate-chill cultivar and the ~660 chill hours observed that winter.

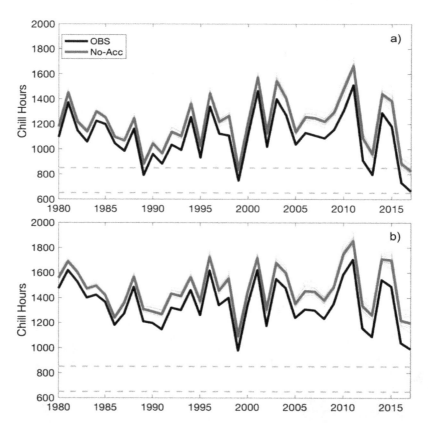

Figure 3. Time series of 1981–2017 chill accumulation for (**a**) the Georgia peach-growing region, and (**b**) the South Carolina peach-growing region. The observed data (OBS) are shown in black, while modeled chill accumulation estimates excluding the influence of anthropogenic climate change (No-Acc) are shown in red, with lighter red lines indicating individual models and the heavy red line indicating the 23-model median. The light pink dashed line indicates the chill requirement for a high-chill peach cultivar and the dashed grey line indicates the chill requirement for a moderate-chill peach cultivar.

The reduced chill accumulation across the SEUS was modeled relative to contemporary 1981–2017 averages for the early and mid-21st centuries, with multi-model mean SEUS declines of ~100 h, and ~185 h, respectively (Figure 4a,b). The geographic patterns of reductions in chill hours were similar to those shown for the influence of modeled climate change for the 1981–2017 period. Over Georgia (South Carolina) peach-growing regions, the average declines in chill were calculated as ~110 (~135) hours by the early 21st century and ~210 (~250) hours by the mid-21st century. In addition to declines in the average chill accumulation, the probability of experiencing a year with low winter chill accumulation increased. Averaged across the SEUS and across all models, approximately 20% of years by the early 21st century and 40% of years by the mid-21st century experienced low winter chill, with the greatest increases across western and northern Alabama, northern and central Georgia, and northern and central South Carolina (Figure 4c,d). By the early and mid- 21st century, Georgia (South Carolina) peach regions saw ~15% (30%) and 32% (52%) of years having low winter chill, respectively.

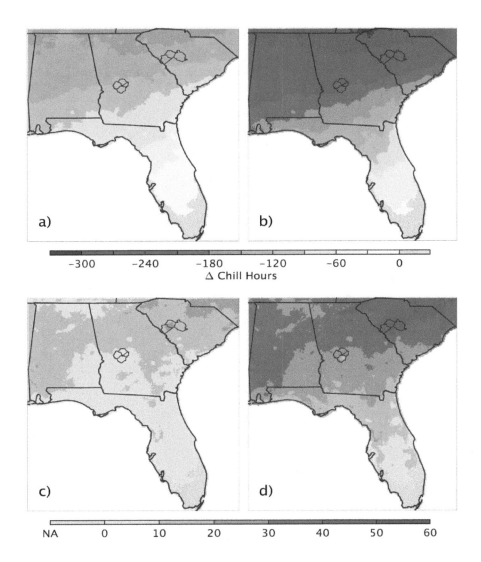

Figure 4. The difference in climatological chill hours for **(a)** the early 21st century (2010–2039) and **(b)** mid-21st century (2040–2069), relative to the observed 1981–2017 period. Panels **(c)** and **(d)** show differences in the probability of a low-chill winter for 2010–2039, and 2040–2069, respectively, relative to the observed 1981–2017 period. For all panels, the areas with <100 chill hours for the 1981–2017 observed climatology are masked in grey.

With respect to peach cultivar-specific chill requirements, 23% (4%) percent of years showed insufficient chill for the high- (moderate-) chill cultivar in prime peach-growing counties in Georgia by the early 21st century, rising to 43% (11%) percent of years by the mid-21st century (Figure 5a). The peach-growing regions in South Carolina, which did not see chill accumulations below established thresholds from 1980–2017, had 5% (0.25%) percent of years with insufficient chill for the high-chill (moderate-chill) cultivar by the early 21st century, and 12% (1.5%) percent of years by the mid-21st century (Figure 5b). Notably, there was substantial inter-model variability in the risk of winters with insufficient chill. For example, the percent of winters with insufficient chill for the moderate-chill cultivar in Georgia ranged from 0–16% for the early 21st century, and 0–30% for the mid-21st century.

By contrast, chill accumulation was sufficient for the low-chill peach cultivar under both future time periods in both states' peach-growing regions.

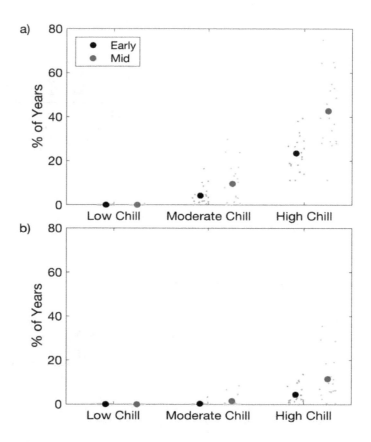

Figure 5. (**a**) For the Georgia peach-growing region, the percent of years with insufficient chill for a low-, moderate-, and high-chill peach cultivar under early 21st (black/grey) and mid-21st (red/pink) century conditions. Small grey and pink dots indicate the percent years for individual models, while the larger black and red dots indicate the 20-model average (**b**) As in (**a**) but for the South Carolina peach-growing region.

4. Discussion

Recent studies have shown that extreme events around the globe would not have been possible without the influence of human-induced warming [35–37]. Temperatures in the SEUS during the October-February chill accumulation periods of 2015–2016 and 2016–2017 were the 2nd and 3rd warmest since 1895, with the 1931–1932 winter being the warmest [38], suggesting that such warm winters are possible within the bounds of natural variability and can occur without significant contributions from anthropogenic climate change. While we do not undertake a detailed attribution analysis, our modeling exercise provides support that recent insufficient chill accumulation in the SEUS peach regions, such as in 2017, would not likely occur under the same synoptic conditions in the absence of climate change. Further, our results showing an increased probability of low-chill winters due to climate change add to the growing body of literature defining the contribution of anthropogenic climate change to observed adverse climate impacts [27,39,40].

Although insufficient chill accumulation is not a principle cause of loss for federally-insured crops in the SEUS [41,42], previous work has postulated that projected declines in chill may reduce suitability for perennial crop production [19,43]. Similarly, the projected future declines in chill accumulation across the SEUS complement previous work showing increases in the average and coldest winter minimum temperatures [16], and declines in chill accumulation in regions around the globe [21]. While this warming may offer range expansion for cold-intolerant crops, the related reduction in the winter chill accumulation in subtropical climates like the SEUS is projected to have negative impacts on warm-region fruit and nut crops, particularly those with moderate and high chill requirements [20,21]. However, the degree to which these declines may impact crop yield is unclear as uncertainties remain regarding the chill requirements that are physiologically needed for production, and the overall effect of marginal chill accumulation on crop yield and quality [44,45]. For example, while a common commercial peach cultivar grown in central Georgia has a stated chill requirement of 850 h, Georgia peach specialists have suggested that only 800 h are needed for a suitable crop [4]. Consequently, we underscore that this work is not predictive of yield impacts related to reduced chill accumulation.

Compounding the problem of crop chill requirements is the questionable accuracy of the chilling model. While this model has been widely used for quantifying crop chill requirements, it may be overly sensitive to warming, potentially overestimating the impact of climate change [43]. However, while it is acknowledged that previous studies have shown that the Dynamic Model may provide a more accurate representation of chill accumulation [21], the 20-model mean changes in the average chill accumulation show an agreement of declines across the SEUS and other warm-winter regions, regardless of the chilling model (Figure 6). Further, we recognize that familiarity with chill portions (the units of the Dynamic Model) may be lacking among extension agents and fruit industry professionals (Pamela Knox, University of Georgia Agricultural Climatologist, personal communication), and that regionally-defined chill portion thresholds do not yet exist for SEUS peach cultivars (Dario Chavez, personal communication). Finally, we acknowledge the limitations of using temporally disaggregated daily data [46], and that the microclimates of orchard sites and orchard management practices may augment or abate the projected changes and impacts.

Despite research suggesting that declines in crop suitability due to climate change may not be as severe as shown in our results [45], it is worth noting that we examined changes in chill accumulation under a conservative, moderate warming scenario. Provided that some degree of reductions in suitability are anticipated for peach crops across the SEUS—as well as for other crops with similar chill requirements—adaptive measures may be warranted to maintain production. These measures may include altering orchard management practices and selective planting. For existing orchards, the application of chemicals such as hydrogen cyanamide may effectively break dormancy in insufficiently-chilled peach crops [47], overhead irrigation to encourage evaporative cooling may aid chill accumulation, and orchard management practices such as controlling tree vigor may help to lower the chill needed for successful bud break [48]. For future orchards, site selection with preferential planting in sites with cooler microclimates, such as low-lying cool-air sinks, may provide an opportunity to increase exposure to chilling temperatures.

Orchard managers may also consider specific scion and rootstock combinations that may help mitigate the negative impacts of low chill [49]. Moreover, a transition to crop cultivars with lower chill requirements (e.g., Gulfcrest or other varieties developed for warmer climates) may reduce or eliminate the negative impacts of declining chill accumulation under climate change, as evidenced by the minimal impact of future warming to the low-chill cultivar examined in this study. However, it is noted that orchards planted in cool-air microclimates may be at increased risk of frost damage, and lower-chill cultivars may be more susceptible to early bloom and subsequent frost damage. While quantifying the complex relationships between chill accumulation, bloom, and the relative risks of insufficient chill and spring frost damage are beyond the scope of this work, the interactions between these physiological and climatic conditions highlight the need to consider a broader suite of environmental and economic considerations in planning for future orchard management.

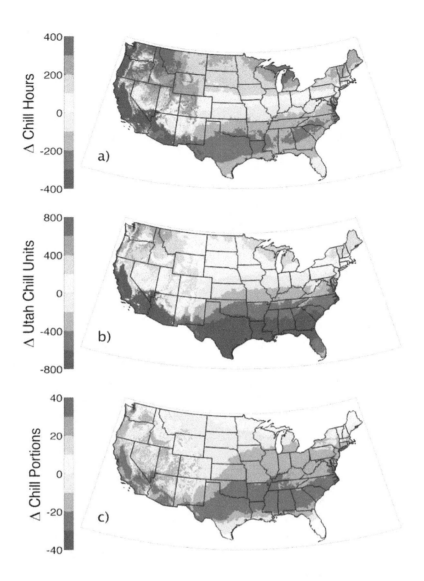

Figure 6. The 20-model average difference in annual accumulated chill between the modeled historical period (1971–2000) and the mid-century (2040–2069) period under RCP 4.5, where chill was accumulation was calculated over the October 1—April 30 cool season using (**a**) the Modified Chill Hour Model as chill hours 32–45 °C, (**b**) the Utah Model as chill units, and (**c**) the Dynamic Model as chill portions. The red shades indicate a reduction in chill accumulation under RCP 4.5, while the blue shades indicate an increase in chill accumulation. The white regions in (**c**) indicate areas with no chill accumulation under historical conditions.

As has been suggested for perennial crop adaptation in other regions [19,50], the translocation of crops to cooler climates may also provide an adaptive measure for maintaining peach cultivation in the SEUS, particularly for those cultivars with higher chilling requirements. Historically, peach cultivation in Georgia extended into the northern portion of the state, but the favorability of that region declined over time due to frequent freeze damage [51]. If climate change reduces the freeze risk in northern Georgia, the area may provide a refuge within the state for continued cultivation of high-chill peach cultivars and other similarly at-risk perennials. However, any future translocation would require significant capital and be contingent upon economic viability, which is likely to be predicated on factors such as topography and soils, the costs associated with the purchase of farmland and the packing or processing facilities, competing land use, and market forces.

5. Conclusions

Quantifying the potential consequences of warming winters on chill accumulation may have implications for long-term orchard management and land use planning and may provide insights useful for climate-informed decision making for a variety of perennial crops that require winter chill. Our results show that anthropogenic climate change has negatively affected chill accumulation in the SEUS over the observed 1981–2017 period, and that ongoing climate change is likely to continue to reduce chill accumulation, with notable impacts on high- and moderate-chill peach cultivars in Georgia. We also highlight the importance of examining interannual variability when assessing climate change risks to agriculture, be that impacts to crop climatic niche or crop yield [19,52]. The adaptation measures (e.g., investments in lower-chill varieties) may be necessary in order for the SEUS, particularly Georgia, to continue to cultivate the crop that has historically been central to its cultural identity. Further, given the relationship between mild winter temperatures, early bloom, and damages due to a false spring—as also seen in 2017—we recommend future work consider the interaction between multiple agro-climatic variables to provide a more complete assessment of future crop suitability and identify the most appropriate adaptive efforts. Finally, as our study employs a methodology that is applicable across other geographic locations, perennial crop cultivars, and agro-climatic metrics, we recommend that similar work be undertaken across agricultural systems and regions to help identify potential crop-specific risks and adaptation opportunities.

Author Contributions: Conceptualization, L.E.P. and J.T.A.; Methodology, L.E.P. and J.T.A.; Formal Analysis, L.E.P. and J.T.A.; Writing—Original Draft Preparation, L.E.P. and J.T.A.; Writing—Review & Editing, L.E.P. and J.T.A.; Visualization, L.E.P. and J.T.A.; Supervision, J.T.A.; Project Administration, J.T.A.; Funding Acquisition, L.E.P. and J.T.A.

Acknowledgments: The authors wish to acknowledge the feedback and local context provided by University of Georgia Assistant Professor and Extension Specialist Dario Chavez, and University of Georgia Agricultural Climatologist Pamela Knox. The authors also wish to acknowledge Katherine Hegewisch for her efforts in incorporating chill accumulation data into the ClimateToolbox. Finally, the authors wish to thank four anonymous reviewers and the journal editor for their comments, which improved the quality of the manuscript.

References

1. Okie, T. Under the Trees: The Georgia Peach and the Quest for Labor in the Twentieth Century. *Agric. Hist.* **2011**, *85*, 72–101. [CrossRef] [PubMed]
2. USDA-NASS. *2016 State Agricultural Overview: California, Georgia, South Carolina*; USDA-NASS: Washington, DC, USA, 2016. Available online: https://www.nass.usda.gov/Statistics_by_State/Ag_Overview/ (accessed on 29 June 2017).
3. Wolfe, K.; Stubbs, K. *Georgia Farm Gate Value Report*; CAES: Athens, GA, USA, 2016; Available online: https://www.caes.uga.edu/content/caes-subsite/caed/publications/farm-gate.html (accessed on 29 June 2017).
4. Thompson, C. Georgia's peach farmers hoping for colder winter this year. Available online: https://newswire.caes.uga.edu/story.html?storyid=6416 (accessed on 29 June 2017).
5. SERCC. *Annual 2017 Climate Report for the Southeast Region*; SERCC: Chapel Hill, NC, USA, 2017.
6. Saure, M.C. Dormancy release in deciduous fruit trees. *Hortic. Rev.* **1985**, *7*, 239–300.
7. Janick, J.; Paull, R.E. *The Encyclopedia of Fruit and Nuts*; CABI: Oxfordshire, UK, 2008; ISBN 0851996388.
8. Atkinson, C.J.; Brennan, R.M.; Jones, H.G. Declining chilling and its impact on temperate perennial crops. *Environ. Exp. Bot.* **2013**, *91*, 48–62. [CrossRef]
9. Weinberger, J.H. Effects of high temperatures during the breaking of the rest of Sullivan Elberta peach buds. *Proc. Am. Soc. Hortic. Sci.* **1954**, *63*, 157–162.
10. Medellín-Azuara, J.; Howitt, R.E.; MacEwan, D.J.; Lund, J.R. Economic impacts of climate-related changes to California agriculture. *Clim. Chang.* **2011**, *109*, 387–405. [CrossRef]
11. Peterson, T.C.; Heim, R.R., Jr.; Hirsch, R.; Kaiser, D.P.; Brooks, H.; Diffenbaugh, N.S.; Dole, R.M.; Giovannettone, J.P.; Guirguis, K.; Karl, T.R. Monitoring and understanding changes in heat waves, cold waves, floods, and droughts in the United States: State of knowledge. *Bull. Am. Meteorol. Soc.* **2013**, *94*, 821–834. [CrossRef]

12. Abatzoglou, J.T.; Barbero, R. Observed and projected changes in absolute temperature records across the contiguous United States. *Geophys. Res. Lett.* **2014**, *41*, 6501–6508. [CrossRef]

13. Walsh, J.; Wuebbles, D.; Hayhoe, K.; Kossin, J.; Kunkel, K.; Stephens, G.; Thorne, P.; Vose, R.; Wehner, M.; Willis, J. Our changing climate. In *Climate Change Impacts in the United States: The Third National Climate Assessment*; U.S. Government Printing Office: Washington, DC, USA, 2014; pp. 19–67.

14. Deser, C.; Terray, L.; Phillips, A.S. Forced and internal components of winter air temperature trends over North America during the past 50 years: Mechanisms and implications. *J. Clim.* **2016**, *29*, 2237–2258. [CrossRef]

15. Sillmann, J.; Kharin, V.V.; Zwiers, F.W.; Zhang, X.; Bronaugh, D. Climate extremes indices in the CMIP5 multimodel ensemble: Part 2. Future climate projections. *J. Geophys. Res. Atmos.* **2013**, *118*, 2473–2493. [CrossRef]

16. Parker, L.E.; Abatzoglou, J.T. Projected changes in cold hardiness zones and suitable overwinter ranges of perennial crops over the United States. *Environ. Res. Lett.* **2016**, *11*, 34001. [CrossRef]

17. Meehl, G.A.; Arblaster, J.M.; Branstator, G. Mechanisms contributing to the warming hole and the consequent US east–west differential of heat extremes. *J. Clim.* **2012**, *25*, 6394–6408. [CrossRef]

18. Schwartz, M.D.; Ault, T.R.; Betancourt, J.L. Spring onset variations and trends in the continental United States: Past and regional assessment using temperature-based indices. *Int. J. Climatol.* **2013**, *33*, 2917–2922. [CrossRef]

19. Parker, L.E.; Abatzoglou, J.T. Shifts in the thermal niche of almond under climate change. *Clim. Chang.* **2018**, *147*, 211–224. [CrossRef]

20. Baldocchi, D.; Wong, S. Accumulated winter chill is decreasing in the fruit growing regions of California. *Clim. Chang.* **2008**, *87*, 153–166. [CrossRef]

21. Luedeling, E. Climate change impacts on winter chill for temperate fruit and nut production: A review. *Sci. Hortic.* **2012**, *144*, 218–229. [CrossRef]

22. Carbone, G.J.; Schwartz, M.D. Potential impact of winter temperature increases on South Carolina peach production. *Clim. Res.* **1993**, *2*, 225–233. [CrossRef]

23. Abatzoglou, J.T. Development of gridded surface meteorological data for ecological applications and modelling. *Int. J. Climatol.* **2013**, *33*, 121–131. [CrossRef]

24. Abatzoglou, J.T.; Brown, T.J. A comparison of statistical downscaling methods suited for wildfire applications. *Int. J. Climatol.* **2012**, *32*, 772–780. [CrossRef]

25. Mitchell, T.D. Pattern scaling: An examination of the accuracy of the technique for describing future climates. *Clim. Chang.* **2003**, *60*, 217–242. [CrossRef]

26. Abatzoglou, J.T.; Williams, A.P. Impact of anthropogenic climate change on wildfire across western US forests. *Proc. Natl. Acad. Sci. USA* **2016**, *113*, 11770–11775. [CrossRef]

27. Williams, A.P.; Seager, R.; Abatzoglou, J.T.; Cook, B.I.; Smerdon, J.E.; Cook, E.R. Contribution of anthropogenic warming to California drought during 2012–2014. *Geophys. Res. Lett.* **2015**, *42*, 6819–6828. [CrossRef]

28. Boryan, C.; Yang, Z.; Mueller, R.; Craig, M. Monitoring US agriculture: The US department of agriculture, national agricultural statistics service, cropland data layer program. *Geocarto Int.* **2011**, *26*, 341–358. [CrossRef]

29. Luedeling, E.; Zhang, M.; Girvetz, E.H. Climatic changes lead to declining winter chill for fruit and nut trees in California during 1950–2099. *PLoS ONE* **2009**, *4*, e6166. [CrossRef] [PubMed]

30. Parker, L.E.; Abatzoglou, J.T. Comparing mechanistic and empirical approaches to modeling the thermal niche of almond. *Int. J. Biometeorol.* **2017**, *61*, 1593–1606. [CrossRef] [PubMed]

31. Weinberger, J.H. Chilling requirements of peach varieties. *Proc. Am. Soc. Hortic. Sci.* **1950**, *56*, 122–128.

32. Linvill, D.E. Calculating chilling hours and chill units from daily maximum and minimum temperature observations. *HortScience* **1990**, *25*, 14–16.

33. Okie, W.R.; Blackburn, B. Increasing chilling reduces heat requirement for floral budbreak in peach. *HortScience* **2011**, *46*, 245–252. [CrossRef]

34. CAES. College of Agriculture and Environmental Sciences (CAES). Available online: http://www.caes.uga.edu/extension-outreach/commodities/peaches/cultivars.html (accessed on 21 March 2017).

35. Knutson, T.R.; Kam, J.; Zeng, F.; Wittenberg, A.T. CMIP5 model-based assessment of anthropogenic influence on record global warmth during 2016. *Bull. Am. Meteorol. Soc.* **2018**, *99*, S11–S15. [CrossRef]

36. Imada, Y.; Shiogama, H.; Takahashi, C.; Watanabe, M.; Mori, M.; Kamae, Y.; Maeda, S. Climate change increased the likelihood of the 2016 heat extremes in Asia. *Bull. Am. Meteorol. Soc.* **2018**, *99*, S97–S101.

[CrossRef]

37. Walsh, J.E.; Thoman, R.L.; Bhatt, U.S.; Bieniek, P.A.; Brettschneider, B.; Brubaker, M.; Danielson, S.; Lader, R.; Fetterer, F.; Holderied, K.; et al. The high latitude marine heat wave of 2016 and its impacts on Alaska. *Bull. Am. Meteorol. Soc.* **2018**, *99*, S39–S43. [CrossRef]

38. NOAA. *Climate at a Glance: Regional Time Series*; NOAA: Silver Spring, MD, USA, 2018. Available online: https://www.ncdc.noaa.gov/cag (accessed on 29 June 2017).

39. Fischer, E.M.; Knutti, R. Anthropogenic contribution to global occurrence of heavy-precipitation and high-temperature extremes. *Nat. Clim. Chang.* **2015**, *5*, 560–564. [CrossRef]

40. Stott, P.A.; Christidis, N.; Otto, F.E.L.; Sun, Y.; Vanderlinden, J.; Van Oldenborgh, G.J.; Vautard, R.; Von Storch, H.; Walton, P.; Yiou, P.; et al. Attribution of extreme weather and climate-related events. *Wiley Interdiscip. Rev. Clim. Chang.* **2016**, *7*, 23–41. [CrossRef]

41. Reyes, J.; Elias, E. Spatio-temporal variation of crop loss in the United States from 2001 to 2016. *Environ. Res. Lett.* **2019**, *14*, 074017. [CrossRef]

42. AgRisk Viewer. Agricultural Risk in a Changing Climate: A Geographical and Historical View of Crop Insurance. Available online: https://swclimatehub.info/rma (accessed on 31 March 2019).

43. Luedeling, E.; Brown, P.H. A global analysis of the comparability of winter chill models for fruit and nut trees. *Int. J. Biometeorol.* **2011**, *55*, 411–421. [CrossRef] [PubMed]

44. Campoy, J.A.; Darbyshire, R.; Dirlewanger, E.; Quero-García, J.; Wenden, B. Yield potential definition of the chilling requirement reveals likely underestimation of the risk of climate change on winter chill accumulation. *Int. J. Biometeorol.* **2019**, *63*, 183–192. [CrossRef] [PubMed]

45. Pope, K.S.; Dose, V.; Da Silva, D.; Brown, P.H.; DeJong, T.M. Nut crop yield records show that budbreak-based chilling requirements may not reflect yield decline chill thresholds. *Int. J. Biometeorol.* **2015**, *59*, 707–715. [CrossRef]

46. Luedeling, E. Interpolating hourly temperatures for computing agroclimatic metrics. *Int. J. Biometeorol.* **2018**, *62*, 1799–1807. [CrossRef] [PubMed]

47. Dozier, W.A.; Powell, A.A.; Caylor, A.W.; McDaniel, N.R.; Carden, E.L.; McGuire, J.A. Hydrogen cyanamide induces budbreak of peaches and nectarines following inadequate chilling. *HortScience* **1990**, *25*, 1573–1575.

48. Erez, A. Means to compensate for insufficient chilling to improve bloom and leafing. *Acta Hortic.* **1995**, *395*, 81–96. [CrossRef]

49. Ghrab, M.; Mimoun, M.B.; Masmoudi, M.M.; Mechlia, N.B. Chilling trends in a warm production area and their impact on flowering and fruiting of peach trees. *Sci. Hortic.* **2014**, *178*, 87–94. [CrossRef]

50. Lobell, D.B.; Field, C.B.; Cahill, K.N.; Bonfils, C. Impacts of future climate change on California perennial crop yields: Model projections with climate and crop uncertainties. *Agric. For. Meteorol.* **2006**, *141*, 208–218. [CrossRef]

51. Taylor, K. Peaches. Available online: http://www.georgiaencyclopedia.org/articles/arts-culture/peaches (accessed on 16 May 2018).

52. Ray, D.K.; Gerber, J.S.; MacDonald, G.K.; West, P.C. Climate variation explains a third of global crop yield variability. *Nat. Commun.* **2015**, *6*, 5989. [CrossRef] [PubMed]

10

Adaptive Effectiveness of Irrigated Area Expansion in Mitigating the Impacts of Climate Change on Crop Yields in Northern China

Tianyi Zhang [1],*, Jinxia Wang [2] and Yishu Teng [3]

[1] State Key Laboratory of Atmospheric Boundary Layer Physics and Atmospheric Chemistry, Institute of Atmospheric Physics, Chinese Academy of Sciences, Beijing 100029, China

[2] School of Advanced Agricultural Sciences, Peking University, Beijing 1000871, China; jxwang.ccap@igsnrr.ac.cn

[3] BICIC, Beijing Normal University, Beijing 1000875, China; tengyishu@bnu.edu.cn

* Correspondence: zhangty@post.iap.ac.cn

† This paper was presented at the Global Land Programme 3rd Open Science Meeting, Beijing, China, 24–27 October 2016.

Academic Editors: Elaine Wheaton and Suren N. Kulshreshtha

Abstract: To improve adaptive capacity and further strengthen the role of irrigation in mitigating climate change impacts, the Chinese government has planned to expand irrigated areas by 4.4% by the 2030s. Examining the adaptive potential of irrigated area expansion under climate change is therefore critical. Here, we assess the effects of irrigated area expansion on crop yields based on county-level data during 1980–2011 in northern China and estimate climate impacts under irrigated area scenarios in the 2030s. Based on regression analysis, there is a statistically significant effect of irrigated area expansion on reducing negative climate impacts. More irrigated areas indicate less heat and drought impacts. Irrigated area expansion will alleviate yield reduction by 0.7–0.8% in the future but associated yield benefits will still not compensate for greater adverse climate impacts. Yields are estimated to decrease by 4.0–6.5% under future climate conditions when an additional 4.4% of irrigated area is established, and no fundamental yield increase with an even further 10% or 15% expansion of irrigated area is predicted. This finding suggests that expected adverse climate change risks in the 2030s cannot be mitigated by expanding irrigated areas. A combination of this and other adaptation programs is needed to guarantee grain production under more serious drought stresses in the future.

Keywords: irrigated area; drought; climate; adaptation; SPEI

---

1. Introduction

Climate change poses serious challenges to Chinese agriculture [1–3]. In recent years, the ability to meet these challenges has been tested by several major extreme climate events. For example, the devastating drought in southwestern China in 2010 critically impaired local agriculture, resulting in an estimated loss of 317 million USD [4]. The average annual total cost of climate disasters is approximately 80 billion USD in China [5]. Climate extremes are anticipated to be aggravated and increasingly influenced by climate change. These conditions will constrain future growth in agricultural sectors; therefore, it is important to take actions to mitigate future climate risks.

Countervailing the current and future adverse climate risks will require adaptation measures. In 2011, the Chinese government announced an important policy requiring that 600 billion USD be invested in agricultural irrigation [6]. The policy set several quantitative targets for improving irrigation over 10 years, starting in 2011. The most important plan is to expand the irrigated area by 2.67 million ha (equivalent to a 4.4% increase). Although the quantitative target is clearly framed, concerns have been raised about the effectiveness of irrigated area expansion in climate risk mitigation [7].

To our knowledge, no study has evaluated the extent that the above irrigated area expansion plan [7] reduces the impacts of future climate change in China. Using process-based crop model and associated assumptions, some studies have evaluated the adaptation effectiveness of potential irrigation in facing climate change risks on Chinese agriculture. For example, assuming no crop water stress was predicted to mitigate 5–15% of the yield reduction in China under future climate scenarios [8,9]. However, the assumption of no water stress is unrealistic and difficult to link with the government plan. Several recent studies were encouraged to the integration of farming management methods into impact assessments as these methods greatly determine the degree of climate impacts on crops. For instance, fertilizer intensive farmers can largely reduce the negative effects of heat stresses in the UK, France and Italy, while the effect is small or even negative in other European countries [10]. On a global scale, vulnerability of key food crops to drought is also greatly dependent on socio-economic conditions and agricultural investments [11]. In China, recent studies have quantified the relationship among climate, crop and irrigation based on statistical data [3,12]. They employed a new data-driven approach, but the major disadvantage is a lack of socioeconomic data with fine spatial-temporal resolution. Furthermore, few of these analyses addressed potential adaptation under future climate scenarios.

Therefore, to understand the adaptation effectiveness of expanding the irrigated area in mitigating climate change impacts, the following objectives are specified in our study: (i) we quantitatively identify crop yield responses to climate and irrigated area based on county-level data during 1980–2011; (ii) we establish a statistical model to assess the adaptive effects of the irrigated areas expansion plan already underway on climate change mitigation; (iii) we explore future climate impacts on crop yields across different irrigated area scenarios.

2. Materials and Methods

2.1. Data and Pretreatment

Our study region is in northern China (Figure 1) because of the increasingly important role the region plays in grain production. The region encompasses 50% of cultivated land and produces 56% of the annual grain production in China. The major grain production areas are the northeastern, northern, and eastern parts of northwestern China. Due to low precipitation and the uneven seasonal distribution of precipitation, crop production in northern China largely depends on irrigation.

Figure 1. Illustration of the study region. The shaded area is northern China, which includes the Northeast, North and Northwest regions of China, as shown in the top-right figure. The number indicates the provinces involved in the study. 1: Heilongjiang; 2: Jilin; 3: Liaoning; 4: Beijing; 5: Hebei; 6: Tianjin; 7: Shandong; 8: Henan; 9: Inner Mongolia; 10: Shanxi; 11: Shaanxi; 12: Ningxia; 13: Gansu; 14: Qinghai; 16: Xinjiang.

Crop data used in the study were obtained from the Chinese Academy of Agricultural Sciences. These data include county-level sown areas and production data for rice, wheat, maize and soybean, which are the four major food crops in our study region, over the period 1980–2011. In addition, we considered county-level irrigated areas and cultivated land areas in our study region and period. Based on the definition by the National Standard of the People's Republic of China [13] and the Food and Agriculture Organization of the United Nations [14], "irrigated area" is the area equipped to be irrigated and it is the most often-used index to quantify irrigation level in earlier studies [3,12]. Percentage of irrigated area (PIA) was calculated based on Equation (1), which represents the irrigated areas relative to sowing areas, an index quantifying irrigated conditions, for each county in each year.

$$PIA_{c,t} = \frac{IRRI_{c,t}}{CulArea_{c,t}} \times 100\% \tag{1}$$

where $PIA_{c,t}$ is the percentage of irrigated area (%), $IRRI_{c,t}$ is the irrigated area (ha), and $CulArea_{c,t}$ is the cultivated land area (ha) of county c in year t.

As crop-specific data for irrigated area are not available, we lump data of the four crops to match the PIA data (Equation (2)).

$$Y_{c,t} = \frac{riceP_{c,t} + wheatP_{c,t} + maizeP_{c,t} + soybeanP_{c,t}}{riceA_{c,t} + wheatA_{c,t} + maizeA_{c,t} + soybeanA_{c,t}} \tag{2}$$

where $Y_{c,t}$ is the yields of the four crops weighted in each county by sown area (ton ha^{-1}); $riceP_{c,t}$, $wheatP_{c,t}$, $maizeP_{c,t}$ and $soybeanP_{c,t}$ (ton) are the production of rice, wheat, maize, and soybean, respectively; and $riceA_{c,t}$, $wheatA_{c,t}$, $maizeA_{c,t}$ and $soybeanA_{c,t}$ (ha) are the sown area of rice, wheat, maize and soybean, respectively, of county c in year t.

Daily temperature and precipitation data in 756 climate stations were downloaded from the China Meteorology Data Sharing Service System [15]. Quality controls and homogenization of these climate data have been executed by the Chinese Meteorological Administration. To derive climate data for each county, we estimated daily climate data using the algorithm presented by Thornton et al. [16]. This algorithm interpolates the abovementioned data of the 756 climate stations into 10 km grid cells and then extracts climatic information for each county from the grid data. The daily grid climatic dataset has been used in a previous study [17]. Subsequently, we calculated the daily climate data for each county by zonal averaging, and then aggregated the daily data into monthly climate data. To represent drought severity, we calculated the monthly Standardized Precipitation Evapotranspiration Index (SPEI) for each county. SPEI is a multi-scalar drought index calculated based on a climatic water balance model [18] considering the role of both precipitation and evapotranspiration. An R package, "SPEI" (https://cran.r-project.org/web/packages/SPEI/), was used to calculate the index with the lag set to 1 month to quantify the monthly moisture conditions due to the climate of the same month. Next, we calculated the mean-growing-season average temperature (Tavg) and SPEI for each crop. The growing season period for each crop was derived from the *Chinese Agricultural Phenology Atlas* [19] (Table 1). Finally, to match the PIA data, Tavg and SPEI were aggregated as weighted by the sown area of the four crops in each year (Equations (3) and (4)).

$$Tavg_{c,t} = \frac{riceTavg_{c,t} \times riceA_{c,t} + wheatTavg_{c,t} \times wheatA_{c,t} + maizeTavg_{c,t} \times maizeA_{c,t} + soybeanTavg_{c,t} \times soybeanA_{c,t}}{riceA_{c,t} + wheatA_{c,t} + maizeA_{c,t} + soybeanA_{c,t}} \tag{3}$$

where $Tavg_{c,t}$ is the mean-growing-season average temperature weighted by sown area; $riceTavg_{c,t}$, $wheatTavg_{c,t}$, $maizeTavg_{c,t}$ and $soybeanTavg_{c,t}$ are the mean growing season average temperature for rice, wheat, maize and soybean, respectively; and $riceA_{c,t}$, $wheatA_{c,t}$, $maizeA_{c,t}$ and $soybeanA_{c,t}$ are the sown area for rice, wheat, maize and soybean, respectively, of county c in year t.

$$SPEI_{c,t} = \frac{riceSPEI_{c,t} \times riceA_{c,t} + wheatSPEI_{c,t} \times wheatA_{c,t} + maizeSPEI_{c,t} \times maizeA_{c,t} + soybeanSPEI_{c,t} \times soybeanA_{c,t}}{riceA_{c,t} + wheatA_{c,t} + maizeA_{c,t} + soybeanA_{c,t}} \quad (4)$$

where $SPEI_{c,t}$ is the mean-growing-season average SPEI weighted by sown area; $riceSPEI_{c,t}$, $wheatSPEI_{c,t}$, $maizeSPEI_{c,t}$ and $soybeanSPEI_{c,t}$ are the mean growing season average SPEI for rice, wheat, maize, and soybean, respectively; and $riceA_{c,t}$, $wheatA_{c,t}$, $maizeA_{c,t}$ and $soybeanA_{c,t}$ are the sown area for rice, wheat, maize, and soybean, respectively, of county c in year t.

2.2. Statistical Model

To evaluate the relationship of climate, yield and PIA, we established a fixed-effect regression model, as given in Equation (5).

$$\begin{aligned} \log(Y_{c,t}) = \alpha_1 PIA_{c,t} + \alpha_2 PIA_{c,t}^2 + \alpha_3 Tavg_{c,t} + \alpha_4 Tavg_{c,t}^2 + \alpha_5 SPEI_{c,t} + \alpha_6 SPEI_{c,t}^2 + \alpha_7 PIA_{c,t} Tavg_{c,t} \\ + \alpha_8 PIA_{c,t} SPEI_{c,t} + \alpha_{9,c} County_c + \alpha_{10,c} County_c \times Year_t + \alpha_{11,c} County_c \times Year_t^2 + \varepsilon_{c,t} \end{aligned} \quad (5)$$

where $Y_{c,t}$ is yield for the four crops weighted by sown area in Equation (2) (ton ha^{-1}), $PIA_{c,t}$ is the percentage of irrigated area in Equation (1) (%); $Tavg_{c,t}$ and $SPEI_{c,t}$ are respectively the mean-growing-season temperature (Equation (3)) and SPEI (Equation (4)) weighted by the sown area of the four crops of county c in year t; $County$ is the dummy variable for county; $Year$ denotes time; and ε is the error term. $\alpha_1 - \alpha_{11}$ are the regression coefficients for each term.

In this model, we used quadratic terms for PIA, Tavg, and SPEI to account for the fact that crops perform best under moderate management and climate conditions and are harmed by extreme cold, hot, dry, or wet field conditions. In addition, we considered the potential interactions between irrigation and climate variables in this model, which represent the changes in climate impacts under different irrigation conditions. Unobserved possible nonlinear time trends at the county level were controlled by using county-by-year linear and quadratic terms and unobserved time-constant variations between counties using a county fixed effect. Consistent with other studies based on statistical model [20], CO_2 effects were not considered. Therefore, results in this study reflect the possible largest impacts from climate change.

The accuracy of the model was evaluated using a bootstrap analysis [21]. By constructing a number of re-samples and replacing the observations, this analysis evaluated the model accuracy defined by confidence intervals. More specifically, years were chosen randomly with replacements for 1000 iterations to estimate the regression coefficients of the model. Then, 1000 sets of regression coefficients were derived, which were then used to calculate yield changes by inputting future climate conditions. The confidence interval not spanning zero indicates a significant effect. Here, the median value and 95% confidence interval (95% CI) of those regression coefficients are reported.

2.3. Climate Scenarios

The climate change projections were taken from the Program for Climate Model Diagnosis and Inter-comparison—Coupled Model Inter-comparison Project Phase 5 for two representative concentration pathways (RCP2.6 and RCP8.5) in our study region. This ensemble climate scenarios were simulated by 26 climate models (Supplementary Materials Table S1). RCP2.6 represents a low emission pathway, i.e., greenhouse gas emissions peak between 2010 and 2020 with emissions declining substantially thereafter; RCP8.5 is a high emission pathway, i.e., emissions continue to rise throughout the 21st century.

The baseline period was set to 1980–2011, consistent with our observations, and the future period was 2020–2039 (referred to as 2030s hereafter), the target period for the abovementioned irrigated areas expansion plan. Following the steps for processing observed climate data described in Section 2.1, we derived the mean-growing-season Tavg and SPEI weighted by the sown areas of four crops for each county-year pair based on the future climate scenarios (here, we assume there is no change in the growing areas of the four crops). The difference in anticipated growing-season Tavg and SPEI relative to the baseline climate was input into our statistical model.

Table 1. Growing season of four crops for each province and mean growing season temperature.

Provinces	ID	Rice Sowing	Harvest	Tavg	Wheat Sowing	Harvest	Tavg	Maize Sowing	Harvest	Tavg	Soybean Sowing	Harvest	Tavg
Northeast													
Heilongjiang	1	1 May	30 September	17.5	1 April	31 July	14.8	1 May	30 September	17.5	1 May	30 September	17.5
Jilin	2	1 May	30 September	18.1	1 April	31 July	15.5	1 May	30 September	18.1	1 May	30 September	18.1
Liaoning	3	1 May	30 September	20.4	1 April	31 July	17.8	1 May	30 September	20.4	1 May	30 September	20.4
North													
Beijing	4	1 April	30 September	20.1	1 October	30 Jun.	6.5	1 June	30 September	22.5	1–30 June	30 September	22.5
Hebei	5	1 April	30 September	21.4	1 October	30 June	8.4	1 June	30 September	23.7	1–30 June	30 September	23.7
Tianjin	6	1 April	30 September	22.3	1 October	30 June	9.0	1 June	30 September	24.7	1–30 June	30 September	24.7
Shandong	7	1 April	30 September	22.0	1 October	30 June	10.0	1 June	30 September	24.5	1–30 June	30 September	24.5
Henan	8	1 April	30 September	22.6	1 October	30 June	11.4	1 June	30 September	24.8	1–30 June	30 September	24.8
Inner Mongolia	9	1May	30 September	17.0	1 April	31 July	14.7	1May	30 September	17.0	1May	30 September	17.0
Northwest													
Shanxi	10	1 May	30 September	19.2	1 October	30 June	5.5	1 May	30 September	19.2	1 May	30 September	19.2
Shaanxi	11	1 May	30 September	20.3	1 October	30 June	8.2	1 May	30 September	20.3	1 May	30 September	20.3
Ningxia	12	1 May	30 September	18.1	1 October	30 June	4.4	1 May	30 September	18.1	1 May	30 September	18.1
Gansu	13	1 April	30 September	16.0	1 October	30 June	3.8	1 May	30 September	16.0	1 May	30 September	16.0
Qinghai	14	NA	NA	NA	1 March	31 July	3.2	1 May	30 September	8.1	1 May	30 September	8.1
Xinjiang	15	1 May	30 September	17.3	1 October	30 June	1.5	1 May	30 September	17.3	1 May	30 September	17.3

Note: ID matches with the numbers in Figure 1; NA denotes no such crop in the province.

2.4. Irrigated Area Scenarios

To explore the effectiveness of irrigated area expansion on mitigating climate impacts, we considered four scenarios of irrigated areas: no change in PIA, 4.4% increase in PIA, 10% increase in PIA, and 15% increase in PIA. Note the maximum value of PIA is 100%. So, in cases where the PIA was greater than 100% after the addition, we reset it to 100%. No change in PIA indicates the scenario without adaptation, a 4.4% increase in PIA is consistent with the existing irrigated area expansion plan, and the last two scenarios (increased PIA by 10% and 15%) indicate potential adaptations if irrigated areas are further amplified.

3. Results

3.1. Irrigated Areas and Crop Yields under the Baseline Climate

Figure 2 demonstrates the average observed PIA and crop yields over the 1980—2011. The PIA varies by locations (Figure 2a). Better irrigation conditions are exhibited particularly in northwestern China (more than 70%) compared with northern (10–70%) and northeastern (less than 50%) China. This result is a major reflection of climatic moisture status, with the dryer climate in the Northwest requiring more irrigation to maintain the local agriculture than in northern and northeastern China. In terms of the crop yields, the spatial distribution is less clear (Figure 2b). Crop yields in most counties vary between 3.5 and 6.5 t ha^{-1}. Regions with relatively low yield include the northern region of the North and the southeastern region of northwestern China; yields vary between 2.0 and 3.5 t ha^{-1}.

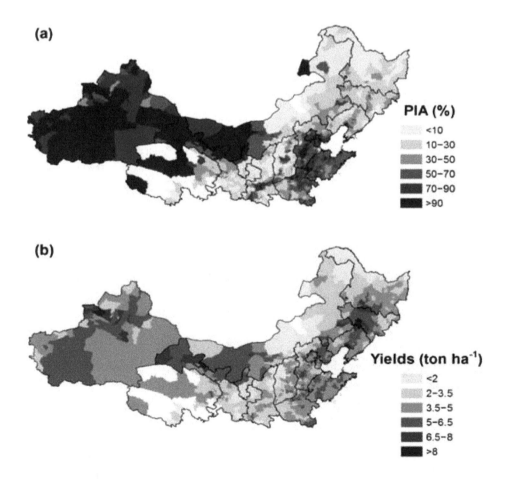

Figure 2. Percentage of irrigated area relative to the area of cultivated land (**a**); Crop yields for the four crops weighted in each county by sown area (**b**).

3.2. Effects of Climate and Irrigation on Crop Yields

A statistical model was established based on our data. Regression results of Tavg, SPEI, and PIA are presented in Table 2. We have also provided a graphical demonstration of the effects of the three individual variables on yields by artificially increasing Tavg by 1 °C, decreasing SPEI by 0.5, and increasing PIA by 10% based on above statistical model.

Table 2. Regression coefficients of the regression model, *t*-statistic and 95% confidence interval estimated using the bootstrap re-sampling approach.

Variables	Regression Coefficients	*t*-Statistic	95% CI
PIA	0.000372	0.41	$(-0.0013, 0.0019)$
PIA^2	-4.68×10^{-6}	-1.07	$(-1.53 \times 10^{-5}, 5.618 \times 10^{-6})$
Tavg	-0.0107	-1.39	$(-0.031, 0.0088)$
Tavg^2	-0.00052 *	-1.84	$(-0.001, -0.0001)$
SPEI	0.260 ***	36.2	$(0.24, 0.28)$
SPEI^2	-0.143 ***	-21.72	$(-0.16, -0.12)$
PIA $\times$ Tavg	6.66×10^{-5} **	-25.88	$(0.00002, 0.00013)$
PIA $\times$ SPEI	-0.00306 **	2.08	$(-0.0033, -0.0027)$
Sample size	28341		
R^2	0.9771		
F-value	330.9		
p-value	<0.001		

* *p*-value < 0.05; ** *p*-value <0.01; *** *p*-value < 0.001.

The full model shows a good agreement, with an R^2 of 0.9771 ($p < 0.001$). The effect of the linear Tavg term on yields is statistically insignificant ($p > 0.05$ with 95% CI between -0.031 and 0.0088). In contrast, there is a significant relationship between the Tavg quadratic term and yields ($p < 0.05$ with 95% CI between -0.001 and -0.0001), as shown in Table 2. Given the present climate, 1 °C further warming would reduce yields by 0–3% in the majority of counties when the Tavg over the growing season is greater than 0°C (Figure 3a). For SPEI, both the linear and quadratic terms on yields are statistically significant ($p < 0.001$ with 95% CI between 0.24 and 0.28 for the linear term and between -0.16 and -0.12 for the quadratic term). With SPEI reduced by 0.5, crops growing above an SPEI of approximately 0.5 in the mean-growing season tend to benefit from the drought, whereas crops grown below this threshold are likely to show a declined yield (Figure 3b).

The effects of PIA and PIA^2 on yields are both statistically insignificant ($p > 0.05$). The 95% CIs vary between -0.0013 and 0.0019 for the linear term and between -1.53×10^{-5} and 5.62×10^{-6} for the quadratic term (Table 2). CIs spanning zero suggest an inconsistent regression coefficient in the sign for each sub-sample generated using the bootstrap analysis. However, significant interaction effects of PIA on climate variables are shown ($p < 0.001$). The 95% CI for PIA $\times$ Tavg is between 0.00002 and 0.00013, while the 95% CI for PIA$\times$SPEI is between -0.0033 and -0.0027 (Table 2). The estimated yield change is approximately 2% when the PIA increases by 10% (Figure 3c). This indicates that irrigated area expansion can alter the magnitude of climate impacts on yields.

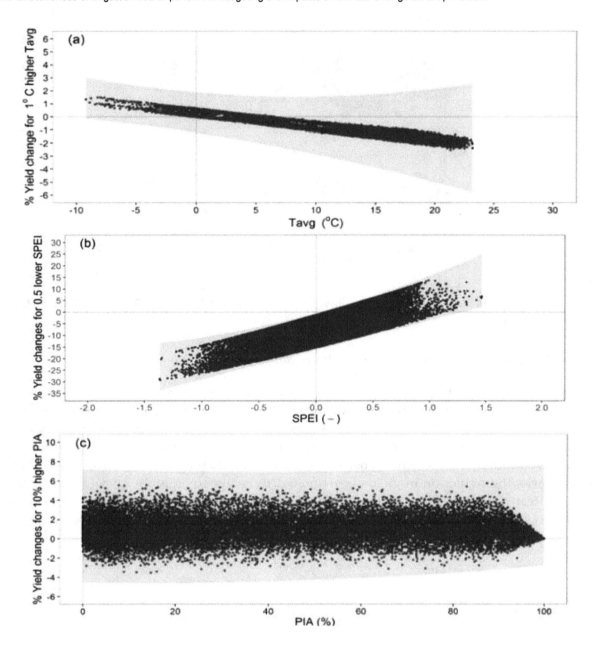

Figure 3. Model-estimated percentage yield changes for (**a**) 1 °C warmer mean-growing-season average temperature (Tavg); (**b**) 0.5 unit lower Standardized Precipitation Evapotranspiration Index (SPEI); and (**c**) 10% higher percentage of irrigated area (PIA). Each of the shaded areas shows the 95% confidence interval in the bootstrap analysis.

3.3. Future Climate Scenario

Based on climate model outputs, a warmer and dryer climate was projected in our study region (Figure 4). Under RCP2.6, it was predicted that Tavg would increase by 1–1.5 °C in northern China as well as the southern region of northeastern China and by more than 1.5 °C in other regions (Figure 4a). In addition, a dryer climate will prevail in most counties: SPEI will experience a 0.0–0.5 reduction in northeastern and most areas of northern China, and more serious decreases in SPEI (0.5–1.5) will occur in northwestern China and parts of northern China (Figure 4b). Under the RCP8.5 scenario, the increase in Tavg will be at least 1.0 °C and most counties will experience a warming with more than 1.5 °C (Figure 4c) relative to the baseline climate. The magnitude of SPEI reduction is also greater (Figure 4d) than in RCP2.6, especially in the central region of northwestern China, where SPEI is estimated to be reduced by approximately 1.0.

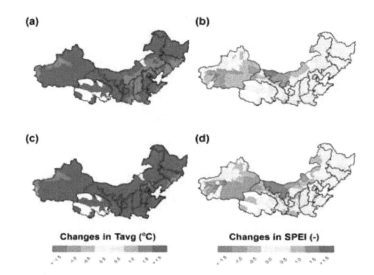

Figure 4. Changes in mean growing season Tavg and SPEI in the 2030s (2020–2039) under RCP2.6 (**a,b**) and RCP8.5 (**c,d**).

3.4. Climate Impacts on Yields under Three Irrigated Area Scenarios

Maintaining the PIA at the baseline climate is anticipated to reduce yields under future climate scenarios, with yields decreasing by 4.7% for the climate under the RCP2.6 scenario averaged over the study region (Table 3). More specifically, we predict 0–5% yield reductions in northeastern and the southern part of northern China, and certain counties in north and northwestern China would experience even more serious reductions, varying between 10% and 20% (Figure 5a). Under the RCP8.5 scenario, the reduction in yields is estimated to be approximately 7.3% (Table 3). Regions with the greatest yield reduction are predicted in the northern part of northern China and central northwestern China, where losses could exceed 20%. In the remaining areas, yields are projected to be reduced by 5–15% (Figure 5e).

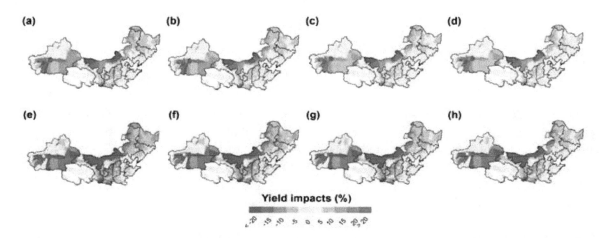

Figure 5. Model-estimated percentage changes in crop yields when (**a**) PIA is constant; (**b**) PIA was increased by 4.4%; (**c**) PIA was increased by 10% and (**d**) PIA was increased by 15% under RCP2.6. The results under RCP8.5 are shown in the bottom panel (**e–h**).

Expanding irrigated areas can alleviate the yield reductions associated with climate impacts. Our model estimates that approximately 0.7% of yields could be saved by a 4.4% increase in PIA under the RCP2.6 climate scenario, and the predicted yield improvement is 0.8% under RCP8.5 (Table 3). Further expansions in PIA are projected to result in greater yield gains. Yield increases are 1.5% under RCP2.6 and 1.8% under RCP8.5 if PIA is expanded by 10%, and the values are 2.2% under RCP2.6

and 2.7% under RCP8.5 if PIA is expanded by 15% (Table 3). However, these yield benefits are still limited relative to adverse climate impacts. The regions with the greatest yield reductions are the northern part of northern China and central northwestern China. Yield could decrease by 5–20% under the RCP2.6 scenario (Figure 5b–d) and by 10–20% under the RCP8.5 scenario (Figure 5f–h). Yield decreases are comparably lower in the northeastern and southern regions of northern China, with 0–15% decreases under RCP2.6 (Figure 5b–d) and 5–15% decreases under RCP8.5 (Figure 5f–h).

Table 3. Projected changes in temperature, SPEI and yields under the four irrigated area scenarios in the study area. The average value has been weighted by sown area.

RCP	RCP2.6	RCP8.5
Temperature change (°C)	1.6	2.0
SPEI change (−)	−0.2	−0.3
Percentage yield change with no change in irrigated area (%)	−4.7	−7.3
Percentage yield change with 4.4% increase in irrigated area (%)	−4.0	−6.5
Percentage yield change with 10% increase in irrigated area (%)	−3.2	−5.5
Percentage yield change with 15% increase in irrigated area (%)	−2.5	−4.6

4. Discussion

4.1. Yield Responses to Climate and Irrigated Areas

Our results demonstrate a significant effect of climate variables on crop yields. Increases in T_{avg} are harmful to yields, with a 0–2% yield reduction per additional 1 °C T_{avg}, because of the associated shorter growing season [14]. Lower yields will be caused by drought in most counties except under very moist climate conditions (i.e., SPEI is approximately 0.5), where yields will be increased. The inverse yield responses to SPEI are associated with less severe water logging and disease under a very wet climate, and hence higher yields, when SPEI is reduced [22,23].

Our model detected a statistically significant effect for the interaction terms of irrigation and climate variables, suggesting that expanding irrigated areas can reduce climate impacts. More specifically, with more irrigated areas, yield reductions caused by heat and drought stresses would be lower. However, our results also indicate that the effects of irrigated area expansion are still very weak on crop yields; the median magnitude is only an approximately 2% yield increase with 10% higher PIA. This finding suggests that the extent to which current irrigation practices will mitigate the negative impacts of climate are quite insufficient in China.

Two primary factors might explain the weak yield response to increased irrigated areas. First and potentially most important, expansions in irrigated areas are not associated with more irrigation water. As noted by theFood and Agriculture Organization of the United Nationsin regard to the definition of irrigated area [14]: "*Due to several reasons (e.g., crop rotation, water shortages, and damage of infrastructure) the area actually irrigated maybe significantly lower than the area equipped for irrigation*". This means that irrigated area data do not reflect the actual accessibility of irrigation water, even though the data of irrigated area is the current primary data to quantify irrigation level. Agricultural water shortage growing in magnitude and frequency in the current [24] and future [25] climate is the main reason. Second, the household contract system was created in 1979, and the use rights of farmland were evenly distributed to the farmers by group farmland ownership [26]. These contracts encouraged farmers to work on their own farmlands but also partially shifted responsibilities previously taken care of by the government to individual farmers, which the farmers could not afford, such as irrigation infrastructure maintenance and repair [27]. Based on a survey conducted by the Ministry of Water Resources of the People's Republic of China in 2006, only 50% of household-based irrigation infrastructures are available to irrigate and 35% of areas categorized as irrigated areas cannot be irrigated [28].

4.2. Future Climate Impacts and Adaptation by Expanding Irrigated Areas

In the 2030s, a warmer and dryer climate is anticipated, posing a serious challenge to agricultural outputs and irrigation water resources over our study region. Expanding irrigated area was projected to save yields from harmful climate impacts. However, such yield benefits are quite limited compared with the negative climate impacts. Similar yield reductions were projected even when irrigated areas are increased by 15%. The scenario experiments demonstrate that yield losses are difficult to avoid under future climate no matter how the irrigated area will be increased.

The model output in our study is not consistent with earlier assessments, which investigated the potential adaptive effects of expanding irrigation using process-based models by assuming different irrigation schemes [29] or assuming no water stress [8]. In theory, there is significant potential for improving irrigation to mitigate the harmful impacts of both of heat and drought on crops [30]. However, in practice, the adaptive potential will not be fully realized. As our study has quantified, the effects of expansion of the irrigated area have little influence on yields due to the aforementioned reasons and thus cannot fundamentally countervail the expected adverse climate change impacts in the 2030s. This finding suggests that the effects of expanding the irrigated area are restricted.

4.3. Implications for the Adaptive Policies for Climate Change in China

Anticipated yield reductions across the irrigated area scenarios suggest that expanding the irrigated area alone cannot achieve our expected climate risk mitigation in the 2030s. Therefore, other solutions are needed. For example, water-saving irrigation technology has been found to reliably increase grain yields while using less water [31]. The adoption of water-saving irrigation technology in sown areas is currently very limited in China [32,33]. According to a farmers' survey across seven provinces in China, only 32% and 4% of sown areas are equipped with household-based and community-based water-saving irrigation technologies, respectively [34]. In northern China, water saving technologies have been reported to show a great potential to reduce water use and improve crop productivities. By using these new technologies, irrigated water reduces by 11.7% and water use efficiency (i.e., yield produced per unit of water) increases by 27.8% for wheat; and the irrigated water saving and water use efficiency improvements are 23.0% and 17.6% for maize, respectively [35]. Therefore, with the low application and substantial potential, water-saving irrigation technology innovation appears to be a more promising approach than establishing more irrigated areas.

Other adaptive measures helpful to improving water use efficiency should not be overlooked. For instance, due to advances in breeding technology, new rice cultivars with high-water efficiency have been bred in China [36], exhibiting a yield advantage of 31–36% under drought [37]. Therefore, policies aimed at climate stress-tolerance cultivars appear to be beneficial to adapting to climate stresses in the future. In addition, the major reason for the future drought is increased evapotranspiration associated with warming [38], thus some technologies that can reduce evapotranspiration, such as plastic sheeting and low-tillage, will be also very helpful. Linking seasonal climate forecasting with crop choice can thus provide another potential climate adaptation. Other adaptation options, including multiple rather than individual adaptive measures, appear more realistic to help reduce future climate risks and should be addressed in future studies.

Finally, it is necessary to develop a new index to represent actual irrigation and water availability at fairly fine resolution. Even though irrigated area is widely used in many earlier works [3,12], its use tends to lack adequate consideration of actual irrigation water and associated adaptive effectsas we showed in this study. Such a new index will prove critical when developing relevant agricultural water use policies.

5. Conclusions

To address the effectiveness of expanding the irrigated area in order to mitigate future climate stresses, this study used county-level data to quantify the adaptive effects of irrigated areas on crop yields and anticipated change in yields under future climate across different irrigated area scenarios. We concluded that expanding irrigated areas cannot countervail future adverse climate impacts on crop yields in northern China. This limitation is primarily attributed to the underutilization of the irrigated area during drought due to water shortage and impaired irrigation infrastructure. Therefore, we hypothesize that the key to improving the resilience of Chinese agriculture under climate impacts is not the size of the irrigated area but, rather, modernizing irrigation. This target change needs to be quantitatively addressed, as it has not been clearly framed in existing policy. Furthermore, the irrigated area expansion plan will require the complementation of other adaptation programs, such as crop breeding and seasonal forecasting. These practices will be particularly useful in regions facing shortages of agricultural water resources. The suitability of different adaptation programs in different regions must be identified in future investigations.

Within the limits of available data, the statistical models used here have been applied to groups of crops and irrigated areas together without crop-specific analysis. These limitations could be overcome with further work by developing an enhanced spatially intensive dataset that further separates agricultural resource inputs for individual crops. In addition, clear knowledge and integrated assessment models to inform farmers' adaptive reactions to climate extremes are in high demand. Such information would enable more accurate predictions of the adaptation potential, costs and benefits, and the agricultural system could be modified based on predicted climate change scenarios.

Acknowledgments: This work was funded by the National Natural Science Foundation of China (41661144006; 31661143012; 41301044). We appreciated the insightful suggestions of anonymous reviewers and Zhi Chen in helping to improve this paper.

Author Contributions: Tianyi Zhang and Jinxia Wang conceived and designed the study; Tianyi Zhang and Jinxia Wang performed the analysis; Tianyi Zhang and Yishu Teng collected data; Tianyi Zhang wrote the paper and all other authors provided comments on the earlier versions of this manuscript.

References

1. Lin, E.; Xiong, W.; Ju, H.; Xu, Y.; Li, Y.; Bai, L.; Xie, L. Climate Change Impacts on Crop Yield and Quality with CO_2 Fertilization in China. *Philos. Trans. R. Soc. B* **2005**, *360*, 2149–2154.

2. Xiong, W.; Lin, E.; Ju, H.; Xu, Y. Climate Change and Critical Thresholds in China's Food Security. *Clim. Chang.* **2007**, *81*, 205–221. [CrossRef]

3. Zhang, T.; Simelton, E.; Huang, Y.; Shi, Y. A Bayesian Assessment of the Current Irrigation Water Supplies Capacity under Projected Droughts for the 2030s in China. *Agric. For. Meteorol.* **2013**, *178*, 56–65. [CrossRef]

4. The Food and Agriculture Organization of the United Nations (FAO). Drought. 2015. Available online: http://www.fao.org/docrep/017/aq191e/aq191e.pdf (accessed on 12 March 2016).

5. Asian Development Bank. *Addressing Climate Change Risks, Disasters, and Adaptation in the People's Republic of China*; Asian Development Bank: Mandaluyong City, Philippines, 2015.

6. Communist Party of China (CPC). Chinese Central Government's Official Web Portal, China's Spending on Water Conservation Doubles During 11th Five-Year Plan. 2011. Available online: http://www.gov.cn/jrzg/2011--01/29/content_1795245.htm (accessed on 12 March 2016).

7. Yu, C. China's water crisis needs more than words. *Nature* **2011**, *470*, 307. [CrossRef] [PubMed]

8. Challinor, A.; Simelton, E.; Fraser, E.; Hemming, D.; Collins, M. Increased crop failure due to climate change: Assessing adaptation options using models and socio-economic data for wheat in China. *Environ. Res. Lett.* **2010**, *5*, 3. [CrossRef]

9. Ju, H.; van der Velde, M.; Lin, E.; Xiong, W.; Li, Y. The impacts of climate change on agricultural production systems in China. *Clim. Chang.* **2013**, *120*, 313–324. [CrossRef]

10. Reidsma, P.; Ewert, F.; Oude Lansink, A.; Leemans, R. Adaptation to climate change and climate variability in European agriculture: The importance of farm level responses. *Eur. J. Agron.* **2010**, *32*, 91–102. [CrossRef]

11. Simelton, E.; Fraser, E.; Termansen, M.; Benton, T.; Gosling, S.; South, A. The socioeconomics of food crop production and climate change vulnerability: A global scale quantitative analysis of how grain crops are sensitive to drought. *Food Secur.* **2012**, *4*, 163–179. [CrossRef]

12. Simelton, E.; Fraser, E.; Termansen, M.; Forster, P.; Dougill, A. Typologies of crop-drought vulnerability: An empirical analysis of the socio-economic factors that influence the sensitivity and resilience of drought of three major food crops in China (1961–2001). *Environ. Sci. Policy* **2009**, *12*, 438–452. [CrossRef]

13. Ministry of Water Resources of China. *Technical Terminology for Irrigation and Drainage*; Ministry of Water Resources of China: Beijing, China, 1993; pp. 56–93.

14. The Food and Agriculture Organization of the United Nations (FAO). Global Map of Irrigated Areas. 2010. Available online: http://www.fao.org/nr/water/aquastat/irrigationmap/index30.stm (accessed on 12 March 2016).

15. China Meteorology Data Sharing Service. Daily climate dataset. Available online: http://cdc.nmic.cn/ (accessed on 12 March 2016).

16. Thornton, P.; Running, S.; White, M.A. Generating surfaces of daily meteorological variables over large regions of complex terrain. *J. Hydrol.* **1997**, *190*, 214–251. [CrossRef]

17. Zhang, T.; Huang, Y.; Yang, X. Climate warming over the past three decades has shortened 20 rice growth duration in China and cultivar shifts have further accelerated the 21 process for late rice. *Glob. Chang. Biol.* **2013**, *19*, 563–570. [CrossRef] [PubMed]

18. Vicente-Serrano, S.; Begueria, S.; Lopez-Moreno, J. A multi-scalar drought index sensitive to global warming: The Standardized Precipitation Evapotranspiration Index-SPEI. *J. Clim.* **2010**, *23*, 1696–1718. [CrossRef]

19. Zhang, F. *Chinese Agricultural Phenology Atlas*; Science Press: Beijing, China, 1987.

20. Liu, B.; Asseng, S.; Müller, C.; Ewert, F.; Elliott, J.; Lobell, D.; Martre, P.; Ruane, A.; Wallach, D.; Jones, J.W.; et al. Similar estimates of temperature impacts on global wheat yield by three independent methods. *Nat. Clim. Chang.* **2016**, *6*, 1130–1136. [CrossRef]

21. Lobell, D.; Burke, M.; Tebaldi, C.; Mastrandrea, M.; Falcon, W.; Naylor, R. Prioritizing climate change adaptation needs for food security in 2030. *Science* **2008**, *319*, 607–610. [CrossRef] [PubMed]

22. Deng, X.; Huang, J.; Qiao, F.; Naylor, R.; Falcon, W.; Burke, M. Impacts of El Nino-Southern Oscillation events on China's rice production. *J. Geogr. Sci.* **2010**, *20*, 3–16. [CrossRef]

23. Zhang, T.; Zhu, J.; Wassmann, R. Responses of rice yields to recent climate changein China: An empirical assessment based on long-term observations at different spatial scales (1981–2005). *Agric. For. Meteorol.* **2010**, *150*, 1128–1137. [CrossRef]

24. Shalizi, Z. *Addressing China's Growing Water Shortages and Associated Social and Environmental Consequences*; World Bank Policy Research Working Paper: No. 3895; World Bank: Washington, DC, USA, 2006.

25. Wang, S.; Zhang, Z. Effects of climate change on water resources in China. *Clim. Res.* **2011**, *47*, 77–82. [CrossRef]

26. Chen, T.; Yabe, M. *Study on the Formation of Household Management in Chinese Agriculture*; Faculty of Agriculture Publications: Fukuoka, Japan, 2009.

27. Xu, K. Why do irrigation infrastructures abandoned? *Coop. Econ. China* **2009**, *2*, 5.

28. Zhang, C.; Li, D. The concept of reinforcing rural irrigation infrastructure constructions in modern China. *China Rural Water Hydropower* **2009**, *7*, 1–3.

29. Chen, C.; Wang, E.; Yu, Q. Modeling wheat and maize productivity as affected by climate variation and irrigation supply in North China Plain. *Agron. J.* **2010**, *102*, 1037–1049. [CrossRef]

30. Zhang, T.; Lin, X.; Sassenrath, G. Current irrigation practices in the central United States reduce drought and extreme heat impacts for maize and soybean but not for wheat. *Sci. Total Environ.* **2015**, *508*, 331–342. [CrossRef] [PubMed]

31. Grassini, P.; Cassman, K. High-yield maize with large net energy yield and small global warming intensity. *Proc. Natl. Acad. Sci. USA* **2011**, *109*, 1074–1079. [CrossRef] [PubMed]

32. Blanke, A.; Rozelle, S.; Lohmar, B.; Wang, J.; Huang, J. Water saving technology and saving water in China. *Agric. Water Manag.* **2007**, *87*, 139–150. [CrossRef]

33. Liu, Y.; Huang, J.; Wang, J.; Rozelle, S. Determinants of agricultural water saving technology adoption: An empirical study of 10 provinces of China. *Ecol. Econ.* **2008**, *4*, 462–472.

34. Cremades, R.; Wang, J.; Morris, J. Policies, Economic incentives and the adoption of modern irrigation technology in China. *Earth Syst. Dyn.* **2015**, *6*, 399–410. [CrossRef]

35. Huang, Q.; Wang, J.; Li, Y. Do water saving technologies save water? Empirical evidence from North China. *J. Environ. Econ. Manag.* **2017**, *82*, 1–16. [CrossRef]

36. Zhang, Q. Strategies for developing Green Super Rice. *Proc. Natl. Acad. Sci. USA* **2007**, *104*, 16402–16409. [CrossRef] [PubMed]

37. Marcaida, M., III; Li, T.; Angeles, O.; Evangelista, G.; Fontanilla, M.; Xu, J. Biomass accumulation and partitioning of newly developed Green Super Rice (GSR) cultivars under drought stress during the reproductive stage. *Field Crop. Res.* **2014**, *162*, 30–38. [CrossRef]

38. Chen, H.; Sun, J. Changes in drought characteristics over China using the standardized precipitation evapotranspiration index. *J. Clim.* **2015**, *28*, 5430–5447. [CrossRef]

11

Possible Scenarios of Winter Wheat Yield Reduction of Dryland Qazvin Province, Iran, based on Prediction of Temperature and Precipitation till the End of the Century

Behnam Mirgol [1] **and Meisam Nazari** [2,3,*]

[1] Department of Water Engineering, Faculty of Engineering and Technology, Imam Khomeini International University, 3414896818 Qazvin, Iran; meisam.nazari1991@gmail.com

[2] Department of Crop Sciences, Faculty of Agricultural Sciences, Georg-August University of Göttingen, Büsgenweg 5, 37077 Göttingen, Germany

[3] Department of Soil Science, University of Kassel, Nordbahnhofstr. 1a, 37213 Witzenhausen, Germany

* Correspondence: meisam.nazari@stud.uni-goettingen.de

Abstract: The climate of the Earth is changing. The Earth's temperature is projected to maintain its upward trend in the next few decades. Temperature and precipitation are two very important factors affecting crop yields, especially in arid and semi-arid regions. There is a need for future climate predictions to protect vulnerable sectors like agriculture in drylands. In this study, the downscaling of two important climatic variables—temperature and precipitation—was done by the CanESM2 and HadCM3 models under five different scenarios for the semi-arid province of Qazvin, located in Iran. The most efficient scenario was selected to predict the dryland winter wheat yield of the province for the three periods: 2010–2039, 2040–2069, and 2070–2099. The results showed that the models are able to satisfactorily predict the daily mean temperature and annual precipitation for the three mentioned periods. Generally, the daily mean temperature and annual precipitation tended to decrease in these periods when compared to the current reference values. However, the scenarios rcp2.6 and B2, respectively, predicted that the precipitation will fall less or even increase in the period 2070–2099. The scenario rcp2.6 seemed to be the most efficient to predict the dryland winter wheat yield of the province for the next few decades. The grain yield is projected to drop considerably over the three periods, especially in the last period, mainly due to the reduction in precipitation in March. This leads us to devise some adaptive strategies to prevent the detrimental impacts of climate change on the dryland winter wheat yield of the province.

Keywords: CanESM2; HadCM3; precipitation; temperature; winter wheat yield

1. Introduction

The temperature of the Earth is increasing more rapidly than during the previous decades, leading to extensive climate change [1]. The Earth's temperature is projected to maintain its upward trend slightly in the next few decades [1]. A significant rise in the concentration of greenhouse gases such as CO_2, CH_4, N_2O, and water vapor, mainly caused by human activities, has intensified this trend [2]. The concentration of greenhouse gases, volume of ozone, aerosols, and sunspots seem to be the most noticeable reason for temperature variations and climate change in the recent century [3].

More than two billion people live in drylands, constituting nearly 40% of the world's population [4]. Cereals are the major crops cultivated in drylands [5]. Crop production in drylands mainly depends on precipitation during the growing season [6]. Moreover, the rise in temperature has led to exacerbating droughts and a considerable loss in crop yields in arid and semi-arid regions [7].

It is necessary to manage drylands in a sustainable way, by which food security is achieved [8]. To do so, there must be some possible measurements and predictions to protect vulnerable sectors such as agriculture and water resources in drylands [9].

General Circulation Models (GCMs) are the most developed tools for the simulation of general responses to the accumulation of greenhouse gases [10]. Studies have shown that the results of GCMs cannot be exploited directly because they are not accurate enough in describing sub-grid data [10]. Therefore, Statistical Downscaling Models (SDSMs) are one of the tools that have been developed to deal with this problem [11]. SDSMs are the most frequently used models in agricultural research, where some independent variables are measured and collected to predict dependent variables [12]. Tatsumi et al. [13] applied the Hadley Centre Coupled Model (version 3; HadCM3) and Coupled Global Climate Model 3 (CGCM3) to forecast the daily minimum, maximum, and average temperature of Shikoku city in Japan, using downscaling techniques. Their results indicated that the temperature is likely to increase in the Shikoku region, Japan, within the period 2071–2099. In a similar study, Ribalaygua et al. [14] used downscaling techniques to simulate the daily minimum and maximum temperature and daily precipitation in a region located in Spain. Their results showed that maximum and minimum temperatures will rise, while precipitation will decrease in the 21st century. Johns et al. [15], by applying the HadCM3 model, predicted that some regions of Central America and Southern Europe might be moister in the future, whereas Australia may experience a type of drier climate.

In recent years, researchers have studied the potential impacts of climate change on plant growth by using different types of simulation models [16,17]. Russell et al. [18] reported that most of the alterations in wheat yield in the United States are related to climate change. Temperature and precipitation, as two important climatic variables for the evaluation of future grain yield, have been investigated by many researchers. For instance, [16] indicated that the changes in temperature and precipitation within the last 30 years in Mexico had positively impacted on the winter wheat yield. In another study, Landau et al. [19], by applying a multiple-regression model, indicated that the temperature increase led to an improvement in the winter wheat crop characteristics, while the precipitation increase could have negative impacts.

The downscaling of GCMs parameters and studying the possible changes in wheat yield due to climatic effects have been distinctly investigated [14,20]. Lhomme et al. [21], for example, studied the potential effect of climate change on durum wheat yield in Tunisia using the downscaled values of some scenarios. Moreover, the efficiency of the IPCC scenarios has rarely been evaluated and compared [22]. In the present study, the downscaling of two important climatic parameters—temperature and precipitation—was done by the Canadian Earth System Model (CanESM2) and HadCM3 models for the province of Qazvin, located in Iran, where the climate is semi-arid and the dryland farming of winter wheat dominates. Then, the most efficient scenario was chosen to predict the dryland winter wheat yield of the province for the next few decades through a multiple-regression model. The efficiency of the fourth and fifth IPCC scenarios in predicting the temperature and precipitation of the region was also compared.

2. Materials and Methods

2.1. Geography, Climate, and Dryland Farming of the Province

The province of Qazvin has an area of 15,821 km^2, located between 48–45 to 50–50 East of the Greenwich Meridian of longitude and 35–37 to 36–45 North latitude of the Equator. Its average altitude is 1278 m above sea level. It has a semi-arid climate with the annual mean precipitation, daily mean temperature, and relative humidity of 301 mm, 14.2 °C, and 51%, respectively. The province is affected by Siberian and Mediterranean winds, which are considerably important factors in controlling the climate of the province. The geographical situation of the studied area is shown in Figure 1.

The total winter wheat yield of the province is 445 million kg, 364 million kg (82%) of which belongs to irrigated farming and 80.7 million kg (18%) to dryland farming. The total cultivated area for winter wheat is nearly 202,497 ha, 95792 ha and 106,704 ha of which are under irrigated and dryland farming, respectively. The average dryland winter wheat yield of the province is estimated to be 1541 kg ha^{-1}.

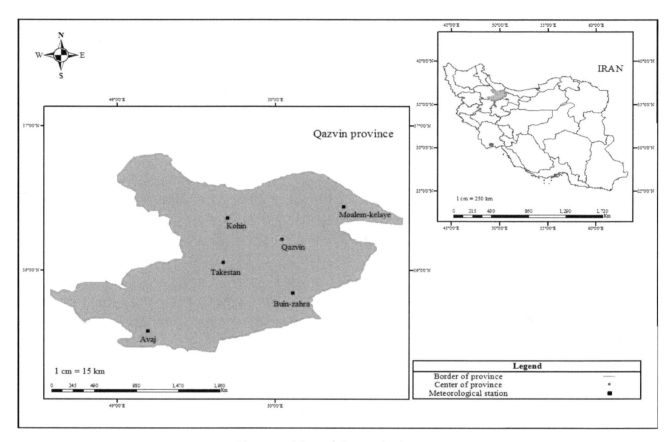

Figure 1. Map of the studied area.

2.2. Methodology

The daily mean temperature and precipitation data for 32 years (1985–2017) were collected from the six meteorological stations in the province (Figure 1). Thereafter, the daily mean temperature and precipitation of all days of all years were calculated separately by the Thiessen polygons method using the software ArcGIS version 10 via Equations (1) and (2):

$$P_a = \frac{\sum p_i A_i}{\sum A_i} \tag{1}$$

$$T_a = \frac{\sum t_i A_i}{\sum A_i} \tag{2}$$

where P_a and T_a are the daily mean precipitation and temperature of the province, respectively; p_i and t_i are the daily mean precipitation and temperature in the station i, respectively; and A_i is the area of the province.

The HadCM3 and CanESM2 models were used to compare the scenarios. HadCM3 has a spatial resolution of 2.5° × 3.75° (latitude by longitude) and the representation produces a grid box resolution of 96 × 73 grid cells. This produces a surface spatial resolution of about 417 km × 278 km, reducing to 295 km × 278 km at 45 degrees North and South. In CanESM2, the long-term time series of standardized daily values are extracted into a one column text file per grid cell. The 128 × 64 grid cells cover global domain according to a T42 Gaussian grid. This grid is uniform along the longitude with a horizontal resolution of 2.81° and is nearly uniform along the latitude of roughly 2.81°. The calibration of the stations (points) against the grid-cells (pixels) was done by the downscaling of the SDSM linear regression model. Data from the years 2006–2015 and 2016–2017 were used for the calibration and validation of both models, respectively. Figures 2 and 3 show the observed versus the simulated values of the temperature and precipitation for the years 2006–2015. Meanwhile, since 26 synoptic variables are considered as predictor variables in these models, having a unique equation was not logically possible because of the accumulated error. To solve this problem, only the predictor variables, being more correlative with the daily mean precipitation and temperature than others, were chosen. Then, the correlation between the variables was detected by Pearson's correlation test ($p < 0.01$) and the most important variables were selected according to the statistical significance between them and the dependent variables ($p < 0.01$). To analyze the climatic data across the study, it was necessary to apply a Statistical Downscaling Model (SDSM). To do so, SDSM version 5.2 was used. SDSM is a decision support tool for assessing local climate change impacts using a powerful statistical downscaling technique. It has the potential to rapidly develop downscaled climatic data [11]. To make statistical connections between the predictor and predicted variables, some regression equations were acquired to predict the climatic variables for the next few periods under the impact of climate change. After acquiring the regression equations and measuring their accuracy, the scenarios were produced through both models for the periods 2010–2039, 2040–2069, and 2070–2099. The properties of these scenarios are indicated in Table 1.

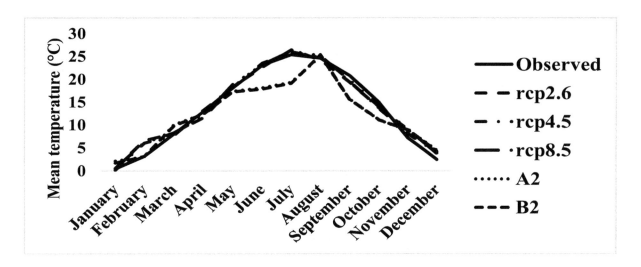

Figure 2. Results of the comparison between the observed and simulated monthly mean temperature values (2006–2015).

Table 1. Properties of the used standard Intergovernmental Panel on Climate Change [10] scenarios.

Models	Scenarios	Properties
CanESM2	rcp2.6	Radiative forcing peaks at 3 W m^{-2} and stabilizes to 2.6 W m^{-2} by the end of 2100; CO_2 concentration is estimated to be 490 ppm by 2100.
	rcp4.5	Radiative forcing is estimated to be 4.5 W m^{-2} by 2100; CO_2 concentration is estimated to be 650 ppm by 2100
	rcp8.5	Radiative forcing is estimated to be 8.5 W m^{-2} by 2100; CO_2 concentration is estimated to be 1370 ppm by 2100
HadCM3	A2	Describes a very heterogeneous world with high population growth, slow economic development, and slow technological change.
	B2	Describes a world with intermediate population and economic growth, emphasizing local solutions to economic, social, and environmental sustainability.

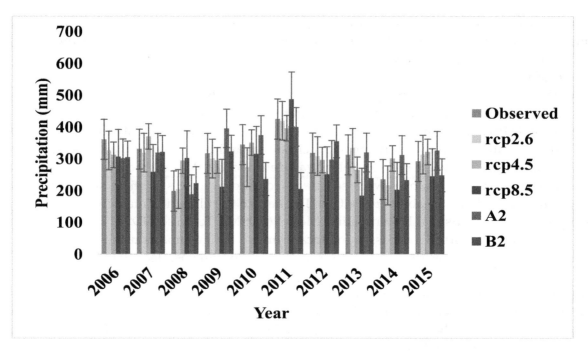

Figure 3. Results of the comparison between the observed precipitation values (2006–2015) and the simulated precipitation values. I = ± SD: standard deviation, the overlapping bars show no significant differences.

The efficiency of the scenarios was compared and the most efficient scenario was recognized through the statistical indicators of Mean Absolute Error (MAE), Root Mean Square Error (RMSE), Nash-Sutcliffe coefficient (NS), Coefficient of Determination (R^2), and Analysis of Variance (at $p < 0.01$) as follows:

$$Z_i = \frac{P_i - \overline{P}}{\sigma_p} \text{ or } Z_i = \frac{O_i - \overline{O}}{\sigma_o} \tag{3}$$

$$MAE = \sum_{i=1}^{n} \left| \frac{P_i - O_i}{n} \right| \tag{4}$$

$$RMSE = \sqrt{\frac{\sum_{i=1}^{n}(P_i - O_i)^2}{n}} \tag{5}$$

$$NS = 1 - \left(\frac{\sum_{i=1}^{n}(O_i - P_i)^2}{\sum_{i=1}^{n}(O_i - \overline{O})^2} \right) \tag{6}$$

$$R^2 = \left[\frac{\frac{1}{n}\sum_{i=1}^{n}(P_i - \overline{P})(O_i - \overline{O})}{\sigma_p \times \sigma_o} \right] \tag{7}$$

where Z_i is the standardized daily mean precipitation or temperature values; O_i and P_i are the observed and simulated daily mean precipitation or temperature values, respectively; $\overline{O}$ is the average of the observed daily mean precipitation or temperature values; $\overline{P}$ is the average of the simulated daily mean precipitation or temperature values; σ_O is the variance of the observed daily mean precipitation or temperature values; σ_P is the variance of the simulated daily mean precipitation and temperature values; and n is the number of data.

Isaaks and Serivastava [23] suggested the MAE and RMSE as statistical indicators able to compare the accuracy of variables. Once the MAE and RMSE values are closer to zero in a scenario, the scenario would be more efficient for predicting climatic variables [24]. When they are exactly 0, it means that there is no error in the predicting task [24]. The Nash-Sutcliffe coefficient (NS) shows to what extent the regression line between the simulated data and measured data can be similar to the regression line 1:1. Its domain is from the negative infinity to 1, and NS = 1 reveals either a complete similarity or a perfect efficiency of a scenario [25]. Meanwhile, R^2 gives information on the correlation between the observed and predicted data and its domain is from 0 to 1 [26]. When R^2 becomes closer to 1, there will be a significant correlation between the data groups [26]. Significant differences between the observed data and values of the predictor scenarios can be distinguished by the analysis of variance [27]. Lack of any significant difference reveals a similarity between the predicted and observed data. In addition, to obtain more appropriate results for the prediction of precipitation, the occurrence of precipitation approach was used. This is a dichotomous method by which the accuracy of whether the occurrence or non-occurrence of precipitation is evaluated. If there is no occurrence of precipitation, then the answer is 'NO', while the answer 'Yes' is a sign of precipitation occurrence [28]. There are four statuses when the observed data are compared with scenario predictions, where a couple of predictions could be true and the remaining predictions could be false. The scenario with a higher percentage of true predictions was selected as the most efficient scenario for predicting the precipitation.

Finally, to predict the dryland winter wheat yield of the province for the next decades and to make a connection between the climatic and yield data for the period 2005–2014, a linear regression model was used. Furthermore, Pearson's correlation test (at $p < 0.01$) between the simulated and observed data, RMSE, and R-square were used to check the regression's validity. All statistical analyses were performed by the software SPSS version 21 (IBM Inc., Chicago, IL, USA).

3. Results

3.1. Temperature Predictions

All three CanESM2 scenarios predicted that the daily mean temperatures would generally increase in the periods 2010–2039, 2040–2069, and 2070–2099 (Table 2). However, the scale of these increases differed by the different scenarios. The scenario rcp2.6 projected that the daily mean temperature of the periods 2010–2039, 2040–2069, and 2070–2099 would be 13.6, 13.9, and 13.9 °C, respectively, which are 0.9, 1.2, and 1.1 °C higher when compared to the observed daily mean temperature. The other scenario rcp4.5 also predicted an increasing trend in the daily mean temperature in the three prospective periods and showed that the mean daily temperature would be 13.4, 14.2, and 14.4 °C in the periods 2010–2039, 2040–2069, and 2070–2099, respectively, each being 0.7, 1.4, and 1.6 °C higher when compared to the observed one. The scenario rcp8.5 predicted the highest temperature trends in comparison with the other two scenarios. It predicted that the mean daily temperature would rise by 13.8, 14.8, and 15.5 °C in the periods 2010–2039, 2040–2069, and 2070–2099, with changes of 1.0, 2.0, and 2.7 °C, respectively, in analogy with the observed value.

Both scenarios (A2 and B2) of HadCM3 generally predicted an increasing daily mean temperature trend for the three future periods in comparison with the observed one, except for scenario B2, which projected a very slightly decreasing trend only for the period 2070–2099 (Table 3). The scenario A2 forecasted that the mean daily temperature would rise to 12.7, 12.8, and 12.8 °C in the periods 2010–2039, 2040–2069, and 2070–2099, being 0.0, 0.1, and 0.2 °C higher, respectively, when compared to the value of the observed period. The mean daily temperatures were projected by the scenario B2 to increase to 12.6 and 12.7 °C in the periods 2010–2039, 2040–2069, respectively. In contrast, it predicted that the mean daily temperature would decrease to 12.6 °C in the period 2070–2099. Accordingly, the predicted temperature changes by scenario B2 are 0.02, 0.05, and −0.04 °C in the periods 2010–2039, 2040–2069, and 2070–2099, respectively, when compared to the observed period.

Table 2. Results of the daily mean temperature predictions of the CanESM2 scenarios for the periods 2010–2039, 2040–2069, and 2070–2099.

Scenarios	Periods	Daily Mean Temperature (°C)
Observed period	1985–2005 (obs)	12.7
rcp2.6	2010–2039 (P1)	13.6
	2040–2069 (P2)	13.9
	2070–2099 (P3)	13.9
	°C change P1 vs. obs	0.9
	°C change P2 vs. obs	1.2
	°C change P3 vs. obs	1.1
rcp4.5	2010–2039 (P1)	13.4
	2040–2069 (P2)	14.2
	2070–2099 (P3)	14.4
	°C change P1 vs. obs	0.7
	°C change P2 vs. obs	1.4
	°C change P3 vs. obs	1.6
rcp8.5	2010–2039 (P1)	13.8
	2040–2069 (P2)	14.8
	2070–2099 (P3)	15.5
	°C change P1 vs. obs	1
	°C change P2 vs. obs	2
	°C change P3 vs. obs	2.7

Table 3. Results of the daily mean temperature predictions of the HadCM3 scenarios for the periods 2010–2039, 2040–2069, and 2070–2099.

Scenarios	Periods	Mean Temperature (°C)
Observed period	1985–2005 (obs)	12.7
A2	2010–2039 (P1)	12.7
	2040–2069 (P2)	12.8
	2070–2099 (P3)	12.8
	°C change P1 vs. obs	0
	°C change P2 vs. obs	0.1
	°C change P3 vs. obs	0.2
B2	2010–2039 (P1)	12.6
	2040–2069 (P2)	12.7
	2070–2099 (P3)	12.6
	°C change P1 vs. obs	0.02
	°C change P2 vs. obs	0.05
	°C change P3 vs. obs	−0.04

3.2. Precipitation Predictions

Overall, the three scenarios of CanESM2 projected a diminishing trend in the annual precipitation for the future periods 2010–2039, 2040–2069, and 2070–2099, when compared to the observed period (Table 4). However, the scenario rcp2.6 projected a less decreasing trend in the annual precipitation for the period 2070–2099. The scenario rcp2.6 predicted that the annual precipitation would drop to 287 and 277 mm in the periods 2010–2039 and 2040–2069, respectively, and decrease to 296 mm in the period 2070–2099. The projected annual precipitation by the scenario rcp4.5 would be 258, 264, and 293 mm in the periods 2010–2039, 2040–2069, and 2070–2099, respectively. The other scenario rcp8.5 forecasted that the annual precipitation would be 283, 278, and 278 mm for the periods 2010–2039, 2040–2069, and 2070–2099, respectively.

Scenario A2 of HadCM3 predicted a decreasing trend in the annual precipitation for the periods 2010–2039, 2040–2069, and 2070–2099, in analogy with the observed period (Table 5). The annual precipitation projected by scenario A2 would be 340, 292, and 276 mm for the periods 2010–2039, 2040–2069, and 2070–2099, respectively. Scenario B2 also forecasted that the annual precipitation for the periods 2010–2039 and 2040–2069 would be 310 and 321 mm, respectively, when compared to the observed period, which conveys a reducing trend. In contrast, it projected an increased annual precipitation of 875 mm for the period 2070–2099, which will be noticeably higher than the observed amount.

Table 4. Results of the annual precipitation predictions of the CanESM2 scenarios for the periods 2010–2039, 2040–2069, and 2070–2099.

Scenarios	Periods	Precipitation (mm)
Observed period	1985–2005 (obs)	346
rcp2.6	2010–2039 (P1)	287
	2040–2069 (P2)	277
	2070–2099 (P3)	296
	% change P1 vs. obs	−18
	% change P2 vs. obs	−21
	% change P3 vs. obs	−15
rcp4.5	2010–2039 (P1)	258
	2040–2069 (P2)	264
	2070–2099 (P3)	293
	% change P1 vs. obs	−29
	% change P2 vs. obs	−26
	% change P3 vs. obs	−16
rcp8.5	2010–2039 (P1)	283
	2040–2069 (P2)	278
	2070–2099 (P3)	278
	% change P1 vs. obs	−20
	% change P2 vs. obs	−21
	% change P3 vs. obs	−21

Table 5. Results of the annual precipitation predictions of the HadCM3 scenarios for the periods 2010–2039, 2040–2069, and 2070–2099.

Scenarios	Periods	Precipitation (mm)
Observed period	1985–2005 (obs)	346
A2	2010–2039 (P1)	340
	2040–2069 (P2)	292
	2070–2099 (P3)	276
	% change P1 vs. obs	−1
	% change P2 vs. obs	−16
	% change P3 vs. obs	−22
B2	2010–2039 (P1)	310
	2040–2069 (P2)	321
	2070–2099 (P3)	875
	% change P1 vs. obs	−10
	% change P2 vs. obs	−7
	% change P3 vs. obs	86

3.3. Comparison of the Scenarios

The variance analysis results showed a higher efficiency for the RCP scenarios than the A and B scenarios in predicting the daily mean temperature of the region (Table 6), because there was no statistically significant difference between the temperature values simulated by the RCPs and the observed values (at $p < 0.01$), while the temperature values simulated by A and B significantly differed from the observed ones (at $p < 0.01$). Among the three scenarios of the model CanESM2, rcp2.6 was selected as the most efficient scenario for predicting the daily mean temperature, as it had the highest Nash-Sutcliffe coefficient and R^2 value and the lowest MAE and RMSE values when compared to scenarios rcp4.5 and rcp8.5.

The results of variance analysis indicated that all scenarios were efficient enough to predict the annual precipitation of the region (Table 7), since no statistically significant difference was found between the simulated and observed values (at $p < 0.01$). The scenario rcp2.6 displayed the lowest values for both MAE and RMSE. Moreover, it showed the highest Nash-Sutcliffe coefficient and R^2 value. Thus, it was selected as the best scenario for predicting the annual precipitation. In addition, the scenarios of CanESM2 simulated closer annual precipitation values to the observed values than the HadCM3 scenarios (Table 8). The CanESM2 scenarios resulted in higher values of true predictions and lower values of false prediction than the scenarios of HadCM3. The indicators provided in Table 8 also, in general, confirmed the excellence of scenario rcp2.6 for predicting the annual precipitation.

Together, these indicators showed a relatively higher efficiency for the CanESM2 scenarios than the HadCM3 scenarios in predicting the daily mean temperature and annual precipitation of the region.

Table 6. Results of the efficiency evaluation of the used scenarios for the daily mean temperature predictions.

Models	Scenarios	MAE	RMSE	Nash-Sutcliffe	R^2	Analysis of Variance
CanESM2	rcp2.6	0.348	0.445	0.808	0.8177	
	rcp4.5	0.355	0.45	0.801	0.8047	0.772 ns
	rcp8.5	0.362	0.461	0.795	0.8174	
HadCM3	A2	0.0529	0.0658	0.707	0.7346	0.000 **
	B2	0.0523	0.0654	0.706	0.7380	

ns: no-significant; **: significant at $p < 0.01$.

Table 7. Results of the efficiency evaluation of the used scenarios for the annual precipitation predictions.

Models	Scenarios	MAE	RMSE	Nash-Sutcliffe	Analysis of Variance
CanESM2	rcp2.6	0.434	1.297	−2.139	
	rcp4.5	0.442	1.298	−3.154	0.279 ns
	rcp8.5	0.45	1.351	−8.576	
HadCM3	A2	0.444	1.33	−7.243	0.453 ns
	B2	0.442	1.299	−3.222	

ns: no-significant.

Table 8. Occurrence of precipitation under the used scenarios.

Occurrences	CanESM2			HadCM3	
	rcp8.5	rcp4.5	rcp2.6	B2	A2
Hit (hit event)	390	395	366	406	425
CN (correct Negative)	1832	1827	1856	1816	1797
Miss (miss event)	1246	1225	1250	1191	1159
FA (false alarm events)	184	205	180	239	271
% true prediction ($\frac{Hit+CN}{n}$)	44.79	44.35	44.25	43.72	43.37
% false prediction ($\frac{Miss+FN}{n}$)	55.2	55.64	55.75	56.27	56.62

3.4. Yield Predictions

The results of the regression analysis and Pearson's correlation test showed that the precipitation in March was the most effective factor for the dryland winter wheat yield of the region (Table 9). The prediction results indicated that the yield would noticeably reduce to 1176, 984, and 890 kg ha^{-1} in the periods 2010–2039, 2040–2069, and 2070–2099, respectively (Table 10). The reduction percentage in the above-mentioned periods is predicted to be −22, −34, and −41%, respectively. These reductions in the yield are consistent with the reductions in the mean precipitation in March during the three prospective periods (Figure 4). The reduction in the yield in the periods 2040–2069 and 2070–2099 will be more severe than that of the period 2010–2039, which is in line with a more severe reduction in the precipitation in March than in the former periods.

Table 9. Regression and correlation results of the yield and precipitation data.

Crop	Regression Model	R	R^2	RMSE (%)	Significance Level	Predictor Model
winter wheat	Forward	0.78	0.62	18.82	0.012 *	Y = 20.883X + 625.846

*: significant at $p < 0.05$ where Y is dryland winter wheat yield; X is the precipitation in March; and the constant numbers are Y-intercepts.

Table 10. Results of the dryland winter wheat yield predictions for the periods 2010–2039, 2040–2069, and 2070–2099.

Crop	Cropping Year	Grain Yield (kg ha^{-1})
Winter wheat	2010–2011 (obs)	1512
	2010–2039 (P1)	1176
	2040–2069 (P2)	984
	2070–2099 (P3)	890
	% change P1 vs. obs	−22
	% change P2 vs. obs	−34
	% change P3 vs. obs	−41

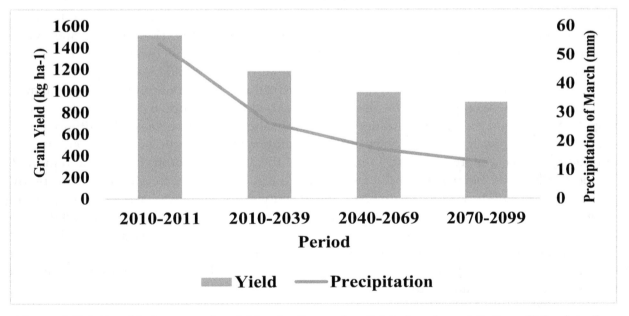

Figure 4. Relationship between the yield reduction and rcp2.6-induced precipitation of March in the three future periods.

4. Discussion

4.1. Temperature Predictions

GCMs have widely been used for predicting future temperature trends. Van Vuuren et al. [29] showed that the mean temperature was likely to increase in the future in many parts of the world. For instance, Basheer et al. [30] claimed that the climate over the Dinder River Basin would be warmer in the upcoming decades. Majhi and Pattnayak [31] also revealed that there would be a gradual temperature increase in Nabarangpur district at the end of the 21st century. Our results also indicated that the temperature would generally increase in the three investigated periods; however, the magnitude of these increases are dependent on the scenarios applied. The CanESM2 scenarios postulated a higher variability in the predicted temperature values than the HadCM3 scenarios. In addition, the temperature changes predicted by CanESM2 were noticeably higher than those predicted by HadCM3. Such different trends have also been observed by [22], who compared some GCMs such as HadCM3 and CanESM2. These diverse trends could have been due to the different scenarios used, as was the case for the study of [32]. Among the CanESM2 scenarios, rcp8.5 and rcp4.5 predicted the highest temperature values, respectively, whilst rcp2.6 projected the lowest ones. These results are in line with the findings of [22]. The greatest temperature values predicted by scenarios rcp8.5 and rcp4.5 seem plausible due to the underlying physical laws to simulate the ongoing increases in the radiative forcing and CO_2 concentrations by the end of the 21st century. In contrast, rcp2.6 simulated a lower radiative forcing towards the end of the 21st century as well as lower CO_2 concentrations.

4.2. Precipitation Predictions

All scenarios, except B2, revealed that there would be a reduction in the annual precipitation in all investigated periods. Scenarios rcp4.5 and rcp8.5 projected the maximum and the minimum reductions in the annual precipitation, respectively, which was a very similar result to what [33] concluded. Scenario B2 projected substantial increases in the annual precipitation for the period 2070–2099. Moreover, scenario rcp2.6 projected a less decreased annual precipitation for the aforementioned period. One study has shown that there is a possibility for a reduction in the rivers' ice thickness in winter and a slight increase in the discharge during the break up from May to June in Siberia [34]. This phenomenon can be caused by extreme warming around Siberia in the period 2070–2099. To confirm this notion, Shiklomanov et al. [35] predicted an increased mean temperature trend for Siberia by the late 21st century. The province of Qazvin is extremely affected by Siberian winds. Therefore, the increased and less decreased annual precipitation projections for the period 2070–2099 by scenarios B2 and rcp2.6 might be logical. Nevertheless, the properties of the scenarios used could be among other reasons for the different precipitation results achieved. Scenarios rcp2.6 and B2 more optimistically simulated the future projections when compared to the other scenarios used. For instance, rcp2.6 predicted a radiative forcing of 3 W m^{-2} and a CO_2 concentration of 490 ppm; and B2 described a world with intermediate population and economic growth, emphasizing local solutions to economic, social, and environmental sustainability. Thus, a more optimistic simulation of the annual precipitation of the region could have been another possible reason for the increased and less decreased precipitation values predicted. Vallam and Qin [22], using a statistical downscaling technique, also showed that scenarios rcp2.6 and B2 could predict either increased or at least lesser decreased rainfall percentage for Frankfurt (Germany), Singapore, and Miami (USA) in the 2080s when compared to the other scenarios used. However, the CanESM2-derived RCP scenarios led to great variabilities in predicting future meteorological variables, especially rainfall in arid regions [22]. This might be another plausible reason for the increase (14%) in the annual precipitation predicted by rcp2.6.

4.3. Yield Predictions

Studies have shown that there is a significant correlation between winter wheat yield and the climatic variables [16]. Thus, the most efficient scenario (rcp2.6) in predicting both temperature and precipitation was applied to predict the dryland winter wheat yield of the province. The results of the Pearson's correlation test indicated that the precipitation in March was the most effective factor on yield (r = 0.78, $p < 0.01$). A study on the effects of precipitation on dryland cereals yield in three provinces of Iran was performed, where the climate is semi-arid [36]. The results of the study showed that the yield of dryland winter wheat was significantly correlated to precipitation, especially the precipitation in April. In the province of Qazvin, dryland winter wheat is at the tillering stage in March (personal communication with the farmers). It seems that the lower precipitation in March could lead to a lower number of head-bearing tillers and lack of the opportunity for their survival, finally resulting in lower grain yields. Karimi [37] investigated the effects of precipitation during the tillering of dryland winter wheat in Iran and reported a significant impact on the final grain yield. Even though agricultural factors such as soil, fertilizers, and other climatic variables like radiation could also be effective, Lobell [16] indicated that precipitation had a more considerable influence on dryland farming. Meanwhile, the value of R^2 between the observed and simulated data was 0.62, meaning that the yield was 62% dependent on the annual precipitation and the other 38% was dependent on other unspecified factors. The percentage of RMSE was about 18% between the observed and simulated data, which was an acceptable value that showed the adequate accuracy of the predictions [38]. Moreover, the observed reductions in the precipitation in March during the three future periods could have been due to shifts in the seasons due to warmer temperatures of the areas by which the studied region is affected. As mentioned earlier, the temperature of Siberia has been projected to rise by the late 21st century [35]. Since the province of Qazvin is extremely affected by Siberian winds, it is plausible that these winds will alter the seasons of this province.

5. Conclusions

In this study, the downscaling of two important climatic variables—temperature and precipitation—was done by the CanESM2 and HadCM3 models for the province of Qazvin, located in Iran. The used scenarios were able to predict the daily mean temperature and annual precipitation for the three different future periods 2010–2039, 2040–2069, and 2070–2099. The CanESM2 scenarios seemed to be more efficient than the HadCM3 scenarios in simulating the future temperature and precipitation trends of the region. Generally, the region's daily mean temperature tended to increase and the annual precipitation tended to decrease in the three prospective periods investigated. However, scenarios rcp2.6 and B2, respectively, predicted that the precipitation would decrease less or even increase in the third period (2070–2099). Scenario rcp2.6 was assumed to be the most efficient to predict the dryland winter wheat yield of the province for the upcoming decades. The grain yield was projected to considerably decrease in the three periods, especially in the last period. The yield reductions are assumed to mainly be due to the decrease in precipitation in March during the investigated periods. Some adaptive strategies to prevent the detrimental impacts of climate change on the province dryland wheat yield include the cultivation of resistant winter wheat varieties to drought as well as earlier sowing dates. The authors would like to recommend the comparative use of the applied CanESM2 and HadCM3 scenarios to predict climatic variables of other semi-arid regions.

Author Contributions: Conceptualization, B.M.; Methodology, B.M. and M.N.; Software, B.M. and M.N.; Validation, B.M. and M.N.; Formal Analysis, B.M. and M.N.; Investigation, B.M. and M.N.; Resources, B.M. and M.N.; Data Curation, B.M. and M.N.; Writing-Original Draft Preparation, M.N.; Writing-Review & Editing, B.M. and M.N.; Visualization, B.M. and M.N.; Supervision, B.M. and M.N.; Project Administration, B.M.; Funding Acquisition, B.M.

Acknowledgments: The authors would like to acknowledge the personnel of the Qazvin Meteorological Organization for providing the meteorological data. Mohammad Eteghadipour is also acknowledged for his useful scientific guides.

References

1. Thomas, R.K.; Zhang, R.; Horowitz, L.W. Prospects for a prolonged slowdown in global warming in the early 21st century. *Nat. Communities* **2016**. [CrossRef]
2. Nozawa, T.; Nagashima, T.; Shiogama, H.; Crooks, S.A. Detecting natural influence on surface air temperature change in the early twentieth century. *Geophys. Res. Lett.* **2005**, 32, L20719. [CrossRef]
3. Santer, B.D.; Taylor, K.E.; Wigley, T.M.; Johns, T.C.; Jones, P.D.; Karoly, D.J.; Mitchell, J.F.B.; Oort, A.H.; Penner, J.E.; Ramaswamy, V.; et al. A search for human influences on the thermal structure of the atmosphere. *Nature* **1996**, 382, 39–46. [CrossRef]
4. White, R.P.; Nackoney, J. Drylands, People, and Ecosystem Goods and Services: A Web-Based Geospatial Analysis. 2003. Available online: http://pdf.wri.org/drylands.pdf (accessed on 17 June 2018).

5.   LADA. *Guidelines for Land Use System Mapping*; Technical Report; FAO: Rome, Italy, 2008.

6.   Wang, X.; Cai, D.; Wu, H.; Hoogmoed, W.B.; Oenema, O. Effects of variation in rainfall on rainfed crop yields and water use in dryland farming areas in China. *Arid Land Res. Manag.* **2016**, *30*, 1–24. [CrossRef]

7.   Andreadis, K.M.; Lettenmaier, D.P. Trends in 20th century drought over the continental United States. *Geophys. Res. Lett.* **2006**, *33*, L10403. [CrossRef]

8.   UNEP. Sourcebook of Alternative Technologies for Freshwater Augmentation in West Asia. 2000. Available online: http://www.unep.or.jp (accessed on 15 June 2018).

9.   Gan, T.Y. Reducing vulnerability of water resources of Canadian Prairies to potential droughts and possible climate warming. *Water Resour. Manag.* **2000**, *14*, 111–135. [CrossRef]

10.   IPCC. *Climate Change: Impacts, Adaptation and Vulnerability*; Cambridge University Press: New York, NY, USA, 2007.

11.   Wilby, R.L.; Dawson, C.W.; Barrow, E.M. SDSM- a decision support tool for the assessment of regional climate change impacts. *Environ. Modell. Softw.* **2002**, *17*, 147–159. [CrossRef]

12.   Gulden, K.U.; Neşe, G. A Study on Multiple Linear Regression Analysis. *Procedia Soc. Behav. Sci.* **2013**, *106*, 234–240.

13.   Tatsumi, K.; Oizumi, T.; Yamashiki, Y. Introduction of daily minimum and maximum temperature change signals in the Shikoku region using the statistical downscaling method by GCMs. *Hydrol. Res. Lett.* **2013**, *7*, 48–53. [CrossRef]

14.   Ribalaygua, J.; Pino, M.R.; Pórtoles, J.; Roldán, E.; Gaitán, E.; Chinarro, D.; Torres, L. Climate change scenarios for temperature and precipitation in Aragón (Spain). *Sci. Total Environ.* **2013**, *463–464*, 1015–1030. [CrossRef] [PubMed]

15.   Johns, T.C.; Gregory, J.M.; Ingram, W.J.; Johnson, C.E.; Jones, A.; Lowe, J.A.; Mitchell, J.F.B.; Roberts, D.L.; Sexton, D.M.H.; Stevenson, D.S. Anthropogenic climate change for 1860 to 2100 simulated with the HadCM3 model under updated emissions scenarios. *Clim. Dyn.* **2003**, *20*, 583–612. [CrossRef]

16.   Lobell, D.B.; Ortiz Monasterio, J.I.; Addams, C.L.; Anser, G.P. Soil, climate and management impacts on regional wheat productivity in Mexico from remote sensing. *Agric. For. Meteorol.* **2002**, *114*, 31–43. [CrossRef]

17.   Lobell, D.B.; Asseng, S. Comparing estimates of climate change impacts from process-based and statistical crop models. *Environ. Res. Lett.* **2017**, *12*, 015001. [CrossRef]

18.   Russell, K.; Chad, L.; Rebecca, M.L.; David, V.S. Impact of Climate Change on Wheat Production in Kentucky. *Plant Soil Sci. Res. Rep.* **2014**, *2*. [CrossRef]

19.   Landau, S.; Mitchell, R.; Barnett, V.; Colls, J.J.; Craigon, J.; Payne, R.W. A parsimonious, multiple- regression model of wheat yield response to environment. *Agric. For. Meteorol.* **2000**, *101*, 151–166. [CrossRef]

20.   Bin, W.D.; Liu, L.; O'Leary, G.J.; Asseng, S.; Macadam, I.; Lines-Kelly, R.; Yang, X.; Clark, A.; Crean, J.; Sides, T.; et al. Australian wheat production expected to decrease by the late 21st century. *Glob. Chang. Biol.* **2018**. [CrossRef]

21.   Lhomme, J.P.; Mougou, R.; Mansour, M. Potential impact of climate change on durum wheat cropping in Tunisia. *Clim. Chang.* **2009**, *96*, 549–564. [CrossRef]

22.   Vallam, P.; Qin, X.S. Projecting future precipitation and temperature at sites with diverse climate through multiple statistical downscaling schemes. *Theor. Appl. Climatol.* **2017**. [CrossRef]

23.   Isaaks, E.H.; Serivastava, R.M. *An introduction to applied Geostatistics*; Oxford University Press: New York, NY, USA, 1989.

24.   Chai, T.; Draxler, R.R. Root mean square error (RMSE) or mean absolute error (MAE)?—Arguments against avoiding RMSE in the literature. *Geosci. Model Dev.* **2014**, *7*, 1247–1250. [CrossRef]

25.   Nash, J.E.; Sutcliffe, J.V. River flow forecasting through conceptual models part I—A discussion of principles. *J. Hydrol.* **1970**, *10*, 282–290. [CrossRef]

26.   Gujarati, D.N.; Porter, D.C. *Basic Econometrics*, 5th ed.; Tata McGraw-Hill Education: New York, NY, USA, 2009; pp. 73–78.

27.   Armstrong, R.A.; Eperjesi, F.; Gilmartin, B. The application of analysis of variance (ANOVA) to different experimental designs in optometry. *Ophthalmic Physiol. Opt.* **2002**, *22*, 248–256. [CrossRef] [PubMed]

28.   Roberts, N.M.; Lean, H.W. Scale-selective verification of rainfall accumulations from high-resolution forecasts of convective events. *Mon. Weather Rev.* **2008**, *136*, 78–97. [CrossRef]

29.   Van Vuuren, D.P.; Meinshause, M.; Plattner, G.K.; Joos, F.; Strassmann, K.M.; Smith, S.J.; Reilly, J.M. Temperature increase of 21st century mitigation scenarios. *Proc. Natl. Acad. Sci. USA* **2008**, *105*, 15258–15262. [CrossRef] [PubMed]

30. Basheer, A.K.; Lu, H.; Omer, A.; Ali, A.B.; Abdelghader, A.M.S. Impacts of climate change under CMIP5 RCP scenarios on the streamflow in the Dinder River and ecosystem habitats in Dinder National Park, Sudan. *Hydrol. Earth Syst. Sci.* **2016**, *20*, 1331–1353. [CrossRef]

31. Majhi, S.; Pattnayak, K.C.; Pattnayak, R. Projections of rainfall and surface temperature over Nabarangpur district using multiple CMIP5 models in RCP 4.5 and 8.5 scenarios. *Int. J. Appl. Res.* **2016**, *2*, 399–405.

32. Mekonnen, D.F.; Disse, M. Analyzing the future climate change of Upper Blue Nile River Basin (UBNRB) using statistical down scaling techniques. *Hydrol. Earth Syst. Sci.* **2016**, *22*, 2391–2408. [CrossRef]

33. Aung, M.T.; Shrestha, S.; Weesakul, S.; Shrestha, P.K. Multi-model climate change projections for Belu River Basin, Myanmar under representative concentration pathways. *J. Earth Sci. Clim. Chang.* **2016**, *7*, L323.

34. Costard, F.; Gautier, E.; Brunstein, D.; Hammadi, J.; Fedorov, A.; Yang, D.; Dupeyrat, L. Impact of the global warming on the fluvial thermal erosion over the Lena River in Central Siberia. *Geophys. Res. Lett.* **2007**, *34*, L14501. [CrossRef]

35. Shiklomanov, N.I.; Streletskiy, D.A.; Swales, T.B.; Kokorev, V.A. Climate change and stability of urban infrastructure in Russian permafrost regions: Prognostic assessment based on GCM climate projections. *Geogr. Rev.* **2017**, *107*, 125–142. [CrossRef]

36. Bannayan, M.; Lotfabadi, S.S.; Sanjani, S.; Mohamadian, A.; Aghaalikhani, M. Effects of precipitation and temperature on crop production variability in northeast Iran. *Int. J. Biometeorol.* **2011**, *55*, 387–401. [CrossRef] [PubMed]

37. Karimi, M. Drought during growing season of 1997–8 and its effects on wheat production in Iran. *Sonbloe J.* **1999**, *30*, 1–7.

38. Rinaldy, M.; Losavio, N.; Flagella, Z. Evaluation of OILCROP-SUN model for sunflower in southern Italy. *Agric. Syst.* **2003**, *78*, 17–30. [CrossRef]

12

Environmental Sustainability of Agriculture Stressed by Changing Extremes of Drought and Excess Moisture

Elaine Wheaton [1] **and Suren Kulshreshtha** [2,*]

[1]   Department of Geography and Planning, University of Saskatchewan, Saskatoon, SK S7N 5A8, Canada; elainewheaton@sasktel.net

[2]   Department of Agricultural and Resource Economics, University of Saskatchewan, Saskatoon, SK S7N 5A8, Canada

*   Correspondence: suren.kulshreshtha@usask.ca

Academic Editor: Iain Gordon

Abstract: As the climate changes, the effects of agriculture on the environment may change. In the future, an increasing frequency of climate extremes, such as droughts, heat waves, and excess moisture, is expected. Past research on the interaction between environment and resources has focused on climate change effects on various sectors, including agricultural production (especially crop production), but research on the effects of climate change using agri-environmental indicators (AEI) of environmental sustainability of agriculture is limited. The aim of this paper was to begin to address this knowledge gap by exploring the effects of future drought and excess moisture on environmental sustainability of agriculture. Methods included the use of a conceptual framework, literature reviews, and an examination of the climate sensitivities of the AEI models. The AEIs assessed were those for the themes of soil and water quality, and farmland management as developed by Agriculture and Agri-Food Canada. Additional indicators included one for desertification and another for water supply and demand. The study area was the agricultural region of the Canadian Prairie Provinces. We found that the performance of several indicators would likely decrease in a warming climate with more extremes. These indicators with declining performances included risks for soil erosion, soil salinization, desertification, water quality and quantity, and soil contamination. Preliminary trends of other indicators such as farmland management were not clear. AEIs are important tools for measuring climate impacts on the environmental sustainability of agriculture. They also indicate the success of adaptation measures and suggest areas of operational and policy development. Therefore, continued reporting and enhancement of these indicators is recommended.

Keywords: environmental sustainability; agricultural sustainability; environmental indicators; climate change; climate extremes; drought; excess moisture; Canadian Prairie Provinces

---

1. Introduction

Considerable changes in climate variables relevant to agriculture and the environment have already occurred and have been documented [1–3]. For the agricultural portion of the Canadian Prairie Provinces, where this study was conducted, these agro-climatic changes include longer growing seasons, more crop heat units, decreasing snow-cover area, changes in precipitation from snow to rain during winter months, and warmer winters. Future changes for the prairie agricultural region indicate continued and perhaps accelerated trends in these variables and many others [1,4]. An increase in climate extremes, including droughts, excess moisture, and heat waves is expected for the Canadian Prairies [5,6]. These extremes can often have adverse effects on the environmental sustainability of agriculture. More recent work also confirms that future drought characteristics (frequency of droughts, duration, and intensity) show increases over the southern prairies [7]. Increases in such extremes would have adverse effects on the environmental sustainability of agriculture. The effects of droughts and excessive moisture on environmental sustainability are of special concern, and are the subject of this paper.

Agriculture is an important part of the economy of the Canadian Prairie Provinces of Alberta, Saskatchewan, and Manitoba. The agriculture and agri-food system of these provinces consists of several industries including primary agriculture, farm input and service providers, food and beverage processing, food distribution, as well as retail, wholesale, and food service industries. In Canada, this sector contributed CAD$108 billion (or 6.6% of Canadian gross domestic product) and employed 2.3 million workers [8]. Much of this production activity occurs in the Prairie Provinces. Primary production is a key part of this system as it affects the other components of the regional economy [9].

Although agriculture is important to the economy, environmental impacts must also be considered for achieving sustainability. Agriculture has many effects on the environment and these effects determine the environmental sustainability. Environmental sustainability is defined as sustainability of ecological services that are provided by the ecosystems [10]. Humans depend upon these services directly or indirectly. A strong environmental sustainability would label any practice unsustainable if the natural ecosystems are put to alternative uses, such as conversion of forest ecosystems to agricultural ecosystems. A more practical definition of environmental sustainability requires that those ecosystems and ecosystem services that are essential to humans be conserved to the point of a minimum safe standard. Examples of effects include those on soil and air quality by the use of different tillage and cropping systems, and those on water quality related to the use of fertilizer and pesticides. Climate trends and extremes are expected to affect air, land, and water resources, and knowledge of these effects are crucial to achieving sustainable agricultural production and food and water security. The effects of excess moisture and drought are especially important, as they can have more pronounced impacts on the environmental sustainability of agriculture than gradual increases in temperature. For example, droughts reduce the protection of soil moisture and vegetation, and erosion can result. Excess moisture and flooding can result in water run-off leading to erosion of soil and damage to vegetative cover that protects the soil. Flooding can also damage the storage areas of fertilizer, manure, and pesticides, releasing them as contaminants into the environment. In this paper, we explore the effects of climate change on the environmental sustainability of agricultural systems, with emphasis on the extreme events of droughts and excess moisture. Therefore, our main objective is specifically to assess the effects of future drought and excess moisture on selected agri-environmental indicators. No other investigations have addressed this topic, to our knowledge.

By 1999, the member countries of the Organization for Economic Cooperation and Development (OECD), including Canada, noted that establishing a key set of agri-environmental indicators (AEIs) that could be useful for member countries was important [11]. In Canada, Agriculture and Agri-Food Canada (AAFC) reports on a set of science-based AEIs using mathematical models showing the interactions between agriculture and the environment [9,12]. These two reports are the latest in the series of Canada's agri-environmental reporting. Therefore, they are the basis for the AEIs we have selected for use, as well as associated trend information. This reporting series and their AEIs are not intended for use with climate change scenarios, but their use may be for strategic adaptation to drought and excess moisture.

Wall and Smit [13] noted that agricultural sustainability and climate change adaptation strategies support one another and that ecosystem integrity is needed for sustaining agricultural production. However, Wheaton et al. [14–16] were the first, according to the authors' knowledge, to assess the possible changes in agricultural sustainability (using AEIs) as expected under climate change, and this study builds upon and expands that work. They found several of the AEIs to be sensitive to climate change and reported a possible decline in the performance of AEIs with climate change for soil erosion, contamination, soil salinization, and water quality categories.

2. Data and Methods

Environmental sustainability indicators of agriculture considered here were based on changes in the set of science-based AEIs developed by Agriculture and Agri-Food Canada [9,12]. The AEIs report agri-environmental performance under four main categories: soil quality, water quality, air quality, and farmland management. Each category has set of indicators addressing sub-themes within the category (e.g., from the soil health theme, sub-indices include soil erosion, soil organic matter, trace elements, and salinity). Many of the indicators can be integrated within climate change studies either directly or indirectly. Directly, these indices can be calculated using climate change scenario data and compared with values obtained under the observed climate record.

Our study methods included literature reviews, development of a conceptual framework, examination of the possible relationships, sensitivities and responses of selected AEIs to climate by examining their mathematical structures. These approaches were used to suggest possible directions of future trends in AEIs with increases of drought and excess moisture under continued climate change. In the remaining sections, the selected AEIs are described, along with an assessment of their future status.

The AEIs selected for this study were those for the soil quality, water quality, and farmland management themes. We added two more indicator types because of their relevance, one for desertification and another for water supply and demand. AEIs were selected for their utility in assessing the possible effects of current and future drought and excess moisture. We examined the mathematical models (factors affecting the relationship among stimulus that causes a change in the AEI level) of each AEI as the first step in choosing them [15]. The AEIs that contain climate variables in their mathematical models are the most clearly sensitive to climate, and therefore either are directly driven by and may have strong relationships with climate change. We determined the nature of the relationships by the types of climate variables used (e.g., temperature, precipitation) in the indicator and whether the relationship with climate was linear or more complex and direct or inverse. Some AEIs for the category of soil quality are good candidates for exploring the direct effects of drought and excess moisture on the environment. Examples include the wind and water erosion, salinity, and particulate emission models as they include climate variables. Where it was not possible to assess indices due to a lack of direct use of climate variables in the models, climatic effects were indirectly implied from ecosystem assessments of changes in vegetation, insects, and diseases, for example.

Although drought and excess moisture affect most aspects of environmental sustainability, we focused on AEI categories and their indicators for soil quality, water quality, farmland management, and water supply and demand, as guided by our conceptual framework (Figure 1), expertise and available literature. From the conceptual framework, we analyzed how the four AEI themes would be affected by changing climate extremes starting with the knowledge of the main characteristics of future possible drought and excess moisture events, as summarized from the literature.

Drought and excess moisture events are expected to become more common in the future on the Canadian Prairies [5,7]. The frequency, intensity, and extent of moderate to extreme droughts are projected to increase. At the other extreme, the review also found agreement that the frequency of severe storms and unusually wet periods is also projected to increase, leading to the conclusion that wet times will become wetter and dry times will become drier, with several driving forces supporting this finding.

Regarding drought, four main characteristics of future possible droughts in the Canadian Prairies were found: (1) increased intensity of dryness, driven by increased evaporation potential with higher temperatures and longer warm seasons; (2) droughts of 6–10 months and longer become more frequent by the 2050s; (3) the frequency of long duration droughts of five years and longer more than doubles in the future to 2100; and (4) decade-long and longer droughts increase by triple in frequency to 2100 [5,6]. The finding of future possible increase in droughts is confirmed by other work that finds increases in drought characteristics in the Canadian Prairie Provinces, especially over the southern study region [7].

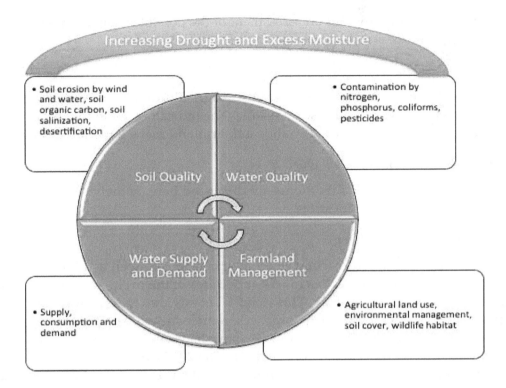

Figure 1. Framework for integrating the effects of changing climate extremes increasing droughts and excess moisture with selected main categories and sub-components of environmental sustainability of agriculture.

Shifting of climate zones poleward with higher temperatures also indicates the occurrence of drought in areas farther north of their usual positions in the study area. Worst-case scenarios should also be considered because of the severe and multiple effects of droughts. Mega droughts have occurred in the past in the Canadian Prairies [17], and it is therefore expected that droughts will be pushed to greater severity with climate warming.

In the context of future trends in the AEIs, it is important to ask: what are the future projections of extreme precipitation events and associated excessive moisture conditions? The IPCC (Intergovernmental Panel on Climate Change) [18] has reported on managing the risks of extreme climate events globally. The report indicates that the frequency of heavy precipitation will increase in the 21st century over many parts of the world. They gave this projection a 66–100% chance of occurring and found that this trend is particularly the case in the high latitudes.

The Canadian Prairie agricultural area has experienced extremely wet conditions in the past and these are projected to increase. Saskatchewan holds Canada's record wettest hour under the current climatic conditions when 250 mm rainfall occurred at Buffalo Gap in the south central area [19]. The largest area eight-hour event in the Canadian Prairies was the rainstorm of 3 July 2000 around Vanguard in southwest Saskatchewan. This storm brought about 375 mm of rainfall, exceeding the average annual precipitation of 360 mm, and caused severe flooding [20]. The projected changes to precipitation amounts in Canada for 2041–2070 show an increase in maximum precipitation in the range of about 10–20% for the prairies for the 20-year return period of one-day precipitation [21]. This means an increase from 40–60 mm (1941–1970) to 48–72 mm for the 20% increase.

Although the work of [17] for the prairies focused on droughts, the climate indices (i.e., Palmer Drought Severity Index and Standardized Precipitation Index) over the future period to 2100 show some very high values, indicating wet periods for a range of Global Climate Model results. For example, some of the future wet periods appear to be as excessive as the wet period of the 1970s. The review of future possible extremes suggested that the overall prairie climate would become drier, but with substantial year-to-year variability, including an increased chance of heavy precipitation and very wet periods [5,6]. The next section provides an assessment of the possible changes in AEIs with projected increases in drought and excess moisture.

3. Results

Descriptions of the environmental sustainability of agriculture as affected by drought and excess moisture are provided in this section for several AEIs. The indicators are in four main categories, soil quality, water quality, water supply and demand, and farmland management (Figure 1).

3.1. Possible Future Trends in AEIs for Soil Quality

Four main AEIs for the soil quality theme are discussed here, namely soil erosion, soil organic carbon, soil salinization, and desertification.

3.1.1. Soil Erosion by Wind and Water

Soil erosion occurs through the action of wind and water, as well as tillage. The soil erosion AEIs had overall improved performances in recent decades in the prairies, indicating reductions in erosion risk between 1981 and 2011 [12]. This trend is mainly due to improved land management, such as adoption of minimum to no-tillage practices, reduced use of summer fallow, and increased forage and cover crops [9]. Recent decades, however, have had severe and extensive droughts, such as in 1999 to 2004 [22], 2008 to 2010 [23], and 2015 [24]. The 2015 drought was found to be likely an outcome of human-influenced warm spring conditions and naturally forced dry weather from May to July. Droughts can result in considerable soil erosion by wind. At least 32 incidents of blowing dust were documented between April and September 2001. This number of incidents was high as it was exceeded only once during the 1977–1988 period of dust storms. Although the wind erosion was severe, it would have been much worse without the increase in soil conservation practices [22]. These events make it clear that drought can result in soil erosion even with the adoption of improved land management practices.

The greater evaporation rates, lower soil moisture, and decreased vegetation cover under droughts result in increased risk of soil erosion (Table 1). This means that management practices to reduce the soil erosion risk would become even more important in the future. Descriptions of future wind speed changes are rare, but Price et al. [25] project little change in wind speed, on average for the prairie semiarid region, with slight reductions in mean summer wind speed of 0.14 m/s for the medium emissions scenario for the 2040–2069 period. However, they did find increases of mean spring wind speed of 0.11 m/s for this scenario and time. Spring is an important time for increased wind erosion risk, as the vegetation cover is not yet well established and the soil is more exposed. Wind speed was a very important factor in the wind erosion component of the AEI as the relationship is direct and cubic [15,26]. The risk of future wind erosion in the province of Saskatchewan, Canada, was estimated to continue to increase with rising temperature and potential evapotranspiration [26].

Agriculture and Agri-Food Canada's assessment of the environmental sustainability of Canadian agriculture found that higher rainfall in eastern provinces, such as Ontario and Quebec, contributed to the lower performances of soil quality indicators [9]. This relationship between higher rainfall and soil quality is useful in assessing future effects in the prairies. Here, drought conditions can shift very quickly to wet conditions. Recent years have shown intense rainfall and severe flooding in several areas of the Prairie Provinces [27]. There were very wet conditions, especially in some parts, in 2010, 2011, and 2012. Spring 2010 was the wettest among the 1948–2012 period, at 64% greater than the areal average for the prairie climate region. Spring 2012 was the third wettest spring, at 52% higher than average. Summer 2010 was also very wet, with a total precipitation amounting to the fourth highest on record at 40% above average. Summer 2012 had the sixth highest areal average precipitation [28]. The heavy rainstorms and high amounts of accumulated precipitation resulted in many excess moisture problems, including more agricultural land being under water than ever recorded. Problems of excess moisture were persistent, lasting from October 2010 to July 2011 for many areas [27]. Intense rainfall events tend to contribute to runoff and increasing soil erosion (Table 1) and do not ease drought conditions as much as gentler rains.

Table 1. Potential effects of increasing droughts and excessive moisture on trends of soil quality indicators.

Soil Health Indicator	Climate Linkage (Direct and Indirect)	Effects of Increased Droughts	Effects of Increased Excess Moisture	Comments Regarding Other Factors
Soil erosion by wind	Wind, temperature and precipitation, soil moisture, vegetation cover	Reduced soil moisture and vegetation cover which increase erosion risk	Increased precipitation intensity can destabilize soil particles	Decreasing snow cover increases exposure to erosion
Soil erosion by water	Precipitation intensity, vegetation cover	Water erosion risk decreases	Increased heavy rainfall increases potential for soil erosion	Heavy rainfall on frozen soil increases erosion risk
Soil organic carbon	Temperature, precipitation, vegetation cover	Reduced vegetation production reduces carbon	Run-off increases carbon losses	Temperature increases tend to increase carbon losses
Soil salinization	Aridity (temperature and precipitation balance), vegetation cover	Evaporation concentrates salts. Reduced vegetation cover can increase salinization	Elevated water tables can increase salinization	Increased variability with drought/wet shifts increases salinization risk
Contamination by trace elements	Precipitation intensity	Possible increased concentrations may occur	Increases	Climate effects estimations require further investigation

The summer of 2011 in Southeastern Saskatchewan provided a good example of several heavy rainfall events [29]. April–June in 2011 had 150 to over 200% of normal precipitation amounts. Multiple rainfall events of 20 mm or greater occurred, and a severe 1:100 year rainfall event occurred on 17 June 2011. These events resulted in unprecedented floods in the Souris River Watershed, causing state of emergency declarations in communities of Weyburn and Estevan. The community of Roche Percee had to evacuate almost every home [30]. Therefore, these extreme precipitation times not only resulted in the flooding of agricultural land with several implications for environmental sustainability, but they also resulted in a loss of homes and other infrastructure (e.g., roads, culverts, and bridges). The damage to infrastructure meant that soil erosion also occurred, though this was more difficult to assess. The impacts to environmental sustainability are discussed with the specific indicator addressed in the following sections. Examples include soil quality (e.g., water erosion of soil) and water quality (e.g., run-off contaminants).

Further changes to snow cover are also expected and have already occurred. Snow cover protects the soil from erosion risk. Northern Hemisphere spring snow cover extent has significantly decreased over the past 90 years and the rate of decrease has accelerated over the past 40 years. An 11% decrease in April snow cover extent has occurred for the 1970–2010 period compared with pre-1970 values. These trends are mainly a result of increasing temperatures [31]. This means that the soil was exposed to wind and water erosion for an increasing length of time in the recent past, and this trend is projected to continue with further warming. Alternatively, snow-melt contributes to overland run-off and can result in water erosion of soil. Recent work finds extensive decreasing trends of snow-water equivalent in Canada related to increasing temperature. The mean size of the decreasing trend for December–April is −0.4 to −0.5 mm/y [32]. Estimates of future possible snow cover changes are challenging because of the complex response of snow cover to warming, but widespread decreases in snow cover duration are projected across the Northern Hemisphere [33]. Implications for the soil erosion AEI are very uncertain because of this complexity, but effects of the continued trend of decreasing snow cover extent should be considered in measures to protect the soil against wind and water erosion.

Intense rainfall events and higher total precipitation accumulations result in greater run-off, with eroded and flooded land (Table 2). The increased water erosion and the flooded land can result in many problems for environmental sustainability of agriculture, including contamination by pollutants of various types resulting in water quality problems.

Table 2. Selected agri-environmental indicator (AEI) categories, indicators, their relationships with climate, and possible future climatic effects related to drought and excess moisture.

Group	Indicator	Measure	Sensitivity to Climate	Links with Climate-Related Changes
Soil Quality	Risk of soil erosion by water	Surface run-off	Strong	Climate change may result in aridity in some parts of the prairies which would increase the probability of surface run-off
				Higher variability in precipitation and incidence of wet events would lead to higher incident of soil erosion
	Risk of wind erosion	Soil loss through wind events	Strong	Future increases are expected with simulated increases in spring wind speed
	Soil organic carbon	Organic carbon level in soil	Medium	Future changes with climate change are not clear because of the interacting effects of management practices
	Risk of soil salinization	Degree of soil salinity	Strong	Climate change may increase salinity from variations of precipitation and dry events
	Contamination by trace elements		Strong	Increased wet and dry periods affect contamination
Water Quality and Quantity	Risk of water contamination by nitrogen	Nitrogen level released by farms into water bodies	Weak	Water run-off containing nitrogen associated with soil erosion is affected by variable precipitation
	Risk of water contamination by phosphorus	Phosphorus level released by farms into water bodies	Weak	Water run-off containing phosphorus associated with soil erosion is affected by variable precipitation
	Water supply and use	Water availability and use	Strong	Climate change would likely impart a reduction in supply, but an increase its demand
Farmland management	Soil cover by crops and residue	Duration of exposed soil	Strong	Vegetative cover is affected by climate change
	Management of farm nutrients and pesticide inputs	Application of organic and inorganic nutrients and pesticides	Medium	Favorable wetter conditions may lead to increased nutrient use. Climate change may lead to increased pest and diseases and the need for their management

3.1.2. Soil Organic Carbon

Other AEI components of soil quality include the tillage erosion risk indicator, soil organic carbon change (SOCC) indicator, the risk of soil salinization indicator, and the risk of soil contamination by trace elements. As indicated earlier, only selected indicators can be considered. The SOCC indicator is affected by land management changes, including the effects of tillage practices, summer fallow frequency, cropping types, and land-use changes. The current trend of the SOCC indicator showed an improved performance from average to a good status as most of the cropland had increasing soil organic carbon from 1981 to 2011. Spatial patterns over the Prairie Provinces to 2011 ranged from no change to large increases in Alberta, large increases over much of Saskatchewan, and mostly moderate increases in Manitoba. The use of reduced tillage practices and reduced summer fallow area was an important influence in this change in the prairies. The Century model was used to predict the rate of change in organic carbon content in soils [9,12].

An important aspect of soil is its carbon storage capacity, which can affect the atmospheric concentrations of carbon dioxide. The level of soil organic carbon would be susceptible to climate change extremes. For example, under a drought period, organic biomass is low which would affect the level of soil organic carbon. Similarly, if land management under climate extremes includes more permanent cover, soil organic carbon would tend to increase because of plant-derived inputs to soils. Vegetation cover is adversely affected by both droughts and flooding and may result in at least short-term decreases in soil organic carbon. Moreover, depending on other constraints, temperature increases would tend to increase soil decomposition loss of carbon [34].

One of the developments needed for the SOCC indicator is to include soil erosion aspects in the model. Even low rates of soil erosion can decrease soil organic carbon [9]. Soil erosion risk increases during droughts and heavy rainfall, so the effects of these extremes should be incorporated into their modeling. Another limitation of the SOCC indicator is that the effects of past (and future) temperature increases do not appear to be assessed and discussed in the reporting by [9]. The Century model does use monthly temperature and precipitation data, so this assessment is possible and is recommended.

3.1.3. Soil Salinization

The risk of soil salinization in the Prairie Provinces decreased from 1981 to 2011, and over this period the land area in the very high-risk class decreased by 2%. The spatial patterns of risk of soil salinization (RSS) on the prairies showed no change to decreased risk in Alberta and Manitoba and large areas of decreased risk in Saskatchewan. The improvements were mostly related to land management, including decreased summer fallow area, and increased area under permanent cover [35]. Again, as for the SOCC, the impact of changing land use practices on the risk of soil salinization dominated, and climate change did not appear to be considered in the modeling for this AEI and results. However, growing season moisture deficits were a factor in the calculations, but the significant yearly variation in the risk of soil salinization was not considered in the indicator [35]. Such sensitivity of RSS to changes in moisture deficits could be determined. Our early estimate of the possible effects of future increased droughts and excessive moisture is the reduced performance of the RSS indicator (Table 2).

3.1.4. Desertification

A risk of desertification indicator is under development [36], and is included here because of its relevance to the topic. This indicator is not included in the most recent AEI reporting [12]. Desertification is the degradation of land in arid to dry sub-humid regions. The preliminary results indicate that average soil erosion rates were usually below the soil tolerance level, meaning that desertification risk due to erosion was low as of 2006 in the Prairie Provinces [36].

Desertification risk increases with soil erosion, losses of soil organic matter, and fluctuating soil salinity [36]. Climatic extremes have the potential to increase all of these factors as discussed previously, and therefore can increase the risk of desertification. Research indicates that the area of land at risk of desertification in the Prairies could increase by about 50% between conditions of 1961–1990 and the 2050s [37]). The World Meteorological Organization [38] states that climate change may exacerbate desertification and soil salinization through alteration of spatial and temporal patterns in temperature, rainfall, solar radiation, and winds. The threat means that the indicator development is recommended to be completed (including climate drivers) and implemented as led by Agriculture and Agri-Food Canada.

The agricultural area of the Canadian Prairies has a large semi-arid climate zone in southwest to west central Saskatchewan and corresponding regions in Alberta. The remainder of the area is mostly classified as dry sub-humid. These climate classifications are based on the Thornthwaite method using a moisture index with the input of monthly mean temperature and precipitation data [39]. These are the climate zones targeted in the desertification definition [37]. A warming climate is expected to expand these dry zones northward in the prairies to cover even greater areas.

3.2. Possible Future Trends for Water Quality

Climate change can be a major instrumental factor affecting water quality, both for surface as well as ground water (Table 2). These changes would occur as a result of two types of developments, both related to climate change. (1) Climate change would likely reduce water quantity (as described later), which would result in changes in flow regimes influencing the chemistry, hydro-morphology and ecology of regulated water bodies [40]. (2) Agricultural activity would face longer growing seasons combined with reduced water availability, with new crops suited to drier [41], warmer conditions. In addition, wetlands that play an important role in water purification may also dry up during such heat events and longer evaporative seasons. The longer growing season and cropping changes could increase the use of fertilizers with subsequent leaching to watercourses, rivers, and lakes, increasing the risk of eutrophication and loss of biodiversity [42]. Many information gaps exist regarding the effects of climate change (e.g., cyclical variability between wet and dry periods) and these are important to quantify to meet the needs of flood control and water quality improvements, for example [43].

Changes in water quality during storms, snowmelt, and periods of elevated air temperature or drought can cause conditions that exceed thresholds of ecosystem tolerance and, thus, lead to water quality degradation [44]. Such precipitation extremes can pose significant risks to water quality outcomes, resulting in a degradation trend of drinking water quality and potential health impacts [45]. At the same time, the impacts of drought and excess moisture are superimposed onto other pressures on water resources [46] and can exacerbate the other pressures. Such pressures may include market pressures, pest, and disease infestations, and effects on producer incomes from other bottlenecks in the agricultural and food complex [8].

3.3. Possible Future Trends in Water Supply and Demand

Water supply and demand are considered (even though they are not included in Canada's AEI reporting series) because as they are critical for environmental sustainability of agriculture. Water demand for agricultural purposes is expected to increase in the future unless conservation measures are in place. Facing periods of frequent droughts, more farmers would lean towards having irrigation on their farms. However, whether this demand would be met or not depends on water availability and its competing uses [47–49].

Water quantity under climate extremes would be affected through reductions in the water stored in glaciers and snow cover. These water sources are currently declining and this trend is projected to continue, e.g., [50]. This trend reduces water availability especially during warm and dry periods (through a seasonal shift in streamflow, an increase in the ratio of winter to annual flows, and reductions in low flows) in regions supplied by this source [51]. Where storage capacities are not sufficient, much of the winter runoff will be lost to the oceans, and this will create regional water shortages.

In addition to surface water, future changes in climate extremes could affect groundwater. Longer droughts may be interspersed with more frequent and intense rainfall events. These changes in climate may affect groundwater through changes in their recharge and discharge [52]. The aquifers where water withdrawal is already higher than their respective recharge amounts would be even more vulnerable to climate change. Such high levels of withdrawals would reduce available quantities considerably [51].

3.4. Farmland Management

Under climate extremes, land management would be affected through changes in soil moisture, which is directly related to climate extremes. Management of soil moisture and water harvesting would be significant adaptation measures to cope during climate extremes. However, many producers may not anticipate and react appropriately to the occurrence of climate extremes and make appropriate adaptations. A survey by [53] shows that, even during serious drought and flood years, only one third of farmers in China were able to use farm management measures to cope with the extreme weather events. In the Prairie Provinces, a survey of producers regarding the 2001–2002 drought indicated that no producer had made any changes in their cultural practices in anticipation of the drought [54]. However, prairie producers are adaptable, and much adaptation occurred during and after the 2001–2002 drought. In this region, many producers have switched from intensive tillage practices to conservation tillage practices. In 1991, only a third of the cropped area was under conservation tillage methods, but by 2011 this area rose by 157% of the 1991 area, thus constituting 85% of the total area prepared for seeding [55].

The four main AEI categories considered here, along with their indicators, measures, estimates of their sensitivities, and relationships with a changing climate are summarized in Table 2. Several of the indicators are estimated to be fairly sensitive to climate extremes, including soil quality, water supply and demand, and portions of farmland management. The reasons for the indication of weaker relationships with climate may be somewhat related to the lack of understanding of the relation of climate variables with the key parameters describing environmental health.

4. Discussion and Conclusions

This paper was an attempt to explore the possible effects of future drought and excess moisture on the environmental sustainability of agriculture. Methods included examining the possible relationships and responses of AEIs to climate drought and excess moisture using the conceptual framework of Figure 1, by evaluating the relationship of AEI models with climate variables, and by using literature reviews. These approaches were used to suggest possible directions of future trends in AEIs with increased drought and excess moisture. The AEIs assessed were those for soil and water quality, and farmland management as developed by Agriculture and Agri-Food Canada [9], with additions of water supply and demand categories.

The estimation of any future occurrence is difficult with many limitations because of several unknowns. However, the projections using several different methods, including climate indices, climate models, and emission scenarios provide strong agreement of the findings of increased intensity and frequency of both future droughts and extreme precipitation (e.g., [6,7,17]). Measuring, monitoring, modeling, projecting, and communicating the characteristics of wet and dry climate extremes are becoming even more critical as the climate shifts and becomes less stable. Sufficient information is needed to guide planning for and implementation of effective actions to adapt to the impacts of climate extremes.

The critical issues of the effects of climate extremes on environmental sustainability of agriculture include effects on natural resources and their ecosystems, including soil quality, water quality, and water supply and demand. Results indicated the nature of future possible changes in AEIs as affected by trends in climate change and extremes. In order to meet the goal of environmental sustainability of agriculture, climate trends and extremes need to be carefully considered. Much better use of climate information and services are required to meet the goal of environmental sustainability of agriculture. The lack of consideration of climate change reduces the capability to adapt and increases vulnerability.

The possible future effects of climate change extremes examined here are conceptual, but are plausible based on the current data from climate science. Actual results may be lower, but they also might be much higher in terms of worst-case scenarios. Solutions for effects of climate extremes should also considered, especially those with the most serious consequences.

Soil quality, as measured by AEIs in the agricultural region of the Prairie Provinces, has showed an improving trend for the 1981–2011 period [12]. However, these AEIs have strong land management drivers, and the effects of climate trends and extremes are not clear. Results regarding the effects of climate change indicate possible declining performances for soil erosion, salinization, and desertification. Results regarding the effects of climate change for other soil AEIs such as soil organic carbon, contamination by trace elements, and farmland management have even less information. All of these AEIs require more work to fully assess the effects of climate change, especially extremes, such as drought and excessive moisture.

The AEIs are numerous with four main categories containing several indicators apiece as described in [12]. Therefore, many could not be addressed here, including air quality and biodiversity indicators. Alternatively, a critical indicator, that of water supply and demand, is not a part of the AEI indicator series by Agriculture and Agri-Food Canada [9,12]. However, water supply and demand was discussed here as a possible indicator, and we recommend it to be included in the AEI series. Next steps in the AEI assessments are recommended to include additional indicators and their relationships with climate change.

Results indicate that the performance of several indicators would likely decrease in a warming climate with more extremes of droughts and extreme moisture. These indicators include risks of soil erosion, soil salinization, water quality and quantity, and soil contamination. Thresholds of climate extremes, however, may be reached and result in accelerated negative performances of such indicators. The impacts of climate change are more difficult to assess for several indicators because of the effect of other factors, such as land management. AEIs are important tools to measure climate impacts on environmental sustainability of agriculture. They also indicate the success of adaptation measures and of required policy development. The climate change risks to environmental sustainability of agriculture require much more attention.

Acknowledgments: We thank the three anonymous reviewers and the Journal editors for their useful comments for the improved version of this manuscript. We thank the Organization for Economic Cooperation and Development (OECD) for the impetus of our earlier work towards assessing the implications of climate change for Agri-Environmental Indicators (AEIs).

Author Contributions: The authors cooperated on all parts of the manuscript.

References

1.   Kulshreshtha, S.; Wheaton, E. Climate change and Canadian agriculture: Some knowledge gaps. *Int. J. Clim. Chang. Impacts Responses* **2013**, *4*, 127–148. [CrossRef]
2.   Qian, B.; Gameda, S.; Zhang, X.; De Jong, R. Changing growing season observed in Canada. *Clim. Chang.* **2012**, *112*, 339–353. [CrossRef]
3.   Nyirfa, W.; Harron, B. *Assessment of Climate Change on the Agricultural Resources of the Canadian Prairies*; The Prairie Adaptation Research Collaborative, University of Regina: Regina, SK, Canada, 2004; 27p.
4.   Qian, B.; De Jong, R.; Gameda, S.; Huffman, T.; Neilsen, D.; Desjardins, R.; Whang, H.; McConkey, B. Impacts of climate change scenarios on Canadian agroclimatic indices. *Can. J. Soil Sci.* **2013**, *93*, 243–259. [CrossRef]
5.   Wheaton, E.; Bonsal, B.; Wittrock, V. *Possible Future Dry and Wet Extremes in Saskatchewan, Canada*; The Water Security Agency, Saskatchewan Research Council: Saskatoon, SK, Canada, 2013.
6.   Wheaton, E.; Sauchyn, D.; Bonsal, B. Future Possible Droughts. In *Vulnerability and Adaptation to Drought: The Canadian Prairies and South America*; Diaz, H., Hurlbert, M., Warren, J., Eds.; University of Calgary Press: Calgary, AB, Canada, 2016.
7.   Masud, M.; Khaliq, M.; Wheater, H. Future changes to drought characteristics over the Canadian Prairie Provinces based on NARCCAP multi-RCM ensemble. *Clim. Dyn.* **2016**, *48*, 2685–2705. [CrossRef]
8.   Agriculture and Agri-Food Canada. *An Overview of the Canadian Agriculture and Agri-Food System 2016*; Agriculture and Agri-Food Canada: Ottawa, ON, Canada, 2017.
9.   Eilers, W.; MacKay, R.; Graham, L.; Lefebvre, A. *Environmental Sustainability of Canadian Agriculture: Agri-Environmental Indicator Report Series*; Report #3; Agriculture and Agri-Food Canada: Ottawa, ON, Canada, 2010; 235p.
10.   Markandya, A.; Perelet, R.; Mason, P.; Taylor, T. *Dictionary of Environmental Economics*; Earthscan: London, UK, 2002.

11. Organization for Economic Cooperation and Development (OECD). *Environmental Indicators for Agriculture: Concepts and Framework*; OECD: Paris, France, 1999; Volume 1.

12. Clearwater, R.; Martin, T.; Hoppe, T. (Eds.) *Environmental Sustainability of Canadian Agriculture: Agri-Environmental Indicators Report Series—Report #4*; Agriculture and Agri-Food Canada: Ottawa, ON, Canada, 2016; 239p.

13. Wall, E.; Smit, B. Climate change adaptation in light of sustainable agriculture. *J. Sustain. Agric.* **2005**, *27*, 113–123. [CrossRef]

14. Wheaton, E.; Kulshreshtha, S.; Eilers, W.; Wittrock, V. *Trends in the Environmental Performance of Agriculture in Canada under Climate Change*; The Organization for Economic Cooperation and Development (OECD), Saskatchewan Research Council: Saskatoon, SK, Canada, 2010; 10p.

15. Wheaton, E.; Eilers, W.; Kulshreshtha, S.; MacGregor, R.; Wittrock, V. *Assessing Agri-environmental Implications of Climate Change and Agricultural Adaptation to Climate Change*; SRC Publication No. 10432-1E11; The Organization for Economic Cooperation and Development (OECD), Saskatchewan Research Council: Saskatoon, SK, Canada, 2011; 31p.

16. Wheaton, E.; Kulshreshtha, S. Agriculture and climate change: Implications for environmental sustainability indicators. In Proceedings of the Ninth International Conference on Ecosystems and Sustainable Development, Bucharest, Romania, 18–20 June 2013; Marinov, A.M., Bebbia, C.A.B., Eds.; Wessex Institute of Technology, WIT Press: Southampton, UK, 2013; pp. 99–110.

17. Bonsal, B.; Aider, R.; Gachon, P.; Lapp, S. An assessment of Canadian prairie drought: Past, present, and future. *Clim. Dyn.* **2013**, *41*, 501–516. [CrossRef]

18. IPCC (Intergovernmental Panel on Climate Change). Summary for Policymakers. In *Managing the Risks of Extreme Events and Disasters to Advance Climate Change Adaptation*; A Special Report of Working Groups I and II of the IPCC; Cambridge University Press: Cambridge, UK, 2012.

19. Phillips, D. *The Day Niagara Falls Ran Dry! Canadian Geographic*; Key Porter Books: Toronto, ON, Canada, 1993; 226p.

20. Hunter, F.; Donald, D.; Johnson, B.; Hyde, W.; Hanesiak, J.; Kellerhals, M.; Hopkinson, R.; Oegema, B. The vanguard torrential storm. *Can. Water Res. J.* **2002**, *27*, 213–227. [CrossRef]

21. Mladjic, B.; Sushama, L.; Khaliq, M.; Laprise, R.; Caya, D.; Roy, R. Canadian RCM projected changes to extreme precipitation characteristics over Canada. *J. Clim.* **2011**, *24*, 2566–2584. [CrossRef]

22. Wheaton, E.; Kulshreshtha, S.; Wittrock, V.; Koshida, G. Dry times: Lessons from the Canadian drought of 2001 and 2002. *Can. Geogr.* **2008**, *52*, 241–262. [CrossRef]

23. Wittrock, V.; Wheaton, E.; Siemens, E. *More than a Close Call: A Preliminary Assessment of the Characteristics, Impacts of and Adaptations to the Drought of 2008–2010 in the Canadian Prairies*; Saskatchewan Research Council: Saskatoon, SK, Canada, 2010; 124p.

24. Szeto, K.; Zhang, X.; White, R.; Brimelow, J. The 2015 Extreme Drought in Western Canada. In *Explaining Extreme Events of 2015 from a Climate Perspective*; Herring, S., Hoell, A., Hoerling, M., Kossing, J., Schreck, C., III, Stott, P., Eds.; Bulletin of the American Meteorological Society; American Meteorological Society: Boston, MA, USA, 2016; Volume 97, pp. S42–S45.

25. Price, D.; McKenney, D.; Joyce, L.; Siltanen, R.; Papadopol, P.; Lawrence, K. *High-Resolution Interpolation of Climate Scenarios for Canada Derived from General Circulation Model Simulations*; Information Report NOR-X-421; Northern Forestry Center, Canadian Forest Service: Edmonton, AB, Canada, 2011.

26. Williams, G.; Wheaton, E. Estimating biomass and wind erosion impacts for several climatic scenarios: A Saskatchewan case study. *Prairie Forum* **1998**, *23*, 49–66.

27. Phillips, D. Canada's Top Ten Weather Stories for 2011. Available online: http://www.ec.gc.ca/meteo-weather/default.asp?lang=En&n=0397DE72-1 (accessed on 5 March 2013).

28. Environment Canada. Climate Trends and Variations Bulletin, Summer 2012, Spring 2012. Available online: http://www.ec.gc.ca/adsc-cmda/default.asp?lang=En&n=30EDCA67-1 (accessed on 5 March 2013).

29. Hopkinson, R. *Anomalously High Rainfall over Southeast Saskatchewan—2011*; Custom Climate Services; The Saskatchewan Watershed Authority: Regina, SK, Canada, 2011.

30. United States Army Corps of Engineers. 2011 Post-Flood Report for the Souris River Basin. Submitted to The International Souris River Board and The United States Department of the Interior. Available online: http://swc.nd.gov/4dlink9/4dcgi/GetSubContentPDF/PB-2794/Souris%202011%20Post%20Flood%20Report.pdf (accessed on 5 March 2013).

31. Brown, R.D.; Robinson, D.A. Northern Hemisphere spring snow cover variability and change over 1922–2010 including an assessment of uncertainty. *Cryosphere* **2011**, *5*, 219–229. [CrossRef]

32. Gan, T.; Barry, R.; Gizaw, M.; Gobena, A.; Balaji, R. Changes in North American Snowpacks for 1979–2007 detected from the Snow Water Equivalent Data of SMMR and SSM/I Passive Microwave and Related Climatic Factors. *J. Geophys. Res. Atmos.* **2013**, *118*, 7682–7697. [CrossRef]

33. Brown, R.; Mote, P. The response of Northern Hemisphere snow cover to a changing climate. *J. Clim.* **2009**, *22*, 2124–2145. [CrossRef]

34. Davidson, E.; Janssens, I. Temperature sensitivity of soil carbon decomposition and feedbacks to climate change. *Nature* **2006**, *440*, 165–173. [CrossRef] [PubMed]

35. Wiebe, B.; Eilers, W.; Brierley, J. Soil Salinity. In *Environmental Sustainability of Canadian Agriculture: Agri-Environmental Indicator Report Series*; Report #3; Eilers, W., MacKay, R., Graham, L., Lefebvre, A., Eds.; Agriculture and Agri-Food Canada: Ottawa, Ontario, Canada, 2010; p. 66.

36. Townley Smith, L.; Black, M. Desertification. Sidebar. In *Environmental Sustainability of Canadian Agriculture: Agri-Environmental Indicator Report Series*; Report #3; Eilers, W., MacKay, R., Graham, L., Lefebvre, A., Eds.; Agriculture and Agri-Food Canada: Ottawa, ON, Canada, 2010; p. 235.

37. Sauchyn, D.; Wuschke, B.; Kennedy, S.; Nykolyak, M. *A Scoping Study to Evaluate Approaches to Developing Desertification Indicators*; Agriculture and Agri-Food Canada, Prairie Adaptation Research Collaborative: Regina, SK, Canada, 2003; p. 109.

38. World Meteorological Organization (WMO). Climate Change and Desertification. Available online: http://www.wmo.int/pages/prog/wcp/agm/publications/documents/wmo_cc_desertif_foldout_en.pdf (accessed on 11 April 2016).

39. Fung, K.; Barry, B.; Wilson, M.; Martz, L. *Atlas of Saskatchewan*; University of Saskatchewan: Saskatoon, SK, Canada, 1999.

40. Waggoner, P.; Revelle, R. Summary. In *Climate Change and U.S. Water Resources*; Waggoner, P.E., Ed.; John Wiley and Sons: Toronto, ON, Canada, 1990.

41. Whitehead, P.G.; Wilby, R.L.; Battarbee, R.W.; Kernan, M.; Wade, A.J. A review of the potential impacts of climate change on surface water quality. *Hydrol. Sci. J.* **2009**, *54*, 101–123. [CrossRef]

42. Moss, B.; Stephen, D.; Balayla, D.; Bécares, E.; Collings, S.; Fernandez-Alaez, C.; Fernandez-Alaez, C.; Ferriol, C.; Garcia, P.; Goma, J.; et al. Continental-scale patterns of nutrient and fish effects on shallow lakes: Synthesis of a pan-European mesocosm experiment. *Freshw. Biol.* **2004**, *49*, 1633–1649. [CrossRef]

43. Anteau, M.; Wiltermuth, M.; van der Burg, M.P.; Pearse, A. Prerequisites for understanding climate-change impacts on northern prairie wetlands. *Wetlands* **2016**, *36*, 299–307. [CrossRef]

44. Murdoch, P.S.; Baron, J.S.; Miller, T.L. Potential effects of climate change on surface water quality in North America. *J. Am. Water Resour. Assoc.* **2000**, *36*, 347–366. [CrossRef]

45. Delpla, I.; Jung, A.-V.; Baures, E.; Clement, M.; Thomas, O. Impacts of climate change on surface water quality in relation to drinking water production. *Environ. Int.* **2009**, *35*, 1225–1233. [CrossRef] [PubMed]

46. Kundzewicz, Z.W.; Mata, l.J.; Arnell, N.W.; Döll, P.; Jimenez, B.; Miller, K.; Oki, T.; Şen, D.; Shiklomanov, I. The implications of projected climate change for freshwater resources and their management. *Hydrol. Sci. J.* **2008**, *53*, 3–10. [CrossRef]

47. Medellin-Azuara, J.; Harou, L.; Olivares, M.; Madani, K.; Lund, J.; Howitt, R.; Tanaka, S.; Jenkins, M.; Zhu, T. Adaptability and adaptations of California's water supply system to dry climate warming. *Clim. Chang.* **2008**, *87*, S75–S90. [CrossRef]

48. Piao, S.; Ciai, P.; Huang, Y.; Shen, Z.; Peng, S.; Li, J.; Zhou, L.; Liu, H.; YihuiDing, Y.; Friedlingstein, P.; et al. The impacts of climate change on water resources and agriculture in China. *Nature* **2010**, *467*, 43–51. [CrossRef] [PubMed]

49. Bates, B.; Kundzewicz, Z.; Wu, S. *Climate Change and Water*; Intergovernmental Panel on Climate Change; Cambridge University Press: Cambridge, UK, 2008.

50. Barnett, T.; Adam, J.; Lettenmaier, D. Potential impacts of a warming climate on water availability in snow-dominated regions. *Nature* **2005**, *438*, 303–309. [CrossRef] [PubMed]

51. Taylor, R.; Scanlon, B.; Döll, P.; Rodell, M.; van Beek, R.; Wada, Y.; Longuevergne, L.; Leblanc, M.; Famiglietti, J.; Edmunds, M.; et al. Ground water and climate change. *Nat. Clim. Chang.* **2013**, *3*, 322–329. [CrossRef]

52. Rosenberg, N.; Epstein, D.; Wang, D.; Vail, L.; Srinivasan, R.; Arnold, J. Possible impacts of global warming on the hydrology of the Ogallala aquifer region. *Clim. Chang.* **1999**, *42*, 677–692. [CrossRef]

53. Huang, J.; Wang, Y.; Wang, J. Farmer's Adaptation to Extreme Weather Events through Farm Management and Its Impacts on the Mean and Risk of Rice Yield in China. In Proceedings of the Agricultural & Applied Economics Association's 2014 Annual Meeting, Minneapolis, MN, USA, 27–29 July 2014.

54. Kulshreshtha, S.N.; Marleau, R. *Canadian Droughts of 2001 and 2002: Economic Impacts on Crop Production in Western Canada*; Publication No. 11602-34E03; SRC Saskatchewan Research Council: Saskatoon, SK, Canada, 2003.

55. Statistics Canada. Table 004-0010-Census of Agriculture, Selected Land Management Practices and Tillage Practices Used to Prepare Land for Seeding, Canada and Provinces, Every 5 Years (Number Unless Otherwise Noted). CANSIM (Database), 2012. Available online: http://www5.statcan.gc.ca/cansim/a47 (accessed on 3 February 2017).

13

Farmers' Net Income Distribution and Regional Vulnerability to Climate Change

Md. Shah Alamgir [1,2], Jun Furuya [3,*], Shintaro Kobayashi [3], Mostafiz Rubaiya Binte [1] and Md. Abdus Salam [4]

[1] University of Tsukuba, Tsukuba, Ibaraki 305-8577, Japan; salamgir.afb@sau.ac.bd (M.S.A.); ruba_zhumu@yahoo.com (M.R.B.)
[2] Sylhet Agricultural University, Sylhet 3100, Bangladesh
[3] Japan International Research Center for Agricultural Sciences; Tsukuba, Ibaraki 305-8686, Japan; shinkoba@affrc.go.jp
[4] Bangladesh Rice Research Institute, Gazipur 1701, Bangladesh; asalam_36@yahoo.com
* Correspondence: furuya@affrc.go.jp

Abstract: Widespread poverty is the most serious threat and social problem that Bangladesh faces. Regional vulnerability to climate change threatens to escalate the magnitude of poverty. It is essential that poverty projections be estimated while bearing in mind the effects of climate change. The main purpose of this paper is to perform an agrarian sub-national regional analysis of climate change vulnerability in Bangladesh under various climate change scenarios and evaluate its potential impact on poverty. This study is relevant to socio-economic research on climate change vulnerability and agriculture risk management and has the potential to contribute new insights to the complex interactions between household income and climate change risks to agricultural communities in Bangladesh and South Asia. This study uses analysis of variance, cluster analysis, decomposition of variance and log-normal distribution to estimate the parameters of income variability that can be used to ascertain vulnerability levels and help us to understand the poverty levels that climate change could potentially generate. It is found that the levels and sources of income vary greatly among regions of Bangladesh. The variance decomposition of income showed that agricultural income in Mymensingh and Rangpur is the main cause of the total income difference among all sources of income. Moreover, a large variance in agricultural income among regions is induced by the gross income from rice production. Additionally, even in the long run the gradual, constant reduction of rice yield due to climate change in Bangladesh is not a severe problem for farmers. However, extreme events such as floods, flash floods, droughts, sea level rise and greenhouse gas emissions, based on Representative concentration pathways (RCPs), could increase the poverty rates in Mymensingh, Rajshahi, Barisal and Khulna—regions that would be greatly affected by unexpected yield losses due to extreme climatic events. Therefore, research into and development of adaptation measures to climate change in regions where farmers are largely dependent on agricultural income are important.

Keywords: income distribution; cost distribution; vulnerable region; adaptation measures; Bangladesh

1. Introduction

Bangladesh has experienced severe famines [1–3]. However, heavy investments in agriculture following these famines have given rise to enhanced food production and have caused significant increases in domestic rice production [4,5]. Both the cultivation techniques and cropping patterns relating to rice production have gradually changed in terms of yield potential [6,7]. Despite huge population pressures, the country has reached self-sufficiency in rice production [8–10]. Additionally, Bangladesh's economic situation is improving; as such, it is one among a rather small group of countries that have seen remarkable progress in terms of both economic performance and development indicators [11]. However, poverty remains a critical social concern in this country [6,12,13].

Climate change will have a largely adverse impact on agricultural production in Asia [14]. For particular geographical locations and due to other environmental reasons, Bangladesh is one of the world's most disaster-prone countries [15–18]. Given climate change impacts, natural resource constraints and competing demands, agriculture and food systems continue to face considerable challenges. The livelihoods of the poor who are directly reliant on agriculture already face a profound threat due to the current climate change in Bangladesh [19,20], which could lead to increased pauperization. At the household level, climate change significantly affects food production [21] which in turn influences food prices and directly affects the poverty of low-income household [22,23]. Agricultural income and non-farm income are the most significant factors in poverty reduction among rural people [24–27]. However, Chaudhry and Wimer reported that household income plays a vital role in the social and economic development of a community and income from agriculture might result in increasing per capita income [28].

Agriculture is strongly influenced by weather and climate, which in turn have impacts on agricultural production [29]. Over the last three decades, temperature has been increasing in Bangladesh [30,31] and the average daily temperature is predicted to undergo an increase of 1.0 °C by 2030 and 1.4 °C by 2050 [32,33]. The annual rainfall is also unevenly distributed in some areas of Bangladesh. Rainfall patterns might change with increasing temperature and drought occur in some areas; however, total rainfall sometimes increases and heavy rainfall induces floods in Bangladesh. Increasing temperature also enhances extreme events, such as cyclones in coastal areas and adversely affects rice production [7,30,34–36]. Additionally, climate change is projected to affect agriculture and it is very likely that climate change will induce significant yield reduction in the future due to climate variability in Bangladesh [37–39], with a projected decline of 8–17% in rice production by 2050 [33,40]. In Bangladesh, nearly 80% of the total cropped area is dedicated to rice production, accounting for almost 90% of total grain production [39,41–46]. Agricultural production, farm income and food security are significantly affected by seasonal growing temperatures [47].

Some previous studies have projected the impacts of climate change on food production and national food security [48,49], as well as their impact on agricultural production, by collecting information under drought, rainfall, sea level rise, flood and temperature increases [39,43,50] and the impact of coastal flooding on rice [7,51,52]. However, there have been fewer studies from micro or regional points of view based on integrated household survey data or poverty measurements under yield reductions of crops due to climate change vulnerabilities. Farmers' low incomes are the main reinforcing factors in poverty traps, so this context of research is not sufficient. To consider suitable adaptation technologies and policies for farmers, impact projections in terms of regional characteristics and poverty are needed far more. To alleviate the severity of climate change's impact on farm production and poverty, adaptation strategies, such as new crop varieties, changing planting times, homestead gardening, planting trees and migration, are vital approaches [6]. Furthermore, research that projects climate change's impacts on poverty or that pinpoints especially vulnerable regions and the vulnerability of farm household income under the impact of climate change is still needed [53,54]. Using statistical analysis, the current study attempts to derive an understanding of regional characteristics in terms of income and agriculture and to assess the contributions of different components on the observed total variance of income and cost, with an eye towards determining regional vulnerability to climate change and projecting the potential effects of climate change on poverty in Bangladesh. In this study, we used high-quality plot-level agricultural production data from the nationally representative survey by the International Food Policy Research Institute (IFPRI) (Appendix A.1). We used different analytical techniques to evaluate regional characteristics and to assess the potential climate change impacts on farm production and poverty under newly developed representative concentration pathways (RCPs) and other climate scenarios. The objective of this study was to project the poverty under the impacts of climate change on crop production and to provide possible adaptive measures.

The paper is designed as follows: we draw a review of the related literature concerning climate change, vulnerability and poverty in Section 2; Section 3 is the methodology section, in which we describe the data sources, compilation procedures and the analytical approaches of the data; in Section 4, descriptive statistics and empirical results of the analysis with discussion are presented; and in Section 5, we conclude by emphasizing the future research directions and some policy guidelines.

2. Review of the Literature

The research on climate change scenarios and poverty in terms of regional characteristics is outlined concisely in this section. Climate change is a reality that is occurring and will increasingly affect the poor; moreover, it is a serious threat to poverty eradication [55]. Poor agricultural communities are always disrupted by climate change's impact on household food security and poverty [56,57]; climate change impacts could increase household poverty [55]. Poverty as a dynamic and multidimensional condition is characterized by the interaction of individual and community features, socioeconomic and political issues, environmental processes and historical circumstances. Particularly in less developed countries and regions through several direct and indirect channels, climatic variability and change can worsen poverty [58]. Lade et al. reviewed the socio-ecological relationship in rural development concepts, emphasizing the economic, biophysical and cultural aspects of poverty. This study classified the poverty alleviation strategies and developed multidimensional poverty trap models and it stated that interventions that ignore nature and culture can reinforce poverty [59].

A multi-factor impact analysis framework was developed by Yu et al. [39] and using this framework [50] Ruane et al. provided sub-regional vulnerability analyses and quantified key uncertainties in climate and crop production. Climate change impacts increase under the higher emissions scenarios and agriculture in Bangladesh is severely affected by sea level rise [50]. Over the same period, several attempts have been made regarding climate scenario development in Bangladesh, mainly using Global Climate Models (GCMs) and in some cases Regional Climate Models (RCMs) [60–62]. From these studies, the overall conclusions include increases in temperature and rainfall, different drought seasons and impacts on crop production.

The projected future yield of rice cultivars in 2030 and 2050 in different areas of Bangladesh by DSSAT crop modelling showed that Bagerhat, Dinajpur, Gaibandha, Maulvibazar, Panchagarh, Rangpur, Sirajganj and Thakurgaon districts will have high yield losses due to climate change impacts. Rainfall, temperature and CO_2 affect the yield for *aman* rice in Rangpur and Khulna divisions and for *boro* rice in Rajshahi, Barisal and the southwest region [63]. Changing patterns of rainfall and temperature in different regions of Bangladesh are significantly higher, compared to IPCC predictions. For sustainable adaptation, location-specific management of seed, crop and irrigation is needed [21]. Soil tolerance, flood tolerance and shorter varieties of rice and other crops could be used to adapt to climate change impacts [64]. Climate change is likely to have an adverse effect on rice and wheat production [5] and significant yield reductions in the future due to climate variability [38] are also directly associated with extreme weather events [19]; due to population pressures, future food production is a challenge in maintaining food security in Bangladesh [5]. Food demand changes because of urbanization, population structure, among other factors; however, food supply can change due to extreme climate change impacts on agricultural production in Bangladesh. The combined effects on rice of major climatic variables were checked by Karim et al. and they found that rice yield would decrease by 33% in both 2046–2065 and 2081–2100 for Rangpur, Barisal and the Faridpur region [65].

Total annual income of a farm household depends on farm and non-farm income. Farm income is always unstable due to the dependency of weather and even if farm income is high poverty may occur; however, higher non-farm income could reduce the poverty [28]. Farm households in Bangladesh are the most prone to the impacts of climatic hazards. Uncertainty is high in farm income and it depends on the wide fluctuations of yields and prices. Unexpected weather can easily damage crop production, rendering farms more vulnerable [66]. In Bangladesh, farmers are fully dependent on weather for their crop production, resulting in lower farm income if extreme climatic events occur. Unexpected yield reductions cause fluctuating farm income and increase food insecurity and poverty. Agriculture is the main source of income of farmers in Bangladesh [8,21] and it might cause per capita income to increase, which in turn could further reduce poverty. The participation of government programs and off-farm income is significantly important in reducing poverty [24].

There has been much research on climate change impacts, adaptations and projections in agriculture. The IPCC's fifth assessment report showed that food production in Asia will vary and decline in many regions under the impact of climate change [37]. Rajendra et al. focused on climate change impacts on farming in northern Thailand, where the vulnerability of farm households persists under the negative impact of climate change [54]. Yamei et al. assessed the adverse effects of future climate on rice yields and provided potential adaptive measures [67]. Nazarenko et al. examined the climate response under a representative concentration pathway (RCP) for the 21st century [68], while there are fewer comprehensive scenarios for the whole country regarding farm income and poverty projections.

In addition, in-depth empirical research on farm income distribution and regional vulnerability to climate change has been lacking. Furthermore, most of the previous studies of climate change impacts on agricultural production have been for specific regions. However, a comprehensive study of climate change impacts comparing the regions of Bangladesh could be enormously significant. One of the motivations of the study is to summarize the farmers' net income scenarios for all of the regions of Bangladesh, assessing the contributions of different components on the observed total variance in income and costs and possible poverty under climate change impacts on agricultural production. Moreover, understanding farmers' local economic situations and coping strategies with climate change impacts could have immense significance for regional point of view. Based on actual farm income, this study evaluates the projected farm income under the scenario that extreme climatic events occur. It then determines the projected poverty to identify vulnerable regions and to suggest appropriate coping and poverty alleviation strategies.

3. Methodology

3.1. Survey Data

In its empirical analysis, this study uses cross-sectional data drawn from nine administrative regions across Bangladesh. These data were derived from the International Food Policy Research Institute (IFPRI), which adopted a multi-stage stratified random sampling method to collect primary data: first a selection of primary sampling units (325 villages) and then a selection of farm households (20 farms) from each primary sampling unit. Randomly selected villages with probability proportional to size (PPS) sampling using the number of households from the Bangladesh population census data in 2001. Randomly selected 20 farm households in each village from the aforementioned national census list. IFPRI researchers designed the Bangladesh Integrated Household Survey (BIHS) (Appendix A.1), the most comprehensive, nationally representative household survey conducted to date. Plot-wise crop production data were collected via semi-structured questionnaire by the IFPRI from 6503 sample farmers across Bangladesh vis-à-vis cultivated crops; the survey period was from 1 December 2010, to 30 November 2011. The original data were collected in a typical agricultural year according to rice production statistics; there was no severe crop loss in the 2010 or 2011 rice years in Bangladesh [69].

3.2. Data Compilation

This study models the poverty rate change under climate change vulnerability in different regions of Bangladesh. Based on the purpose of this study, to analyze the data we applied descriptive, inferential, statistical and multivariate techniques. Plot-wise raw data were compiled in line with the study objectives. We compiled data pertaining to many income sources for each separate household into some important sectors. In addition, for agricultural activities, we also compiled all types of input cost data into some important cost items and output values for each crop. We then compiled and combined them into one data set of households for all 6503 farms. Bangladesh consists of 30 agro-ecological zones (AEZs) that overlap with each other [69,70]. For the convenience of this research, some homogenous agro-ecological zones were combined into the nine administrative regions with their geographical locations. In this manner, we tried to develop nine mutually exclusive regions for our research. To overcome the resulting challenge in consistency under the same impact of climate change in each region [50], we categorized all the sample farmers per the nine administrative zones of Bangladesh, calling each a division (nine different colors indicating the individual divisions) (Figure 1): Barisal (700 sample farmers), Chittagong (300), Comilla (660), Dhaka (1380), Khulna (1020), Mymensingh (600), Rajshahi (580), Rangpur (543) and Sylhet (720).

Figure 1. Map of the objective regions of Bangladesh.

We estimated the costs and incomes associated with 17 major crops produced by farmers in Bangladesh (each is considered an important crop); other crops (such as pulses, oil seeds, spices except for chili and onion, vegetables, leafy vegetables, etc.) and all types of fruits (such as banana, mango, pineapple, jackfruit, papaya, guava, litchi, orange, etc.) were added to another group, "all other crops." The 18 groups are *aus* (Appendix **??**), rice local, *aus* rice LIV, *aus* rice HYV, *aman* rice local, *aman* rice LIV, *aman* rice HYV, *aman* rice Hybrid, T *aus* rice HYV, *boro* rice HYV, *boro* rice Hybrid, wheat local, wheat HYV, maize, jute, potato, chili, onion and all other crops.

To estimate per-capita income for farm household members in all nine administrative regions of Bangladesh, this study considers all income sources, including income from agriculture. The basic unit of analysis is each farm, while farming is the only significant source of income among other sources, such as employment, small business and so on, for the family in a one-year period. Net income for the farm household from agriculture was calculated by deducting total input costs from gross income:

$$\pi = \sum_i P_i Y_i - \sum_i \sum_j P_{ij} X_{ij} \tag{1}$$

where π is net income, P_i is price of crop i, Y_i is production of crop i, P_{ij} is price of input j for crop i and X_{ij} is input j for crop i.

This analysis used only the accounting costs to estimate net income from agriculture (Appendix B.1); these costs include the so-called explicit costs actually incurred by the farms and in surveys, farmers reported their own cost data. For this reason, this study regards supply of one's own land and family labor as part of agricultural income. The farm gate price of each crop for each household was used to estimate gross income derived from agricultural crops, livestock and poultry and fish production; additionally, actual input prices were used to estimate the production costs cited by each farmer and in-kind payments by crops are deducted for estimating gross income. For farmers with no information about farm gate prices or input prices for their respective crops, we used the average prices from the region. This study crosschecked the farm gate prices and input prices with data pertaining to the average national retail price data of select commodities in Bangladesh [71] during the aforementioned study period. Farmers used farm gate prices to sell their crops and for this reason, there was some divergence between national retail prices and the farmers' prices. To estimate per-capita income for each member of the farm, this study assumes that all negative returns tend towards zero so that we can calculate shares of income sources.

Income data were collected for each household and these data were used to calculate overall household income. Income was broadly classified into seven major sectors, as follows:

(i)   Agricultural crop income: income from all crop types produced by farmers throughout the year;
(ii)   Income from fish/shrimp farming;
(iii) Income from livestock and poultry enterprises;
(iv)   Nonagricultural enterprise income: income from nurseries, food processing, fishing, nonagricultural day labor, retail, wholesale, construction, manufacturing, wooden furniture and other businesses;
(v)   Remittances: remittances from within or outside Bangladesh, with the persons who sent the remittances excluded from their respective households;
(vi)   Employment: both formal and informal employment, income from self-employed and/or owned businesses that are not agricultural, income received from relatives and friends not presently living with the household and so on; and
(vii) Other income: income received from land rent or property rent, income from life and nonlife insurance, profit from shares, gratuities, or retirement benefits, income from lotteries or prizes, interest received from banks, charity assistance, other cash receipts and/or other in-kind receipts.

These seven sectors of household income were used to determine the actual income and income sector shares, both of which reflect income distributions significantly.

3.3. Analytical Approach

This study used four types of statistical analysis.

3.3.1. Analysis of Variance (ANOVA)

After dividing farm households into the nine aforementioned regions, we conducted single-factor analysis of variance (ANOVA) to examine differences among the farm households of the nine regions in Bangladesh in terms of mean per-capita income.

3.3.2. Cluster Analysis

The cluster analysis (CA) technique was used to determine the main and dominant income sources in Bangladesh's various regions. Environmental (i.e., topographical) divergence is a common phenomenon in Bangladesh and it diversifies farm production, although farm households within a certain region do tend to be similar. Ward's hierarchical method and the partitioning method can be used to determine the most appropriate clusters regarding the main income sources in each region. A dendrogram—a graphical representation of the hierarchy of nested cluster explanations—is a manifestation of Ward's method and it provides clues for finding the preferable number of clusters regarding income sources.

3.3.3. Decomposition of Variances

To understand the interregional differences and to assess the contributions of different components to the observed total variance of input cost and income, different crop production data are used [72–75]. These data include per hectare crop yields, prices and all costs at the farm level and we decompose the variances in net cost and net income into different factors using the following relations.

$$V(X \pm Y) = V(X) + V(Y) \pm 2\text{Cov}(X, Y) \tag{2}$$

where X and Y are stochastic variables, such as the costs of inputs or incomes from different sectors; $V(\cdot)$ is variance and $\text{Cov}(\cdot)$ is covariance.

3.3.4. Projections: Log-Normal Distributions

There are different types of probability distributions studied in probability theory. Lognormal distribution is one of the most important one and was established long ago [76–78]. Lognormal distribution is a type of a continuous distribution. It is a probability distribution in which the logarithm of the random variable is distributed normally. This distribution is closely related to the normal distribution. Lognormal distribution is very commonly used in the social sciences, economics and finance [79].

Arata [80] pointed out that the income distribution among individuals is very important and is one of the main themes in economics. Income distribution is widely understood to be well described by a log-normal distribution.

Lognormal distribution has two parameters: mean (μ) and standard deviation (σ). If x is distributed log-normally with parameters μ and σ, then $\log(x)$ is distributed normally with mean μ and standard deviation σ. The log-normal distribution is applicable when the quantity of interest must be positive since $\log(x)$ exists only when x is positive. A positive random variable X is log-normally distributed if the logarithm of X is normally distributed.

$$ln(X) \sim N\left(\mu, \sigma^2\right) \tag{3}$$

Let Φ and φ be, respectively, the cumulative probability distribution function and the probability density function of the $N(0, 1)$ distribution.

The probability density function of the log-normal distribution is;

$$f(x|\mu,\sigma) = \frac{1}{x\sigma\sqrt{2\pi}} exp\left\{\frac{-(lnx - \mu)^2}{2\sigma^2}\right\}; x > 0 \tag{4}$$

If we substitute a poverty line into x and integrate the probability density function up to x, we can obtain a poverty rate. The poverty line, which is estimated by world Bank, is inserted into the equation [12,67].

We estimate the incomes of all sample families on the assumption of climate change impacts and draw the distribution of the estimated incomes, assuming that the distribution follows log normal distribution. To draw log normal distribution, we must find the mean and standard deviation of $ln(x)$ (Appendix B.2). From the actual per-capita income of household members in the study areas, we obtain the actual distribution of per-capita income using the lognormal distribution. Next, we project the crop yield loss from the assumption of the literature reviews and we estimate the projected per-capita income. From projected per-capita income using lognormal distribution, we obtain the estimated distribution of per-capita income. By simulating these two distributions, we find the poverty rate graph.

4. Results and Discussion

4.1. Comparison of Income Levels Among Regions

Agricultural income is a key driver in reducing poverty in Bangladesh, where it accounted for 90% of all poverty alleviation between 2005 and 2010 [81]. In terms of employment, Bangladesh's economy is primarily dependent on agriculture. Approximately 85% of the population is directly or indirectly attached to the agriculture sector [38,69].

Agriculture continues to be the main source of income in the sample households in all regions (Table 1) and this result is consistent with Hossain and Silva (2013) [5]. However, in all regions, nonagricultural profit and employment are important income sources and these results are consistent with Bangladesh Economic Review [45]. The amount of remittances varies by region: that in Sylhet is not the highest nationally but the people there do consider remittances to be the main income source in the region. The agricultural income is higher in Rajshahi than in other regions and the per capita income of this region per the study sample is US\$ 423.6 (Table 2). Diversification of agricultural crops results in this region having highest income from agriculture.

Table 1. Each income sector's share in total household income (%), by region.

	B	CH	CO	D	K	M	RJ	RN	S	BD
Agril. crops	12.71	8.14	5.50	13.55	19.43	20.15	18.72	21.41	9.03	14.32
Main crops	6.08	2.89	2.34	8.25	10.81	11.44	11.72	14.84	6.15	8.36
Other crops	6.63	5.25	3.16	5.30	8.62	8.71	7.00	6.58	2.87	5.96
Fish	9.23	1.54	0.57	2.18	7.93	6.06	2.87	1.14	3.16	3.96
Livestock	2.19	1.17	1.48	3.60	6.15	5.12	4.43	3.10	1.80	3.47
Non-ag. profit	20.76	19.25	14.13	21.22	18.09	17.66	19.61	14.88	20.05	18.80
Remittance	11.04	24.99	41.48	15.68	7.64	9.11	4.48	7.58	17.77	15.22
Employment	38.91	44.35	30.80	41.10	38.52	39.04	38.83	50.54	44.02	40.10
Other income	5.16	0.55	6.04	2.66	2.23	2.86	11.06	1.35	4.18	4.12
Total	100	100	100	100	100	100	100	100	100	100

B = Barisal, CH = Chittagong, CO = Comilla, D = Dhaka, K = Khulna, M = Mymensingh, RJ = Rajshahi, RN = Rangpur, S = Sylhet, BD = Bangladesh, Main crops = *Aus*, *Aman* and *Boro* rice and other crops = Wheat, maize, jute, potato, chili, onion and so on.

Table 2. Mean, median and standard deviation of per-capita income (US\$/yr), by region.

	B	CH	CO	D	K	M	RJ	RN	S	BD
Mean	308.93	336.75	378.35	362.17	369.84	307.63	423.63	308.76	301.63	327.55
Median	289.93	217.83	246.25	242.87	254.11	215.04	283.14	226.99	204.82	232.94
SD	314.75	418.11	314.22	403.66	382.81	278.08	372.71	246.61	301.02	348.64
PR	0.51	0.48	0.46	0.46	0.42	0.51	0.33	0.47	0.49	0.46

B = Barisal, CH = Chittagong, CO = Comilla, D = Dhaka, K = Khulna, M = Mymensingh, RJ = Rajshahi, RN = Rangpur, S = Sylhet, SD = Standard deviation and PR = Poverty rate.

Table 1 shows significant differences in main income sources among farmers in various regions in Bangladesh. Employment is the predominant income source in most regions, followed by nonagricultural profits and agriculture. The share of agriculture in total income varies by region. Among Bangladeshi farming households, the employment share is 40.10%, although the overall share of agriculture in total income is 14.32%. Rangpur has the highest share of agricultural income in total annual income (21.41%), followed by the Mymensingh region (20.15%). Comilla's share of remittances in total annual income was highest (41.48% of total income); in comparison, the share generated by agricultural crops in Comilla was only 5.50%. Currently, overseas workers are more often from the Comilla region than other regions in Bangladesh, with a significant proportion of them sending remittances, becoming a vital source of income in the Comilla region. Rice and other crops were the main sources of income among the sampled farm households in the study areas (Appendix C). Incomes from maize and potato appear to be growing but their respective shares remain small. There are regional land conditions and climate differences among Bangladesh's regions, so wheat, maize, onion and potato production is not familiar to all farmers. Consequently, farmers in all areas of Bangladesh tend to focus on rice cultivation.

Table 2 shows descriptive statistics of income status by region. Poverty rates were estimated by applying the poverty line and the purchasing power parity from the World Bank [22] to log-normal income distributions. The findings presented in Table 2 indicate differences in mean, median and standard deviation of net incomes among the nine regions in Bangladesh; using these findings, one can pinpoint relatively rich and poor regions.

In terms of mean net income, incomes of sampled farm households in Rajshahi are the highest, while those of Barisal, Mymensingh, Rangpur and Sylhet are lower. As some farmers had negative or zero per-capita income, the standard deviation is relatively large in certain regions. The highest standard deviation value is found in Chittagong (US$ 418.1), reflecting a large income gap among the farmers there.

The highest poverty rate (i.e., 0.51) was found in Mymensingh and Barisal (Table 2), while the lowest (i.e., 0.33) was in Rajshahi; overall, the country's upper poverty rate is 0.46. The rates in Chittagong and Sylhet were also relatively low (i.e., 0.49). The officially estimated upper poverty rate and national average poverty rate are both in the vicinity of 0.35 [12,82], which makes sense because the original data were collected from rural, farming-engaged people and excluded affluent or single urban people.

Among regions where the poverty rates were high, Barisal, Mymensingh and Sylhet had the lower mean incomes. In contrast, Chittagong had the highest standard deviation, compared to the other regions. In the regions of Barisal, Mymensingh and Sylhet, it appeared that the mean income level was low; however, in the other regions, the mean income was large. These results show that these low-income regions are vulnerable regions and should be the targets of farmers' support policies.

From results of Table 2, this study found that there are differences in mean, median and standard deviation of net incomes among the nine regions in Bangladesh and for validation of this difference, we perform ANOVA and report the results in Table 3. Analysis of variance (ANOVA) is a statistical test designed to examine means across more than two groups by comparing variances, based upon the variability in each sample and in the combined samples. We analyzed the variance within and between the sample farmers to determine the significance of any differences in per capita income of farm household members among the regions of Bangladesh. The results of the overall F test in the ANOVA summary shows the results regarding the variability of means between groups and within groups. As indicated, the overall F test is significant (i.e., p-value < 0.05), indicating that means between groups are not equal and it is statistically concluded that there have been significant differences among the regions in terms of mean per-capita income.

Table 3. ANOVA mean differences across regions.

Source of Variation	SS	df	MS	F	p-Value	F Crit
Between groups	6.31×10^{10}	9	7.01×10^9	4.757462	2.39×10^{-6}	1.880604
Within groups	1.91×10^{13}	12,996	1.47×10^9			
Total	1.92×10^{13}	13,005				

The first column in ANOVA provides us with the sum of squares between and within the groups and for the total sample farmers. The total sum of squares represents the complete variance on the dependent variable for the total sample. The second column represents the degrees of freedom, $(n - 1)$. The total degrees of freedom represent $13,006 - 1 = 13,005$; degrees of freedom between groups equals the number of groups minus one $(10 - 1 = 9)$. The within groups degrees of freedom equals $13,005 - 9 = 12,996$. The third (mean square) column contains the estimates of variability between and within the groups. The mean square estimate is equal to the sum of the squares divided by the degrees of freedom. The between groups mean square is 7.01×10^9; the within-groups mean square is 1.47×10^9. The fourth column, the F ratio, is calculated by dividing the mean square between groups by the mean square within the groups. The F ratio should be one if the null hypothesis is true, while both mean square estimates are equal. However, as shown in Table 3, larger F values (4.757462) imply that the means of the per capita income groups are greatly different from each other, compared to the variation in the individual sample farmers in each group. The next column is the significance level (p-value) and it indicates that the value of F ratio is sufficiently large to reject the null hypothesis. The significance level is 2.39×10^{-6}, which is less than 0.05. Therefore, the mean per capita incomes of sample households among the regions of the country were significantly different in the study year.

4.2. Regional Characteristics on Income Source

This section intends to classify regions of Bangladesh to determine the regional characteristics of income sources in each administrative region. Sectoral income shares from Table 1 are analyzed by cluster analysis and are shown in Figure 2. Here, a dendrogram depicts the income source relationships among the regions. The horizontal axis of the dendrogram (in Figure 2) represents the distance or dissimilarity between clusters and the vertical axis represents the objects (regions) of clusters. From the cluster analysis, this study attempted to find the similarity and clustering with the dendrogram, which visually displays a certain cluster shape. Regions that are close to each other (have small dissimilarities) are linked near the right side of the plot. In Figure 2, we note that Khulna and Mymensingh are very similar compared to the regions that link up near the left side, which are very different. For example, Comilla appears to be quite different from any of the other regions. The number of clusters formed at a particular cluster cutoff value can be quickly determined from this plot by drawing a vertical line at this value and counting the number of lines that the vertical line intersects. In this study, we can see that, if we draw a vertical line at the value of 18.0, four clusters will result. One cluster contains four regions, one contains three regions and two clusters each contain only one region, as shown in Figure 2, in which Barisal, Mymensingh, Khulna and Rajshahi are more alike than resembling Rangpur. In addition, Chittagong, Dhaka and Sylhet are more alike than resembling Comilla.

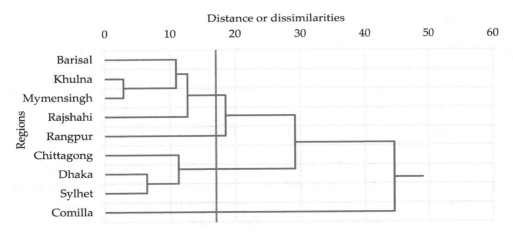

Figure 2. Dendrogram showing clusters for main income sources, by region.

Table 4 summarizes regional characteristics of income sources. Clusters 1 and 2 are largely dependent on agriculture. Clusters 3 and 4 are not largely dependent on agriculture. This result indicates the importance of agricultural research for clusters 1 and 2.

Table 4. Cluster characteristics of main income sources, by region.

Cluster	Region	Main Income Source	Distinction
1	Barisal, Mymensingh, Khulna, Rajshahi	Agricultural. crops, non-agricultural profit, employment	
2	Rangpur		Dominant Employment
3	Chittagong, Dhaka, Sylhet	Non-agricultural profit, remittance, employment	
4	Comilla		Dominant Remittance

Using the dendrogram in Figure 3 (agricultural crop share in total agricultural income analyzed by cluster analysis), four clusters were determined (Table 5) as the clusters suitable for representing agricultural crop income sources among the regions. We followed the same procedure for this dendrogram (Figure 3) that we followed in Figure 2.

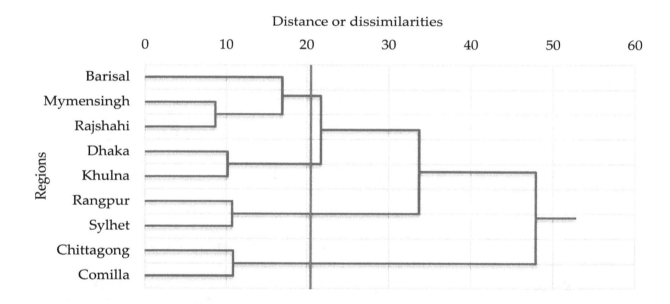

Figure 3. Dendrogram showing clusters for agricultural income sources, by region.

Table 5. Cluster characteristics of agricultural income sources, by region.

Cluster	Region	Main Income Source	Distinction
1	Barisal, Mymensingh, Rajshahi	Rice, other crops	
2	Rangpur, Sylhet		Dominant rice
3	Chittagong, Comilla		Dominant other crops
4	Dhaka, Khulna	Rice, jute, chili, onion, other crops	

The selected clusters show significant differences among the regions. Rice and other crops were identified as the main agricultural income sources of clusters 1–3, whereas rice, jute, chili, onion and other crops were those of cluster 4. The selected clusters produced the significant differences among the regions. In addition, rice predominated in cluster 2, while other crops predominated in cluster 3.

These findings imply, for example, that rice is the main agricultural income source in Rangpur and Sylhet, while other crops are those in Chittagong and Comilla.

4.3. Reasons for Broad Income Distribution within a Region

To grasp the diversity of income for sampled farm households, the income can be decomposed into seven broad components, such as Agriculture, Fish, Livestock and poultry, Nonagricultural enterprise profit, Remittance, Other income and Employment income, in each region. We applied decomposition of variances and the results are shown in Table 6. The decomposition of variances is useful in evaluating how much each source of income contributes to total income variation of farm households. The decomposed variance share was derived from annual per capita income from the seven aforementioned broad income source sectors. Across Bangladesh, differences in remittances, other income and employment are important factors that all contribute the largest share of variation in total income. If a family can find good employment both inside and outside its region, it can become relatively wealthy, although income share from employment does not significantly more contribute in all regions (Table 6).

Table 6. Share of broad income components (%) in total income variation, by region.

	B	CH	CO	D	K	M	RJ	RN	S	BD
V(b)	6.57	1.67	1.94	4.19	8.18	13.87	3.18	20.59	2.49	4.79
V(c)	20.03	0.19	0.03	1.57	35.73	8.17	1.11	0.23	1.98	6.42
V(d)	1.08	0.18	0.17	0.87	1.78	4.58	2.81	0.98	1.05	1.54
V(e)	17.39	13.64	6.33	16.50	13.47	11.90	5.09	7.84	19.73	11.63
V(f)	8.70	40.78	54.36	10.94	10.22	12.99	1.61	30.23	29.95	17.78
V(g)	4.84	0.05	14.76	1.16	0.61	2.38	69.70	0.37	2.82	21.63
V(h)	19.44	27.29	11.61	44.54	17.17	25.26	7.16	38.32	21.01	22.05
2*Cov(e,h)	21.95	15.22	10.81	20.22	12.85	14.22	7.32		20.96	14.16
2*Cov(b,c)								1.43		
2*Cov(c,h)							2.03			
2*Cov(f,g)		0.99								
2*Cov(c,e)						6.63				
Total	100	100	100	100	100	100	100	100	100	100

B = Barisal, CH = Chittagong, CO = Comilla, D = Dhaka, K = Khulna, M = Mymensingh, RJ = Rajshahi, RN = Rangpur, S = Sylhet and BD = Bangladesh; b = Agriculture, c = Fish, d = Livestock and poultry, e = Nonagricultural enterprise profit, f = Remittance, g = Other income and h = Employment income.

We found in Table 6 that agriculture is one of the main contributors to income differences in Mymensingh and Rangpur regions. Figure 4 shows total income distribution by income sources for the whole country, of which 22% of income inequality of total income is explained by inequality of employment income, while 13.87% and 20.59% of income inequality of total income explained by agriculture in Mynemnsingh and Rangpur respectively (Figures 5 and 6). Furthermore, this result indicates that remittance is the most important sector inducing income disparity in Comilla, compared to employment in Dhaka and Rangpur. In addition, other income sources are significant sources of income to confirm the total income disparity in Rajshahi. This finding likely explains that the income inequality of total income makes the larger contribution of inequality in agricultural income for crop farm households in Bangladesh.

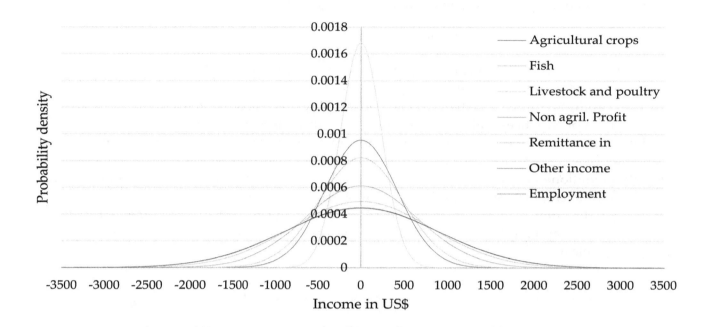

Figure 4. Distribution of total income for farm households in Bangladesh by income sources.

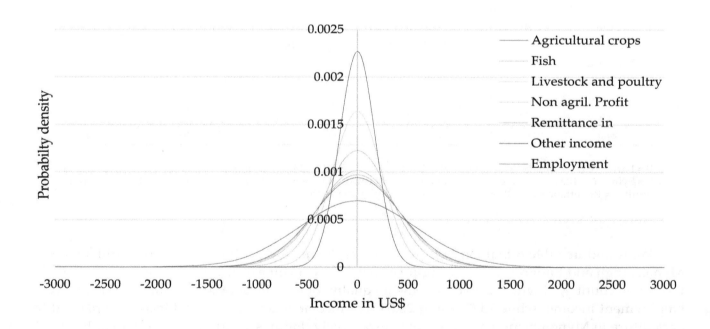

Figure 5. Distribution of total income (US$) for farm households in Mymensingh by income sources.

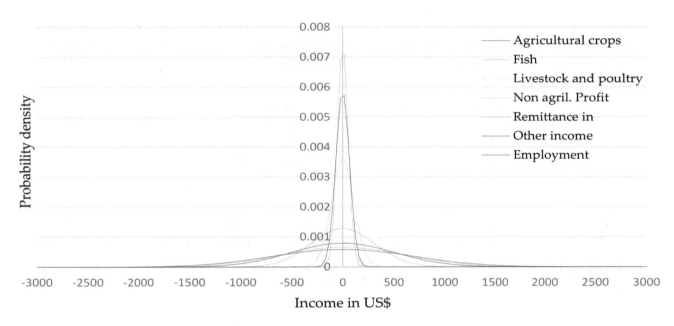

Figure 6. Distribution of total income (US$) for farm households in Rangpur by income sources.

4.4. Factors in Agricultural Income Differences

The main factors of agricultural income differences are shown in Table 7 obtained by the decomposed variance method. We estimate the variance component shares of crops for all farms across nine regions. From Table 6, we identify that agriculture is one of the main reasons for income differences in Mymensingh, Rangpur, Barisal, Khulna and Rajshahi. The empirical estimates of Table 7 indicate that the main variation in agricultural income comes from *aman* HYV (g) and *boro* HYV (j) rice. However, the results also display the contributions of other crop income to total agricultural income variation.

Table 7. Shares of crop income (%) in total agricultural income variation, by region.

	B	CH	CO	D	K	M	RJ	RN	S	BD
V(b)	0.35	0.07	0.03	0.15	0.10	0.00	0.01	0.00	0.36	0.11
V(c)	0.08	0.04	0.03	0.00	0.00	0.06	0.06	0.01	0.04	0.04
V(d)	0.64	0.43	0.01	0.02	1.54	0.06	0.13	0.13	1.06	0.53
V(e)	5.23	0.00	0.36	0.36	0.53	0.50	0.50	0.15	2.06	1.02
V(f)	0.47	0.02	0.16	0.02	0.07	0.06	0.01	0.15	0.00	0.10
V(g)	8.95	7.67	1.12	1.63	10.15	3.84	7.64	12.95	7.88	8.50
V(h)	0.02	0.00	0.00	0.00	0.09	0.09	0.05	0.11	0.00	0.06
V(i)	0.70	0.00	0.06	0.01	0.06	0.00	0.00	0.36	0.16	0.14
V(j)	6.36	4.32	8.13	34.03	17.72	20.89	17.72	14.03	48.26	25.30
V(k)	2.49	2.13	1.26	5.71	3.88	0.69	3.56	3.40	17.82	5.03
V(l)	0.00	0.00	0.00	0.00	0.00	0.00	0.00	0.00	0.00	0.00
V(m)	0.00	0.00	0.01	0.04	0.15	0.00	0.23	0.18	0.00	0.11
V(n)	0.00	0.00	0.27	0.07	0.10	0.00	0.53	0.65	0.00	0.28
V(o)	0.26	0.00	4.28	4.74	2.46	0.04	0.91	0.93	0.14	2.38
V(p)	0.49	0.04	20.77	0.35	0.03	0.08	1.78	6.48	0.16	2.68
V(q)	1.65	0.90	0.81	11.56	12.40	0.98	0.17	0.49	0.08	6.00
V(r)	0.00	0.00	0.00	6.51	0.54	0.00	0.63	0.02	0.00	1.91
V(s)	67.37	75.85	43.55	29.35	44.77	62.62	16.16	24.67	21.98	44.00
2*Cov(o,r)				5.43	0.85		0.81			1.79
2*Cov(g,j)		5.75					9.73	11.64	13.34	
2*Cov(g,k)		2.79			0.37			4.55	7.94	
2*Cov(g,p)							0.02	3.58	11.66	
2*Cov(o,p)			18.45				0.34	6.19	2.33	
2*Cov(g,s)								9.54		
2*Cov(j,s)								13.61		
2*Cov(d,j)	4.95		0.72		4.20					
Total	100	100	100	100	100	100	100	100	100	100

B = Barisal, CH = Chittagong, CO = Comilla, D = Dhaka, K = Khulna, M = Mymensingh, RJ = Rajshahi, RN = Rangpur, S = Sylhet, BD = Bangladesh; b = *Aus* rice local, c = *Aus* rice LIV, d = *Aus* rice HYV, e = *Aman* rice Local, f = *Aman* rice LIV, g = *Aman* rice HYV, h = *Aman* rice Hybrid, i = T *Aus* rice HYV, j = *Boro* rice HYV, k = *Boro* rice Hybrid, l = Wheat Local, m = Wheat HYV, n = Maize, o = Jute, p = Potato, q = Chili, r = Onion, s = All other crops.

Rice is the leading crop in Bangladesh, accounting for more than 90% of total cereal production covering 75% of Bangladesh's total cropped area [45,69]. For Mymensingh and Rangpur, variances in both *aman* HYV and *boro* HYV rice are high. For other regions, variances in *boro* HYV are high.

All other crops(s) are among the main causes (44% variance share) of income differences for all of Bangladesh since all types of pulses, oil seeds, spices, vegetable, leafy vegetables and fruits are included in the group of "all other crops." Moreover, all other crops(s) explain the larger contribution to total agricultural income variation because, in some regions, vegetables and fruits, among others, excluding rice, are important agricultural income sources.

The distribution of crop income among total agricultural income for the whole country is shown in Figure 7, which follows in Figures 8 and 9 for Mymensingh and Rangpur, respectively, with selected crops mainly produced by farmers in these regions. We found that *boro* rice has the widest variation in both the region and the highest inequality of total agricultural income, explained by the inequality of *boro* HYV income.

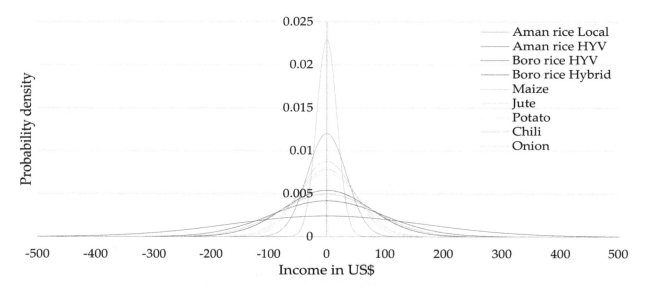

Figure 7. Distribution of agricultural income for farm households in Bangladesh by crop income.

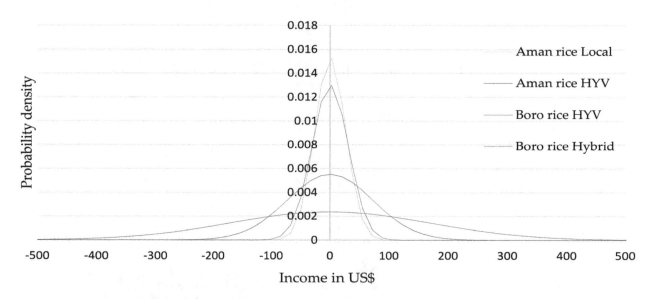

Figure 8. Distribution of agricultural income for farm households in Mymensingh by crop income.

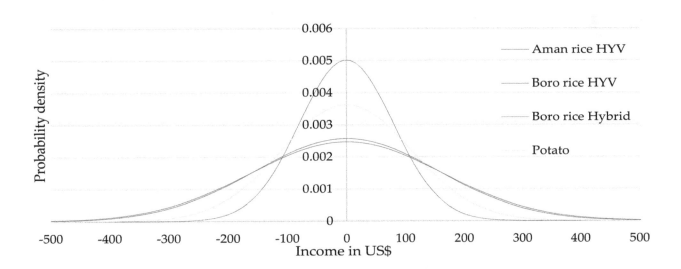

Figure 9. Distribution of agricultural income (US$) for farm households in Rangpur by crop income.

4.5. Factors Contributing to Variations in Income from Aman HYV and Boro HYV Rice Production

According to the results of Table 7, it is important to determine the factor causing the net income differences in *aman* HYV production. From decomposed variance of gross income and gross cost, we find in Table 8 that gross income is the main factor in net income difference, indicating that, although farmers in same region cultivated *aman* HYV rice, their gross incomes were different.

Table 8. Decomposed variances share (%) of GI and GC for *aman* HYV rice, by region.

	B	CH	CO	D	K	M	RJ	RN	S	BD
V(GI)	75.31	74.34	98.38	53.87	76.53	57.17	66.88	74.25	45.49	69.45
V(GC)	80.97	33.57	35.80	91.18	36.13	49.23	55.56	30.27	55.10	45.67
−2*Cov(GI, GC)	−56.27	−7.91	−34.18	−45.06	−12.66	−6.39	−22.44	−4.52	−0.59	−15.11
Total	100	100	100	100	100	100	100	100	100	100

B = Barisal, CH = Chittagong, CO = Comilla, D = Dhaka, K = Khulna, M = Mymensingh, RJ = Rajshahi, RN = Rangpur, S = Sylhet and BD = Bangladesh; GI = Gross income; and GC = Gross cost.

These gross income differences mainly induce the net income disparities in Comilla, Khulna, Chittagong and Rangpur, while gross cost induces the income disparities in Dhaka and Barisal for *aman* HYV rice. Additionally, gross cost also contributes to the total net income disparity of *aman* HYV rice production. To determine the variance in gross cost for *aman* HYV rice production, we estimate the variance component shares of all costs contributing to gross cost and present them in Table 9.

The results show the factors responsible for large variations in cost from *aman* HYV rice production. As shown in Table 9, variances in seed (c) shows in third row, chemical fertilizer (g) in row seven and hired labor costs (k) in row eleven, are high across all regions. In Dhaka, the highest 80% of inequality of gross cost for *aman* HYV rice production is explained by the inequality of hired labor cost (k), while in Barisal, the highest 25% inequality of gross cost is explained by inequality of seed cost. These costs were the main factors inducing the income differences in *aman* HYV rice production. This result indicates the importance of farming knowledge and easy input access to rice cultivation.

Table 9. Decomposed variances share (%) of costs for *aman* HYV rice production, by region.

	B	CH	CO	D	K	M	RJ	RN	S	BD
V(b)	3.64	3.73	3.79	0.97	3.66	5.50	3.72	8.79	4.32	3.24
V(c)	25.01	1.87	24.54	1.47	3.55	5.56	3.12	6.78	3.81	5.15
V(d)	0.53	1.79	1.04	1.32	8.33	2.04	4.15	6.70	0.67	3.69
V(e)	0.07	0.18	0.19	0.08	0.41	0.64	0.77	0.64	0.23	0.33
V(f)	0.54	0.48	0.28	0.07	0.65	0.10	0.65	0.54	0.14	0.35
V(g)	5.32	9.73	6.27	1.54	12.74	6.72	7.57	7.05	3.38	6.42
V(h)	0.98	0.06	0.01	0.04	0.30	2.76	0.05	0.57	1.42	0.50
V(i)	9.49	2.29	1.88	0.35	4.25	1.29	1.31	2.70	1.62	2.10
V(j)	3.47	0.58	1.62	0.10	0.44	0.70	0.15	0.26	3.04	0.69
V(k)	15.16	39.90	45.37	80.58	37.61	70.65	40.88	58.04	74.50	59.53
2*Cov(f,g)	1.72	2.37	1.33	0.33	2.14	0.77	3.05	1.26		1.41
2*Cov(i,f)	2.07		0.59	0.13			1.17	1.03	0.41	0.54
2*Cov(i,g)	11.50		3.88	0.77	5.69	3.26	4.29	4.69	1.94	3.32
2*Cov(k,g)	5.46	20.32		8.55	19.47		18.35			12.74
2*Cov(c,j)	15.04							0.95	4.52	
2*Cov(k,f)		3.79		2.04			4.82			
2*Cov(k,i)		1.90	9.21	1.67	0.75		5.94			
2*Cov(c,k)		11.0								
Total	100	100	100	100	100	100	100	100	100	100

B = Barisal; CH = Chittagong; CO = Comilla; D = Dhaka; K = Khulna; M = Mymensingh; RJ = Rajshahi; RN = Rangpur; S = Sylhet; and BD = Bangladesh; b = Rental cost of land; c = Seed cost; d = Irrigation cost; e = Manure/compost cost; f = Pesticide cost; g = Chemical fertilizer cost; h = Draft animal cost for land preparation; i = Rental cost for tools and machinery; j = Threshing cost; and k = Hired labor cost.

In Table 7, we note that *boro* HYV also had an influence on agricultural income. It is essential to determine the factors affecting the net income variation for *boro* HYV rice cultivation. Table 10 summarizes the decomposed variance of gross income and gross cost from *boro* HYV rice production and shows that gross income is the main factor in net income differences for *boro* HYV rice production, except for in Chittagong and Sylhet. However, gross cost also contributes to the total net income disparity of *boro* HYV rice production.

Next, we want to know which costs are the main factors in income differences in *boro* HYV rice production. To know the variance in gross costs for *boro* HYV rice production, we estimate the variance component shares of all cost expenditures contributing to gross cost and present them in Table 11. We found that the variances in seed (c) shows in third row, irrigation (d) in row four, chemical fertilizer (g) in row seven and hired labor cost (k) in row eleven, are high in all regions, indicating that adaptation strategies, such as low input costs, have priorities for the large gross income variances of *boro* rice cultivation.

Table 10. Decomposed variance share (%) of gross income and cost of *boro* HYV rice, by region.

	B	CH	CO	D	K	M	RJ	RN	S	BD
V(GI)	101.34	46.75	264.6	62.73	79.59	70.15	69.81	80.61	67.68	91.68
V(GC)	43.86	79.49	97.26	41.17	40.46	47.38	60.96	28.25	84.98	54.04
−2*Cov (GI, GC)	−45.20	−26.24	−261.9	−3.90	−20.05	−17.53	−30.77	−8.86	−52.66	−45.72
Total	100	100	100	100	100	100	100	100	100	100

B = Barisal, CH = Chittagong, CO = Comilla, D = Dhaka, K = Khulna, M = Mymensingh, RJ = Rajshahi, RN = Rangpur, S = Sylhet and BD = Bangladesh; GI = Gross income and GC = Gross cost.

These input costs were made the net income differences in this rice production for sample farmers. Based on the findings in Table 11, it is also important to note that, in Chittagong region, the variance in hired labor cost (k) is highest (69.84%) while it is lowest in Comilla region (27.25%). This result implies that 69.84% of inequality of gross cost is elucidated by the inequality of hired labor cost in Chittagong region. As shown in the fourth row, irrigation cost (d) contributes a significant share of the variation of gross cost; the highest 22.93% of inequality of gross cost is explained by the inequality of irrigation cost in Dhaka, compared to the lowest in Chittagong. This result implies that reduction of input cost variances will ensure the low net income differences for this rice production. Farm households are not entirely self-sufficient regard the labor supply for their farming. In peak times of agricultural production, such as transplanting, weeding and harvesting, hired labor demand occurs. However, the labor supply is low in Chittagong due to hill tract areas of Bangladesh [69], resulting in the higher costs of labor.

Table 11. Decomposed variance share (%) of costs for *boro* HYV rice production, by region.

	B	CH	CO	D	K	M	RJ	RN	S	BD
V(b)	2.87	0.66	0.50	1.88	2.66	4.11	1.32	5.32	2.63	2.27
V(c)	4.10	0.71	2.21	3.67	4.78	2.72	1.73	4.34	2.20	3.61
V(d)	8.89	2.70	4.06	22.93	22.39	22.42	10.70	16.00	7.57	18.01
V(e)	0.24	0.05	1.10	0.31	0.76	0.88	0.33	2.56	0.12	0.80
V(f)	0.89	0.09	0.18	0.16	0.48	0.33	0.31	0.60	0.07	0.33
V(g)	7.71	3.31	1.98	6.71	14.76	12.82	4.71	13.54	3.23	8.21
V(h)	0.04	0.03	0.00	0.05	0.79	10.08	0.13	0.38	2.04	1.16
V(i)	2.42	0.89	1.01	0.93	1.47	1.09	0.47	1.68	1.12	1.23
V(j)	0.98	0.20	0.15	1.08	0.75	2.24	0.24	0.39	0.18	0.78
V(k)	38.05	69.84	27.25	42.04	38.45	31.49	51.04	38.17	65.10	51.51
2*Cov(f,g)	3.91	0.73	0.66	0.90	2.15		1.49	3.46	0.50	1.55
2*Cov(d,g)	4.98		1.18				4.35			
2*Cov(f,i)	1.07	1.15	2.62	0.39	0.52		0.52	0.97	0.26	0.61
2*Cov(g,i)	4.68	2.70	1.99	2.87	5.47	3.76	2.14	5.69	1.99	3.43
2*Cov(g,k)	11.72	14.45	6.27	11.25			10.64		11.72	
2*Cov(i,k)	7.46		6.84	4.83	4.58	8.05	3.89			5.90
2*Cov(e,i)		2.50	9.58					1.25	0.22	0.60
2*Cov(f,k)			5.34				5.99			
2*Cov(e,g)			1.50					4.90	0.44	
2*Cov(e,f)			7.04					0.76	0.63	
2*Cov(d,k)			8.70							
2*Cov(e,k)			9.85							
Total	100	100	100	100	100	100	100	100	100	100

B = Barisal, CH = Chittagong, CO = Comilla, D = Dhaka, K = Khulna, M = Mymensingh, RJ = Rajshahi, RN = Rangpur, S = Sylhet and BD = Bangladesh; b = Rental cost of land, c = Seed cost, d = Irrigation cost, e = Manure/compost cost, f = Pesticide cost, g = Chemical fertilizer cost, h = Draft animal cost for land preparation, i = Rental cost for tools and machinery, j = Threshing cost and k = Hired labor cost.

4.6. Future Projections

Production levels in agriculture, fishery and livestock raising are projected to change due to climate change [39,83]. We therefore sought to project the impact of rice yield change on the state of poverty in Bangladesh. If rice is a commercial crop, a price hike due to any damage from climate change could increase Bangladeshi farmers' living standards. However, rice remains a subsistence crop among most Bangladeshi farmers; therefore, we assume that rice yield reduction will lead to a rice consumption reduction.

The effects of climate change on rice yields, as has been estimated and shown by International Food Policy Research Institute [37], are such that, without adaptation to climate change impacts, *aman* HYV and *boro* HYV rice yields will decline by 3.5% and 10.2%, respectively, in Bangladesh. According to the Geophysical Fluid Dynamics Laboratory (GFDL) scenarios, if temperature changes by 4.0 °C, then 17% decline in overall rice will occur in Bangladesh [84].

According to this projection, we assumed that, due to climate change effects on *boro* HYV and *aman* HYV, rice yields will be reduced by 10% and 4%, respectively, as well as a 17% reduction in overall rice among the sample households. We applied log-normal distribution to project the poverty rate due to income reduction by yield loss on the effects of climate change.

Figure 10 shows the annual per-capita income (actual and projected) in US$ of the sample households across Bangladesh. In general, one can see from this figure that the sample population density (i.e., probability density) mostly lies within the low annual per-capita income range, which is less than the poverty line. Additionally, the probability density of the low-income range increases in the projected income distribution when one considers rice yield loss due to climate change.

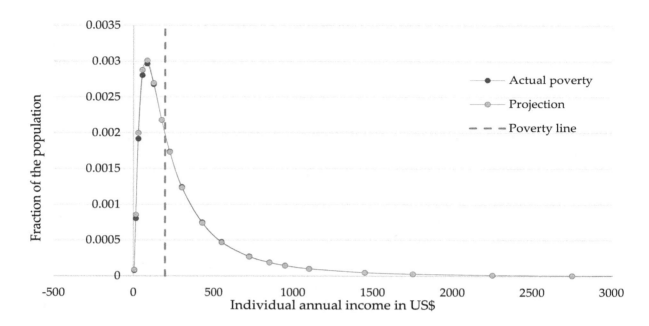

Figure 10. Annual per-capita income (US$) distribution of Bangladesh (17% loss of rice).

From the decomposed variance share of income sources in Table 9, we found that agriculture was the main reason for income differences in Mymensingh and Rangpur. Now, we can examine the effects of climate change on rice production (10% and 17% losses) in these two regions by log-normal distribution.

We analyzed and found that constant reduction of rice yield (10% loss) by climate change in Bangladesh is not such a severe problem for farmers. Because the change in net per-capita income is very small, there is not a dramatic change of poverty rate. However, if unexpected extreme events, such as floods, flash floods, droughts and sea level rise, occur in specific areas of Bangladesh, they create a more vulnerable situation for the farmers' livelihood. In addition, the probability density of low-income range increases (Figures 11 and 12) in both Mymensingh and Rangpur districts, where rice income decreases due to climate change.

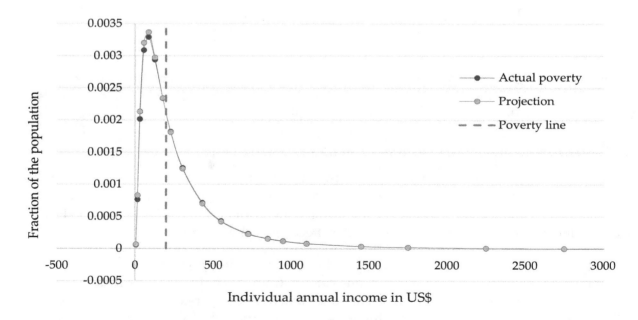

Figure 11. Annual per-capita income (US$) distribution of Mymensingh (17% loss of rice).

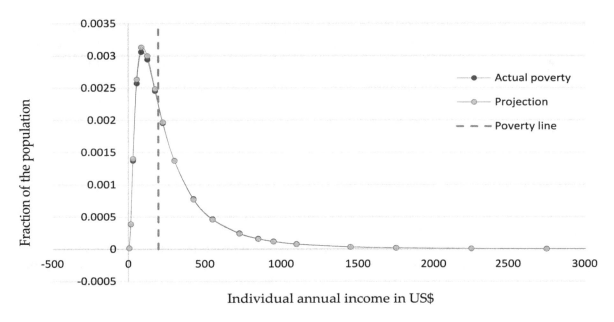

Figure 12. Annual per-capita income (US$) distribution of Rangpur (17% loss of rice).

We also applied the same analysis in Figures 10–12 to all of the regions and Table 12 shows the results of the poverty rate after income changes due to assumed yield losses of *aman* HYV, *boro* HYV rice and overall rice.

Table 12. Change in poverty rate following a loss of rice yield due to climate change.

		B	CH	CO	D	K	M	RJ	RN	S	BD
	Actual	0.507	0.484	0.446	0.455	0.415	0.496	0.323	0.462	0.484	0.454
	Projected	0.508	0.491	0.447	0.458	0.417	0.502	0.330	0.466	0.487	0.457
10% loss	Change	0.001	0.007	0.001	0.003	0.002	0.006	0.007	0.004	0.003	0.003
	Increase (%)	0.197	1.446	0.224	0.659	0.482	1.210	2.167	0.866	0.620	0.661
	Projected	0.513	0.494	0.449	0.460	0.422	0.511	0.335	0.473	0.490	0.461
17% loss	Change	0.006	0.010	0.003	0.005	0.007	0.015	0.012	0.011	0.006	0.007
	Increase (%)	1.183	2.066	0.673	1.099	1.687	3.024	3.715	2.381	1.240	1.542

B = Barisal, CH = Chittagong, CO = Comilla, D = Dhaka, K = Khulna, M = Mymensingh, RJ = Rajshahi, RN = Rangpur, S = Sylhet and BD = Bangladesh.

The estimated results suggest that rice yield loss would reduce the annual per-capita income of the sample farm households and increase the poverty rate in various regions across Bangladesh. It was found that the highest poverty rate increase (3.024%) would occur in Mymensingh, Rajshahi (3.715%) and Rangpur (2.381%). Rajshahi and Rangpur are in northwestern Bangladesh and are prone to drought; climate change would affect rice production specifically in the summer, when *boro* rice is being produced. Mymensingh is affected by floods, flash floods and heavy rainfall each year, owing to the effects of climate change on *aman* and *boro* harvests.

Climate Change Impact Scenario

Extreme events, such as floods, droughts and changes in seasonal rainfall patterns, negatively impact crop yields in vulnerable areas [85–87]. In Bangladesh, the rural poverty rate would be exacerbated [88] as a result of the impacts of extreme events on the yield of rice crop and increases in food prices and the cost of living [89,90]. The impacts of climate change on poverty would be heterogeneous among countries [91]. Due to the impact of climate change, rice production would decrease and some rice exporting countries, such as Indonesia, the Philippines and Thailand, would benefit from global food price rises and reduced poverty, while Bangladesh would experience a net increase in poverty of approximately 15% by 2030 [89,91].

Climate change refers to changes in climate attributed directly as temperature, precipitation, CO_2 concentrations and solar radiation or indirectly as river floods, flash floods and sea level rise that alter the composition of the global atmosphere, as well as to natural climate variability observed over comparable time periods [33,50].

Temperature Increase

Temperature is an important factor for *boro* rice production and the maximum temperature is always more vulnerable with a negative impact on rice yields. In Bangladesh, seasonal temperature suddenly fluctuates, causing drastically declines in the yield of *boro* rice. *Boro* rice yields decrease by a maximum of 18.7% due to an increase in minimum temperature of 2.0 °C–4.0 °C and by 36.0% for 2.0 °C–4.0 °C maximum temperature increases in different location of Bangladesh in 2008 [92]. According to the Intergovernmental Panel on Climate Change (IPCC), SRES emissions scenarios and climate models being considered, global mean surface temperature is projected to rise in the range of 1.8 to 4.0 °C by 2100 [93]. Following the previous assessment, the IPCC concludes in their fifth assessment report (AR5) that it will be difficult to adapt with large-scale warming of approximately 4°C or more, which will increase the likelihood of severe, pervasive and irreversible impacts [91,94,95].

According to the previous projection of temperature fluctuations in Bangladesh, we assume that, due to the maximum and minimum temperature fluctuations, in the future, the overall rice production will decrease by approximately 17% of the sample farmers and results are shown in Table 12. The table shows that maximum 3.7% poverty will increase in Rajshahi and second highest (3.0%) in Mymensingh region and this implies that it is important to adaptation strategies for Rajshahi and Mymensingh for high temperature.

Rainfall Decreases (Drought)

Inadequate rainfall leads to greater drought frequency and intensity, while increased evaporation increases the chance of complete crop failure [96,97]. Drought is the most widespread and damaging of all environmental stresses [35,98]. In South and Southeast Asia, including some states of India, severe drought affects rain-fed rice and yield, with losses as high as 40% and the total area affected measuring 23 million hectares, amounting to $800 million [99]. Bangladesh experienced severe drought in different years and locations in the districts of the northwestern border [100]. Erratic rainfall and drought reduce crop production by 30% and 40%, respectively [84]. *Boro* rice production will decrease due to rainfall in winter [92]. This study noted that, with 5-mm and 10-mm rainfall reductions in the future, *boro* rice will decrease by a maximum of 16.6% and 24.2%, respectively, in the winter. Drought caused 25% to 30% crop reduction in the northwestern part of Bangladesh based on from 2008 [101]. Due to the high rainfall variability and dryness, the northwestern region is the most drought-prone area in Bangladesh [102,103]. Rajshahi, Chapai-Nawabganj, Naogaon, Natore, Bogra, Joypurhat, Dinajpur and Kustia districts are drought prone areas in Bangladesh because of their moisture-retention capacity and infiltration rate characteristics [104].

According to the previous projection of drought, we assume that, if rainfall decreases and drought occur in the future, the overall rice production will decrease by approximately 20% of the sample farmers in northwestern districts of Bangladesh. By using log-normal distribution, we project the poverty rate due to income reduction by yield loss because of drought.

Table 13 shows the results of the poverty rate (Figure 13) after income changes due to assumed yield losses of overall rice by drought in the northwestern region in Bangladesh, while the Dinajpur (10.175% poverty increase), Rajshahi (5.670% poverty increase) and Naogaon (11.245% poverty increase) districts are most vulnerable to poverty. Dependency on agriculture with high variability of annual rainfall has made the northwestern regions highly susceptible to droughts and high poverty rates, compared to other parts of the country. Conservation of water could play an important role in reducing the impact of drought and alleviating poverty in this area [103].

Table 13. Poverty rate in drought-prone districts on rainfall decrease.

	BG	CN	DI	KU	NG	NT	RJ	JT
Actual	0.242	0.354	0.285	0.447	0.249	0.448	0.388	0.268
Projected	0.263	0.361	0.314	0.452	0.277	0.452	0.410	0.282
Change	0.021	0.007	0.029	0.005	0.028	0.004	0.022	0.014
Increase (%)	8.678	1.977	10.175	1.119	11.245	0.893	5.670	5.224

BG = Bogra, CN = Chapai-Nawabganj, DI = Dinajpur, KU = Kustia, NG = Naogaon NT = Natore, RJ = Rajshahi and JT = Joypurhatr.

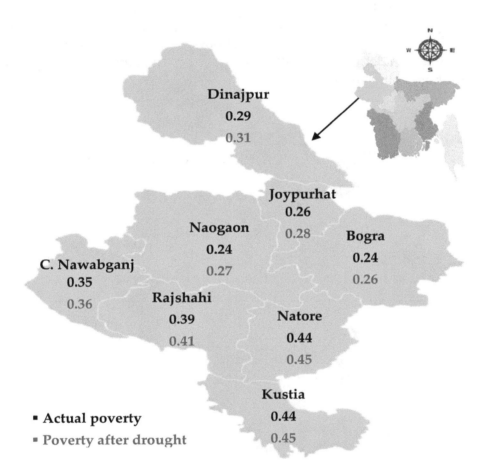

Figure 13. Changing poverty rates caused by drought in northwestern regions.

Flood From the GBM basins, the monsoonal discharge of water causes seasonal floods and affects most of the areas of Bangladesh, with extent varying by year [50]. Floods occur almost every year and in 1998, floods covered almost 70% of total land area in Bangladesh, causing the maximum damage by floods in Bangladesh [105]. According to the IPCC's fourth assessment report, the intensity and frequency of floods and cyclones will increase in the near future [33]. Moreover, the IPCC's fifth assessment report (AR5) predicts that greater risks of flooding will increase on the regional scale [91,94–99]. In addition, extreme flood events will reduce crop production by 80% in Bangladesh [37,84].

Mymensingh, Sylhet, Dhaka, Comilla, some parts of Rangpur and Khulna regions are the mainly river-flooded areas in Bangladesh [50]. We assume that, if extreme floods, as in 1998 (the magnitude of the 1998 flood was the maximum in Bangladesh), occur, farm production will decrease by 80% in the flood-prone regions of Bangladesh. By log-normal distribution we project the poverty rate due to income reduction by yield loss due to the effects of extreme floods. The results are shown in Table 14.

Table 14. Poverty rate due to yield loss by flood in Bangladesh.

	CO	D	K	M	RN	S
Actual	0.446	0.455	0.415	0.496	0.462	0.484
Projected	0.465	0.502	0.479	0.554	0.529	0.519
Change	0.019	0.047	0.064	0.058	0.067	0.035
Increase (%)	4.260	10.330	15.422	11.694	14.502	7.231

CO = Comilla, D = Dhaka, K = Khulna, M = Mymensingh, RN = Rangpur and S = Sylhet.

The estimated results in Table 14 suggest that rice yield loss would reduce the annual per-capita income of the sample farm households and increase the poverty rate in various regions across Bangladesh (Figure 14). It was found that the highest poverty rate increases would occur in Rangpur (14.502%) and Khulna (15.422%). This result implies that coping strategies to highly flood affected areas of crops loss should have priority.

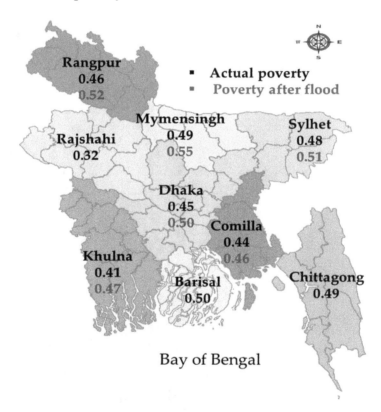

Figure 14. Changing poverty rates caused by floods in different regions.

Flash Floods

The northeastern parts of Bangladesh—mostly Sunamganj, Kishorganj, Netrokona, Sylhet, Habiganj and Maulvibazar—are prone to flash floods during the months of April to November and these areas are covered by many haors, where water remains stagnant [106]. Farmers of these districts produced *boro* rice in almost 80% of their land, while only approximately 10% of the area is covered by transplanted *aman* production [107]. In 2017, flash floods affected these areas and damaged almost 90% (maximum) of *boro* rice [108]. According to this scenario, we assumed that if in the future this extreme event occurs in haor areas, *boro* rice yields will be reduced by a maximum of 90% of the sample households. We applied log-normal distribution to project the poverty rate due to income reduction by yield loss due to the effects of flash floods on *boro* rice yields by a maximum of 90%.

Table 15 shows the results of the poverty rate after incomes changed due to assumed yield loss of *boro* rice in flash flood regions in Bangladesh, while Kishorganj district is most vulnerable to poverty

(19.214% increase) if flash floods occur (Figure 15). The projected results are treated as flash flood to be changed the poverty in northern-eastern parts of Bangladesh and this region are vulnerable on flash flood. Therefore, ex-ante coping strategies are important to the damages of flash flood.

Table 15. Poverty rate in flash flood region in Bangladesh.

	HB	KI	MV	NT	SU	SY	TH
Actual	0.354	0.458	0.624	0.585	0.511	0.427	0.354
Projected	0.381	0.546	0.637	0.628	0.550	0.452	0.381
Change	0.027	0.088	0.013	0.043	0.039	0.025	0.027
Increase (%)	7.627	19.214	2.083	7.350	7.632	5.855	7.627

HB = Habiganj, KI = Kishorganj, MV = Maulvibazar, NT = Netrokona SU = Sunamganj, SY = Sylhet and TH = Total Haor.

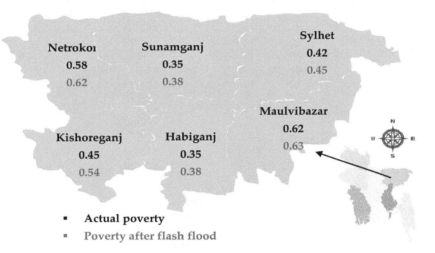

Figure 15. Changing poverty rate caused by flash floods in northeastern regions.

Sea Level Rise

Approximately 80% of the land of Bangladesh is flatlands, while 20% is 1 m or less above sea level, which is the coastal area (southern 19 districts beside the Bay of Bengal) and particularly vulnerable to sea level rise [109]. The coastal area covers approximately 20% of the country (including 19 districts beside the Bay of Bengal), which is approximately 30% of the net cultivable area and 25.7% of the population of Bangladesh [110,111]. Sea level rise will directly result in increased coastal flooding, which will increase in the event of storm surges. IPCC's fourth assessment report [33] reports that a 1-m sea level rise will displace approximately 14,800,000 people by inundating a 29,846-sq. km. coastal area [112]. Nicholls and Leatherman in 1995 [113] predicted that a 1-m sea level rise would result in a 16% of national rice production loss in Bangladesh [114].

In terms of number of people affected with respect to sea level rise, Bangladesh has been rated as the third most vulnerable country in the world. By 2050, approximately 33 million people would be suffering from surging, assuming a sea level rise of 27 cm. A full 18% of the total land area in Bangladesh would submerge with a 1-m rise in sea level [115]. Based on the IPCC fifth annual report (AR5), across all representative concentration pathways (RCPs), global mean temperature (°C) is projected to rise by 0.3 to 4.8 °C by the late-21st century and global mean sea level (m) is projected to increase by 0.26 to 0.82 m [91]. The Global Circulation Model (GCM) predicts an average temperature increase of 1.0 °C by 2030, 1.4 °C by 2050 and 2.4 °C by 2100; the study revealed that the sea level will rise by 14 cm, 32 cm and 62 cm, respectively. A rise in temperature would cause significant decreases in production of 28 % and 68 % for rice and wheat, respectively [84].

According to this scenario, we assumed that, due to sea level rise in the southern part of Bangladesh, *boro* rice yields will be reduced by 30% of the sample households. We applied log-normal distribution to project the poverty rate due to income reduction with yield loss based on the effects of sea level rise.

Table 16 shows the results of the poverty rate after income changes due to assumed yield loss of rice in coastal regions due to sea level rise, while Khulna district is the most vulnerable to poverty and poverty will increase by 6.752% (Figure 16). Changing continuous sea level rise in the coastal region result in no significant loss reduction for rice.

Table 16. Poverty rate in sea level rise regions in Bangladesh.

	SK	KH	BT	PR	JL	BG	BS	PT	BL	LK	NK	FN	CT	CX
Actual	0.599	0.295	0.363	0.388	0.640	0.532	0.419	0.628	0.491	0.529	0.438	0.481	0.505	0.462
Projected	0.609	0.315	0.370	0.390	0.650	0.545	0.431	0.636	0.493	0.533	0.440	0.487	0.515	0.464
Change	0.010	0.020	0.007	0.002	0.011	0.013	0.013	0.008	0.002	0.004	0.002	0.007	0.010	0.002
Increase (%)	1.688	6.752	1.924	0.527	1.674	2.388	3.081	1.255	0.491	0.770	0.410	1.361	1.901	0.367

SK = Satkhira, KH = Khulna, BT = Bagerhat, PR = Pirozpur, JL = Jhalakati, BG = Barguna, BS = Barisal, PT = Patuakhali, BL = Bhola, LK = Lakshmipur, NK = Noakhali, FN = Feni, CT = Chittagong and CX = Cox's Bazaar.

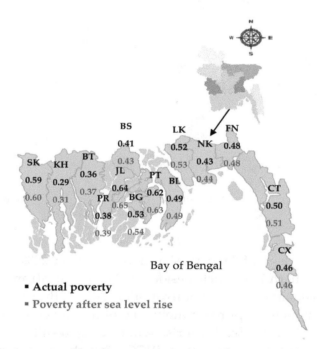

Figure 16. Changing poverty rate caused by sea level rise in southern regions.

Representative Concentration Pathways (RCPs)

In assessing future climate change, the fifth assessment report (AR5) of the IPCC selected four RCPs, –RCP 2.6, RCP 4.5, RCP 6.0 and RCP 8.5 [91], with RCP 4.5 and RCP 8.5 covering both medium and extreme scenarios. These four RCPs describe four probable climate futures depending on how much greenhouse gasses are emitted over the next 85 years.

According to the IPCC's fifth annual report (AR5), across all representative concentration pathways (RCPs), global mean temperature (°C) is projected to rise by 0.3 to 4.8 °C by the late-21st century [68]. Increasing temperatures will increase the number of growing days over time. Heat stress is a major issue for crop production and reduces yields.

Climate change will certainly continue in coming decades and affect agricultural production. Yamei Li et al. worked on simulating total climate change impacts on rice production under RCP scenarios and projected that average rice yields during the 2020s, 2050s and 2080s would decrease by 12.3%, 17.2% and 24.5% under RCP 4.5 and by 14.7%, 27.5% and 47.1% under RCP 8.5, respectively [67].

According to this scenario, we assumed that, due to total climate change impacts, rice yields would be reduced by a maximum of 47% based on RCP 8.5 among the sample households. We applied log-normal distribution to project the poverty rate due to income reduction by yield loss. Table 17 shows that, under RCP 4.5 and RCP 8.5, the poverty rate will increase in all of the regions because of rice income reductions.

Additional increases in average poverty occur in Rajshahi, Mymensingh, Rangpur, Khulna and Sylhet region under both RCP 4.5 and RCP 8.5 with variations in the total climate change impacts on rice production. The yield of rice is predicted to decrease more under RCP 8.5 than RCP 4.5, resulting in per-capita income decreases. Under RCP 8.5, this study predicts a maximum increase in poverty of 10.526% in Rajshahi and the lowest of 3.139% in Comilla (Table 17). It is possible that our predicted rice yield declines by RCP scenario and relatively drought prone areas, such as Rajshahi, will be more vulnerable (Figure 17). The results from our drought scenarios are comparable to the results for RCP 8.5 and it is consistent that Rajshahi region is more vulnerable under climate change impacts. In both scenarios, our predicted yield decline and resulting per-capita income decline increase poverty. Climate change forces a decline in rice yield [116], suggesting that the predicted decreases in heat stress yield can be mostly attributed to an increased drought tolerant variety.

Table 17. Changes in poverty rates following a loss of rice yield due to RCPS.

		B	CH	CO	D	K	M	RJ	RN	S	BD
	Actual	0.507	0.484	0.446	0.455	0.415	0.496	0.323	0.462	0.484	0.454
25% loss of rice under RCP 4.5	Projected	0.516	0.490	0.455	0.462	0.424	0.510	0.345	0.471	0.497	0.463
	Change	0.009	0.006	0.009	0.007	0.009	0.014	0.022	0.009	0.013	0.009
	Increase (%)	1.775	1.240	2.018	1.538	2.169	2.823	6.811	1.948	2.686	1.982
47% loss of rice under RCP 8.5	Projected	0.524	0.500	0.460	0.470	0.438	0.526	0.357	0.488	0.507	0.474
	Change	0.017	0.016	0.014	0.015	0.023	0.030	0.034	0.026	0.023	0.020
	Increase (%)	3.353	3.306	3.139	3.297	5.542	6.048	10.526	5.628	4.752	4.405

B = Barisal, CH = Chittagong, CO = Comilla, D = Dhaka, K = Khulna, M = Mymensingh, RJ = Rajshahi, RN = Rangpur, S = Sylhet and BD = Bangladesh.

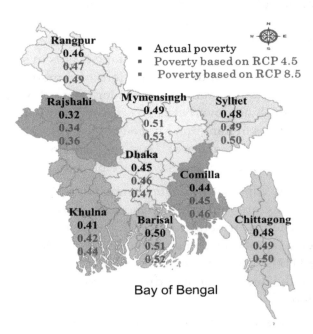

Figure 17. Changing poverty rate caused by total climate change impact based on RCP 4.5 and 8.5.

5. Conclusions

This paper has focused on the agrarian sub-national regional analysis of climate change vulnerability in Bangladesh under various climate change scenarios and its potential impact on poverty. It has drawn some significant evidence of regional vulnerability to climate change from regional characteristics, per-capita income, total income disparity, cost of production and poverty, based on statistical analysis of farm survey data. Our findings indicated that some regions are vulnerable to climate change impact on agricultural production among the administrative regions of Bangladesh, where coping strategies and techniques are important.

Bangladeshi farmers are producing crops, although there is much uncertainty due to associated risks of climate change. The results of our study show that, from the income shares of income source sectors, farmers in Mymensingh and Rangpur are largely dependent on agriculture. Of these regions, Mymensingh is one of the regions with the highest poverty rates. The income share in income sources revealed that income category shares across the various regions of Bangladesh are far from uniform. Income share comparisons and cluster analysis classified the regions into three groups as follows. (1) In some regions, namely Rajshahi, Khulna and Dhaka, income from agriculture is important and these regions receive relatively high income. (2) In other regions, namely Mymensingh, Rangpur and Barisal, agriculture income is important but the regions receive relatively low income. (3) The other regions, which are Comilla, Chittagong and Sylhet, are not strongly dependent on agriculture and Comilla region strongly relies on income from remittances. The principal targets of agricultural research for poverty reduction are considered to be in group (2).

Variance decomposition of income showed that agricultural income in Mymensingh and Rangpur is the main cause of income differences. Moreover, large variances in agricultural income in the regions are induced by gross incomes from rice production, indicating that rice yield can have large impacts on income levels. Therefore, research and development and technical support for farmers to realize high and stable rice yields in these regions are important.

This paper used modelling to predict crop yield changes by different aspects of climate change under droughts, floods, flash floods, sea level rise and RCP scenarios. We account for some uncertainty in crop yields and the resulting reduction in per-capita income of farm households. The proposed lognormal distribution projected the poverty rate and examined the vulnerable regions. The key is to understand the future projections of poverty rates on assumptions of *boro* HYV and *aman* HYV rice yield decreases on each farm due the climate change impacts and climate volatility subjecting the poor to poverty rate increases in different regions. Current climate change impacts are not the same in different regions; in particular, different extreme climatic events in specific regions often result in irreversible losses. One of the examples of the interventions of climatic events is that dependency on agriculture with high variability in annual rainfall has render the northwestern parts highly vulnerable to droughts and has increased the high poverty rates, compared to other parts of the country. Extreme floods can increase the poverty rates in Rangpur, Mymensingh and Khulna regions. Kishorganj district is the most vulnerable on poverty (8.8% increase) if sudden flash floods occur in the northeastern part of the country. Due to sea level rise, coastal areas will face poverty.

Strategies and techniques to cope with climate change for regions where small-scale farmers are largely dependent on agriculture are important challenges. Among the negative consequences of climate change impacts, subsistence farmers are suffering more from vulnerabilities such as extreme poverty or hunger. However, adaptation techniques in agriculture are a vital tool to avoid the adverse impacts of climate change [117]. Given the complex nature of droughts, floods, flash floods and sea level rise as phenomena, the development of drought-tolerant, short-maturing and salt-tolerant varieties is critically important.

More generally, our results are focused on farm income and poverty, including regional vulnerability due to climate change impacts on agricultural production. In recent years, climate change impacts have played a vital role in increasing the poverty rate and income variability among farm households in Bangladesh. Extreme environmental hazards are faced by farmers in this country and their net farm production decreases drastically, increasing the poverty rate while changes in weather conditions are a less severe problem for farmers due to their involvement in other income activities. We actually performed this study focusing on revealing the comprehensive impact of climate change on farm production and the crops are that the most important for per capita income differences across the country and that enhance the poverty rate, using the covariance and lognormal distribution methods.

This study has attempted to bridge the gap between academic research and professional practices in the context of potential climate change impacts on crop production and poverty. Because of the relatively large sample size, compilation and manipulation of the data were challenging. With the assessment of poverty and regional vulnerability due to climate changes, it is hoped that the study in general will assist in guiding authorities in terms of interventions aimed at climate change risk reduction in Bangladesh. Therefore, we believe that this research will help to reveal the mechanisms behind the per capita income differences and projected poverty rates of farm households based on different climate change impact scenarios across Bangladesh. Future work might also be more micro level for policy making to test root-level poverty and to further evaluate the impact of climate change on different crops and it should include the model for poverty determinants to confirm the relationships studied and their adaptations.

Author Contributions: M.S.A. conceived the research, compilation and analyze the data, drafted, edited and revised the manuscript; J.F. modified the methodology of the research, checked, edited and revised the manuscript; S.K. designed, compilation and analyze the data, edited and revised the manuscript; M.R.B. checked the statistical tools and maps of the objective regions; and M.A.S. helps to compilation of data and first draft of the manuscript.

Acknowledgments: The authors are grateful to three anonymous reviewers for their constructive comments and suggestions. We would like to thank the International Food Policy Research Institute (IFPRI) for providing us with the primary data. We also acknowledge the support from JIRCAS under the project "Climate Change Measures in Agricultural Systems" and University of Tsukuba.

Appendix A

Appendix A.1

In this study, we used the primary data from Bangladesh Integrated Household Survey (BIHS 2011–2012) by IFPRI, https://dataverse.harvard.edu/dataset.xhtml?persistentId=hdl:1902.1/21266.

Appendix A.2

"aus" is former rainy season, *"aman"* is rainy season and *"boro"* is dry season irrigated rice.

Appendix B

Appendix B.1

$$Net\ accounting\ cost_i = \sum_i \sum_j P_{ij} X_{ij}$$
$$= c_{i,\ Rental\ cost\ of\ land} + c_{i,\ seed\ cost} + c_{i,\ irrigation\ cost} \tag{A1}$$
$$+ c_{i,\ manure\ or\ compost\ cost} + c_{i,\ pesticides\ cost} + c_{i,\ fertilizer\ cost}$$
$$+ c_{i,\ draft\ animal\ cost} + c_{i,\ machinery\ cost} + c_{i,threshing\ cost} + c_{i,\ hired\ labor\ cost}$$

$$Production\ value_i = \sum_i P_i Y_i \tag{A2}$$

$$Gross\ income_i = \sum_i P_i Y_i - Inkind\ payment_i$$
$$= \sum_i P_i Y_i - (C_{i,\ irrigation\ cost\ paid\ by\ crop} + C_{i,\ labor\ cost\ paid\ by\ crop}) \tag{A3}$$

$$Net\ income(\pi)_i = Gross\ income_i - Net\ accounting\ cost_i \tag{A4}$$

Appendix B.2

We estimate per-capita incomes (US$) of all sample families on assumption of climate change impacts and draw the distribution of the estimated incomes assuming that the distribution follows log normal distribution. To draw log normal distribution, we have to find mean and standard deviation of $ln(x)$. Firstly, we divide the per capita income in different class and make the average (x) of each class and we find the frequency of household (n) in each per-capita income class. Then we find the log of average per-capita class, log (x); and multiplied by the frequency of household in each class, $n * \log(x)$. Next average,

$$u = \frac{\sum n\{\log(x)\}}{\sum n} \tag{A5}$$

Then we estimate, $\log(x) - u$, $\{\log(x) - u\}^2$ and $n\{\log(x) - \mu\}^2$
Next standard deviation,

$$\sigma = \sqrt{\frac{\sum n\{\log(x) - u\}^2}{\sum n}} \tag{A6}$$

Returns the lognormal distribution of x, where ln (x) is normally distributed with parameters Mean and Standard deviation. Use this function to analyze data that has been logarithmically transformed.

$$f_X(x) = \frac{1}{dx}Pr(X \leq x) = \frac{1}{dx}Pr(lnX \leq lnx) = \frac{1}{dx}\Phi\left(\frac{lnx-\mu}{\sigma}\right) = \varphi\left(\frac{lnx-\mu}{\sigma}\right)\frac{1}{dx}\left(\frac{lnx-\mu}{\sigma}\right)$$
$$= \varphi\left(\frac{lnx-\mu}{\sigma}\right)\frac{1}{\sigma x} = \frac{1}{x}\cdot\frac{1}{\sigma\sqrt{2\pi}}exp\left(-\frac{(lnx-\mu)^2}{2\sigma^2}\right) \tag{A7}$$

Syntax: LOGNORM.DIST(x, mean, standard deviation and cumulative)

Appendix C

Table A1. Household income (US$/yr.) from different sources, by region.

	B	CH	CO	D	K	M	RJ	RN	S	BD
Agril. crops	159.35	124.17	82.83	194.67	273.63	225.23	322.78	246.71	131.77	200.28
Main crops	76.23	44.11	35.22	118.52	152.25	127.87	202.10	170.95	89.86	116.89
Other crops	83.13	80.06	47.61	76.16	121.39	97.36	120.69	75.76	41.92	83.39
Fish	115.70	23.47	8.54	31.34	111.73	67.72	49.43	13.14	46.17	55.45
Livestock	27.43	17.81	22.35	51.76	86.61	57.25	76.48	35.67	26.20	48.60
Non-Ag. profit	260.29	293.63	212.95	304.83	254.71	197.39	338.22	171.49	292.70	262.92
Remittance	138.41	381.12	624.89	225.28	107.64	101.84	77.30	87.37	259.51	212.90
Employment	487.70	676.42	464.06	590.46	542.42	436.33	669.77	582.29	642.59	560.94
Other income	64.65	8.41	90.96	38.22	31.36	32.01	190.70	15.53	60.98	57.61
Total	1253.53	1525.04	1506.60	1436.53	1408.12	1117.77	1724.70	1152.23	1459.92	1398.71

B = Barisal, CH = Chittagong, CO = Comilla, D = Dhaka, K = Khulna, M = Mymensingh, RJ = Rajshahi, RN = Rangpur, S = Sylhet, BD = Bangladesh, Main crops = *Aus*, *Aman* and *Boro* rice and other crops = Wheat, Maize, Jute, Potato, Chili, Onion etc.

Table A2. Each agricultural crop's share in total net agricultural income (%), by region.

Crops	B	CH	CO	D	K	M	RJ	RN	S	BD
Rice	45.51	33.66	32.99	37.39	43.52	55.62	51.27	57.72	67.05	47.22
Aus	6.37	2.89	1.51	0.64	3.03	0.84	1.11	1.39	5.19	2.24
Aman	24.36	17.83	6.42	5.22	15.55	15.37	17.27	22.12	18.45	14.96
Boro	14.78	12.95	25.06	31.54	24.95	39.42	32.89	34.21	43.41	30.02
Wheat	0.00	0.00	0.19	0.22	0.70	0.07	1.32	0.96	0.00	0.48
Maize	0.00	0.00	0.84	0.30	0.26	0.00	1.40	2.01	0.00	0.56
Jute	0.61	0.00	3.03	10.53	5.85	0.44	2.80	2.96	0.11	4.37
Potato	0.66	0.37	5.49	0.53	0.18	0.36	4.04	4.68	1.00	1.62
Chili	1.82	2.17	2.69	6.85	5.72	1.54	0.67	1.20	0.53	3.40
Onion	0.00	0.00	0.01	5.79	1.01	0.00	1.81	0.32	0.00	1.70
Other crops	51.39	63.80	54.77	38.38	42.76	41.96	36.67	30.16	31.31	40.65
Total	100	100	100	100	100	100	100	100	100	100

B = Barisal, CH = Chittagong, CO = Comilla, D = Dhaka, K = Khulna, M = Mymensingh, RJ = Rajshahi, RN = Rangpur, S = Sylhet and BD = Bangladesh.

Table A3. Costs and income (US$/ha) associated with aman and boro HYV rice production, by region.

	Aman HYV										Boro HYV									
	B	CH	CO	D	K	M	RJ	RN	S	BD	B	CH	CO	D	K	M	RJ	RN	S	BD
b	53.77	74.83	76.13	53.84	30.12	45.34	46.93	38.08	57.28	47.08	64.59	82.41	75.69	50.94	32.39	45.13	49.55	42.71	54.49	49.14
c	64.29	38.14	72.11	79.90	64.27	33.96	45.27	30.77	45.13	47.80	60.51	46.47	70.82	71.22	66.36	42.49	73.53	40.41	43.01	58.24
d	1.33	4.58	8.04	34.37	11.12	27.00	37.48	12.43	5.65	19.52	63.70	60.16	135.28	165.63	114.87	122.83	116.16	93.95	61.48	113.42
e	1.19	1.54	2.87	1.55	1.73	2.81	7.00	2.45	3.84	3.22	2.40	5.36	9.24	4.22	10.59	8.17	8.65	25.41	1.92	7.98
f	5.98	11.33	8.48	6.22	3.34	9.16	9.36	9.31	4.81	7.49	14.01	14.25	13.96	7.34	9.24	13.41	11.12	13.73	3.65	9.72
g	26.33	45.58	60.39	50.65	40.65	63.05	49.46	50.75	27.61	47.88	59.67	92.28	92.46	90.84	97.05	106.66	73.24	107.18	45.80	84.34
h	9.08	0.61	0.43	0.67	2.57	3.96	1.61	1.84	6.60	3.22	1.55	1.58	0.30	1.02	2.82	5.72	2.55	2.06	5.54	3.05
i	26.59	43.06	37.71	33.86	25.06	25.65	22.36	26.46	31.04	27.43	42.48	46.83	36.75	33.65	28.94	26.73	21.83	30.41	25.77	29.51
j	17.58	17.31	9.34	6.11	9.51	8.45	4.04	3.36	5.89	7.64	15.92	29.29	14.55	16.23	19.54	10.05	5.96	9.59	4.27	11.99
k	85.80	155.19	133.77	171.81	113.27	115.80	134.27	106.25	107.67	120.55	152.40	305.40	237.84	242.40	151.19	157.81	190.60	125.47	227.20	192.16
TC	291.93	392.18	409.27	438.98	301.63	335.18	357.78	281.70	295.53	331.82	477.24	684.02	686.89	683.49	533.00	539.01	553.19	490.92	473.14	559.55
TP kg/ha	3573	3655	1913	3131	2515	2776	3650	3500	2572	3023	4659	4821	5136	6181	5122	4950	6025	5733	4218	5304
GI	734.65	710.58	387.39	614.66	477.69	577.30	661.75	669.42	476.78	585.58	841.58	964.00	999.64	1169.99	1009.64	1082.65	1115.88	1115.55	749.11	1023.34
GI-TC	442.72	318.40	-21.88	175.69	176.07	242.12	303.96	387.72	181.25	253.75	364.35	279.98	312.75	486.49	476.65	543.65	562.69	624.64	275.96	463.80

B = Barisal, CH = Chittagong, CO = Comilla, D = Dhaka, K = Khulna, M = Mymensingh, RJ = Rajshahi, RN = Rangpur, S = Sylhet and BD = Bangladesh; b = Rental cost of land, c = Seed cost, d = Irrigation cost, e = Manure/compost cost, f = Pesticide cost, g = Chemical fertilizer cost, h = Draft animal cost for land preparation, i = Rental cost for tools and machinery, j = Threshing cost, k = Hired labor cost, TC = Total cost, TP = Total production and GI = Gross income.

References

1. Schendel, W.V. *A History of Bangladesh*; Cambridge University Press: Cambridge, UK, 2009.
2. Ravallion, M. The Performance of Rice Markets in Bangladesh during the 1974 Famine. *Econ. J.* **1985**, *95*, 15–29. [CrossRef]
3. Gisella, K. *Long Run Impacts of Famine Exposure: A Study of the 1974–1975 Bangladesh Famine*; University of Colorado: Denver, CO, USA, 2012.
4. Dorosh, P.A.; Rashid, S. *Bangladesh Rice Trade and Price Stabilization*; IFPRI Discussion Paper 01209; International Food Policy Research Institute: Washington, DC, USA, 2012; 2p.
5. Hossain, A.; Silva, J.A.T. Wheat and rice, the epicenter of food security in Bangladesh. *Songklanakarin J. Sci. Technol.* **2013**, *35*, 261–274.
6. Alam, G.M.; Alam, K.; Mushtaq, S. Climate change perceptions and local adaptation strategies of hazard-prone rural households in Bangladesh. *Clim. Risk Manag.* **2017**, *17*, 52–63. [CrossRef]
7. Alamgir, M.S.; Furuya, J.; Kobayashi, S. Determinants of Early Cropping of Rice in Bangladesh: An Assessment as a Strategy of Avoiding Cyclone Risk. *Jpn. J. Agric. Econ.* **2017**, *19*. [CrossRef]
8. Israt, J.S.; Misuzu, T.N.; Mana, K.N.; Mohammad, S.H.; Yoshiaki, I. Rice Cultivation in Bangladesh: Present Scenario, Problems, and Prospects. *J. Int. Cooper Agric. Dev.* **2016**, *14*, 20–29.
9. Bell, A.R.; Bryan, E.; Ringler, C.; Ahmed, A. Rice productivity in Bangladesh: What are the benefits of irrigation? *Land Use Policy* **2015**, *48*, 1–12. [CrossRef]
10. Chen, Y.; Lu, C. A Comparative Analysis on Food Security in Bangladesh, India and Myanmar. *Sustainability* **2018**, *10*, 405. [CrossRef]
11. World Bank. *World Development Report 2013: Jobs*; World Bank: Washington, DC, USA, 2012; p. 197.
12. Poverty and Inequality in Bangladesh. *Journey towards Progress (2014–2015)*; Macroeconomic Wing, Finance division, Ministry of Finance, Government of the People's Republic of Bangladesh: Dhaka, Bangladesh, 2015.
13. Sulaiman, M.; Misha, F. *Comparative Cost–Benefit Analysis of Programs for the Ultra-Poor in Bangladesh, Bangladesh Priorities*; Copenhagen Consensus Center: Copenhagen, Denmark, 2016; 29p.
14. Hijioka, Y.; Lin, E.; Pereira, J.J.; Corlett, R.T.; Cui, X.; Insarov, G.E.; Lasco, R.D.; Lindgren, E.; Surjan, A. *Climate Change 2014: Impacts, Adaptation, and Vulnerability. Part B: Regional Aspects. Contribution of Working Group II to the Fifth Assessment Report of the Intergovernmental Panel on Climate Change*; Barros, V.R., Field, C.B., Dokken, D.J., Mastrandrea, M.D., Mach, K.J., Bilir, T.E., Chatterjee, M., Ebi, K.L., Estrada, Y.O., Genova, R.C., et al., Eds.; Cambridge University Press: Cambridge,UK; New York, NY, USA, 2014; pp. 1327–1370.
15. Choudhury, A.M. Managing Natural Disasters in Bangladesh. Presented at the Dhaka Meet on Sustainable Development in Bangladesh: Achievements, Opportunities, and Challenges at Rio10, Bangladesh Unnayan Parishad, Dhaka, Bangladesh, 16–18 March 2002.
16. Shimi, A.C.; Parvin, G.R.; Biswas, C.; Shaw, R. Impact and adaptation to flood—A focus on water supply, sanitation and health problems of rural community in Bangladesh. *Disaster Prev. Manag.* **2010**, *19*, 298–313. [CrossRef]
17. World Bank. *Natural Disaster Hotspots: A Global Risk Analysis*; Disaster Risk Management Series. No. 5; World Bank: Washington, DC, USA, 2005.
18. World Bank. *Disaster Risk Management in South Asia: A Regional Overview*; World Bank: Washington, DC, USA, 2012.
19. Wassmann, R.; Jagadish, S.V.K.; Sumfleth, K.; Pathak, H.; Howell, G.; Ismail, A.; Serraj, R.; Redona, E.; Singh, R.K.; Heuer, S. Regional vulnerability of climate change impacts on Asian rice production and scope for adaptation. *Adv. Agron.* **2009**, *102*, 91–133.
20. World Bank. *Economics of Adaptation to Climate Change: Bangladesh*; World Bank: Washington, DC, USA, 2010; 79p.
21. Titumir, R.A.M.; Basak, J.K. Effects of climate change on crop production and climate adaptive Techniques for Agriculture in Bangladesh. *Soc. Sci. Rev.* **2012**, *29*, 215–232.
22. Ferreira, F.H.G.; Chen, S.; Dabalen, A.; Dikanov, Y.; Hamadeh, N.; Jolliffe, D.; Narayan, A.; Prydz, E.B.; Revenga, A.; Sangraula, P.; et al. *A Global Count of Extreme Poor in 2012: Data Issues, Methodologies and Initial Results*; World Bank: Washington, DC, USA, 2015.

23. Hasan, S.A. The impact of the 2005–2010 rice price increase on consumption in rural Bangladesh. *Agric. Econ.* **2016**, *47*, 423–433. [CrossRef]

24. El-Osta, H.S.; Morehart, M.J. Determinants of Poverty among U.S. Farm Households. *J. Agric. Appl. Econ.* **2008**, *40*, 1–20. [CrossRef]

25. Mat, S.H.C.; Jalil, A.Z.A.; Harun, M. Does Non-Farm Income Improve the Poverty and Income Inequality among Agricultural Household in Rural Kedah? *Proc. Econ. Financ.* **2012**, *1*, 269–275. [CrossRef]

26. Yamano, T.; Sserunkuuma, D.; Otsuka, K.; Omiat, G.; Ainembabazi, J.H.; Shimamura, Y. The 2003 REPEAT Survey in Uganda: Results, Foundation for Advanced Studies on International Development (FASID) and Makerere University. 2004. Available online: http://www3.grips.ac.jp/~globalcoe/j/data/repeat/REPEATinUgandaReport.pdf (accessed on 17 May 2018).

27. Janvry, A.D.; Sadoulet, E.; Zhu, N. *The Role of Non-Farm Incomes in Reducing Rural Poverty and Inequality in China*; CUDARE Working Papers; University of California: Berkeley, CA, USA, 2005; pp. 1–31.

28. Chaudhry, A.; Wimer, C. Poverty is Not Just an Indicator: The Relationship Between Income, Poverty, and Child Well-Being. *Acad. Pediatr.* **2016**, *16*, S23–S29. [CrossRef] [PubMed]

29. Gornall, J.; Richard, B.; Eleanor, B.; Robin, C.; Joanne, C.; Kate, W.; Andrew, W. Implications of climate change for agricultural productivity in the early twenty-first century. *Philos. Trans. R. Soc. B* **2010**, *365*, 2973–2989. [CrossRef] [PubMed]

30. GOB (Government of Bangladesh); UNDP (United Nations Development Program). *The Probable Impacts of Climate Change on Poverty and Economic Growth and Options of Coping with Adverse Effects of Climate Change in Bangladesh*; Policy Study: Dhaka, Bangladesh, 2009.

31. Sarker, M.A.R.; Alam, K.; Gow, J. Exploring the relationship between climate change and rice yield in Bangladesh: An analysis of time series data. *Agric. Syst.* **2012**, *112*, 11–16. [CrossRef]

32. FAO (Food and Agriculture Organization). *Livelihood Adaptation to Climate Variability and Changes in Drought-Prone Areas of Bangladesh*; Food and Agriculture Organization: Rome, Italy, 2006.

33. IPCC (Intergovernmental Panel on Climate Change). *Climate Change 2007: Impacts, Adaptation and Vulnerability: Contribution of Working Group II to the Fourth Assessment Report of the Intergovernmental Panel on Climate Change*; Cambridge University Press: Cambridge, UK, 2007.

34. Alauddin, M.; Hossain, M. *Environment and Agriculture in a Developing Economy: Problems and Prospects for Bangladesh*; Edward Elgar: London, UK, 2001.

35. FAO (Food and Agriculture Organization). Climate Change and Food Security: Risks and Responses. 2016. Available online: http://www.fao.org/3/a-i5188e.pdf (accessed on 27 April 2018).

36. UNDP. *Fighting Climate Change: Human Solidarity in a Divided World*; Human Development Report; Oxford University Press: Oxford, UK, 2008.

37. IFPRI (International Food Policy Research Institute). *Agriculture and Adaptation in Bangladesh, Current and Projected Impacts of Climate Change*; Discussion Paper 01281; International Food Policy Research Institute: Washington, DC, USA, 2013.

38. Islam, M.B.; Ali, M.Y.; Amin, M.; Zaman, S.M. *Climatic Variations: Farming Systems and Livelihoods in the High Barind Tract and Coastal Areas of Bangladesh*; Lal, R., Sivakumar, M., Faiz, S., Mustafizur, Rahman, A., Islam, K., Eds.; Climate Change and Food Security in South Asia; Springer: Dordrecht, The Netherlands, 2011.

39. Yu, W.H.; Alam, M.; Hassan, A.; Khan, A.S.; Ruane, A.C.; Rosenzweig, C.; Major, D.C.; Thurlow, J. *Climate Change Risk and Food Security in Bangladesh*; EarthScan: New York, NY, USA, 2010.

40. BBS (Bangladesh Bureau of Statistics). *Compendium of Environment Statistics of Bangladesh*; Government of the People's Republic of Bangladesh: Dhaka, Bangladesh, 2005.

41. Alauddin, M.; Tisdell, C. Trends and projections of Bangladeshi food production: An alternative viewpoint. *Food Policy* **1987**, *12*, 318–331. [CrossRef]

42. Alauddin, M.; Tisdell, C. *The 'Green Revolution' and Economic Development: The Process and Its Impact in Bangladesh*; Macmillan: London, UK, 1991.

43. Asaduzzaman, M.; Ringler, C.; Thurlow, J.; Alam, S. *Investing in Crop Agriculture in Bangladesh for Higher Growth and Productivity, and Adaptation to Climate Change*; Bangladesh Food Security Investment Forum: Dhaka, Bangladesh, 2010.

44. BBS. *Yearbook of Agricultural Statistics of Bangladesh*; Government of the People's Republic of Bangladesh: Dhaka, Bangladesh, 2009.

45. BER. *Bangladesh Economic Review, Government of the People's Republic of Bangladesh*; Ministry of Finance: Dhaka, Bangladesh, 2017.

46. Karim, Z.; Hussain, S.G.; Ahmed, M. Assessing Impact of climate Variation on food grain production in Bangladesh. *J. Water Air Soil Pollut.* **1996**, *92*, 53–62.

47. Battisti, D.S.; Naylor, R.L. Historical warnings of future food insecurity with unprecedented seasonal heat. *Science* **2009**, *323*, 240–244. [CrossRef] [PubMed]

48. Kobayashi, S.; Furuya, J. Comparison of climate change impacts on food security of Bangladesh. *Stud. Reg. Sci.* **2011**, *41*, 419–433. [CrossRef]

49. Salam, M.A.; Furuya, J.; Alamgir, M.S.; Kobayashi, S. *Policy Adaptation Cost for Mitigation of Price Variation of Rice under Climate Change in Bangladesh*; Center for Environmental Information Science: Tokyo, Japan, 2016; Volume 30, pp. 1–7.

50. Ruane, A.C.; Major, D.C.; Yu, W.H.; Alam, M.; Hussain, S.G.; Khan, A.S.; Rosenzweig, C. Multi-factor impact analysis of agricultural production in Bangladesh with climate change. *Glob. Environ. Chang.* **2013**, *23*, 338–350. [CrossRef]

51. Ali, A. Climate change impacts and adaptation assessment in Bangladesh. *Clim. Res.* **1999**, *12*, 109–116. [CrossRef]

52. Sarwar, M.G.M. Impacts of Sea Level Rise on the Coastal Zone of Bangladesh. Master's Thesis, Lund University, Lund, Sweden, 2005.

53. Abid, M.; Schilling, J.; Scheffran, J.; Zulfiqar, F. Climate change vulnerability, adaptation and risk perceptions at farm level in Punjab, Pakistan. *Sci. Total Environ.* **2016**, *547*, 447–460. [CrossRef] [PubMed]

54. Rajendra, P.S.; Nuanwan, C.; Sunsanee, A. Adaptation to Climate Change by Rural Ethnic Communities of Northern Thailand. *Climate* **2017**, *5*, 57. [CrossRef]

55. African Development Bank; Asian Development Bank; Department for International Development, United Kingdom; Directorate-General for Development, European Commission; Federal Ministry for Economic Cooperation and Development, Germany; Ministry of Foreign Affairs—Development Cooperation, The Netherlands; Organization for Economic Cooperation and Development; United Nations Development Programme; United Nations Environment Programme; and The World Bank: Poverty and Climate Change, Reducing the Vulnerability of the Poor through Adaptation. 2002. Available online: http://www.oecd.org/env/cc/2502872.pdf (accessed on 19 May 2018).

56. Ali, A.; Erenstein, O. Assessing farmer use of climate change adaptation practices and impacts on food security and poverty in Pakistan. *Clim. Risk Manag.* **2017**, *16*, 183–194. [CrossRef]

57. Menikea, L.M.C.S.; Arachchi, K.A.G.P. Adaptation to Climate Change by Smallholder Farmers in Rural Communities: Evidence from Sri Lanka. *Proc. Food Sci.* **2016**, *6*, 288–292. [CrossRef]

58. Leichenko, R.; Silva, J.A. Climate change and poverty: Vulnerability, impacts, and alleviation strategies. *WIREs Clim. Chang.* **2014**, *5*, 539–556. [CrossRef]

59. Lade, S.J.; Haider, L.J.; Engstrom, G.; Schulter, M. Resilience offers escape from trapped thinking on poverty alleviation. *Sci. Adv.* **2017**, *3*, e1603043. [CrossRef] [PubMed]

60. Ahmed, A.U.; Alam, M. *Development of Climate Change Scenarios with General Circulation Models*; Huq, S., Karim, Z., Asaduzzaman, M., Mahtabs, F., Eds.; Vulnerability and Adaptation to Climate Change for Bangladesh; Kluwer Academic Publishers: Dordrecht, The Netherlands, 1998; pp. 13–20.

61. Agrawala, S.; Ota, T.; Ahmed, A.U.; Smith, J.; van Aalst, M. *Development and Climate Change in Bangladesh: Focus on Coastal Flooding and the Sundarbans*; Organisation for Economic Co-Operation and Development: Paris, France, 2003.

62. Tanner, T.M.; Hassan, A.; Islam, K.M.N.; Conway, D.; Mechler, R.; Ahmed, A.U.; Alam, M. *ORCHID: Piloting Climate Risk Screening in DFID Bangladesh*; Detailed Research Report; Institute of Development Studies, University of Sussex: Brighton, UK, 2007.

63. CEGIS (Center for Environmental and Geographic Information Services). *Vulnerability to Climate Induced Drought: Scenario and Impacts*; Center for Environmental and Geographic Information Services: Dhaka, Bangladesh, 2013.

64. Sikder, R.; Xiaoying, J. Climate Change Impact and Agriculture in Bangladesh. *J. Environ. Earth Sci.* **2014**, *4*, 35–40.

65. Karim, M.R.; Ishikawa, M.; Ikeda, M.; Islam, M.T. Climate Change model predicts 33% rice yield decrease in 2100 in Bangladesh. *Agron. Sustain. Dev.* **2012**, *32*, 821–830. [CrossRef]

66. Key, N.; Prager, D.; Burns, C. *Farm Household Income Volatility: An Analysis Using Panel Data from a National Survey*; Economic Research Report, No.26; United States Department of Agriculture, Economic Research Service: Washington, DC, USA, 2017.

67. Li, Y.; Wu, W.; Ge, Q.; Zhou, Y.; Xu, C. Simulating Climate Change Impacts and Adaptive Measures for Rice Cultivation in Hunan Province, China. *Am. Metrol. Soc.* **2016**, 1359–1376. [CrossRef]

68. Nazarenko, L.; Schmidt, G.A.; Miller, R.L.; Tausnev, N.; Kelley, M.; Ruedy, R.; Russell, G.L.; Aleinov, I.; Bauer, M.; Bauer, S.; et al. Future climate change under RCP emission scenarios with GISS ModelE2. *J. Adv. Model. Earth Syst.* **2015**, *7*, 244–267. [CrossRef]

69. BBS. *Yearbook of Agricultural Statistics*; Government of the People's Republic of Bangladesh: Dhaka, Bangladesh, 2016.

70. Quddus, M.A. Crop production growth in different agro-ecological zones of Bangladesh. *J. Bangladesh Agric. Univ.* **2009**, *7*, 351–360. [CrossRef]

71. Department of Agricultural Marketing. *The Average Nationwide Retail Price of Selected Commodities in Bangladesh from 1 December 2010 to 30 November 2011*; Government of the People's Republic of Bangladesh, Ministry of Agriculture: Dhaka, Bangladesh, 2017.

72. Schmit, T.M.; Boisvert, R.N.; Tauer, L.W. Measuring the Financial Risks of New York Dairy Producers. *J. Dairy Sci.* **2001**, *84*, 411–420. [CrossRef]

73. Chang, H.-H.; Schmit, T.M.; Boisvert, R.N.; Tauer, L.W. *Quantifying Sources of Dairy Farm Business Risk and Implications for Risk Management Strategies*; Working Paper WP 2007-11, July 2007; Department of Applied Economics and Management, Cornell University: Ithaca, NY, USA, 2007.

74. Wolf, C.A.; Black, J.R.; Hadrich, J.C. Upper Midwest dairy farm revenue variation and insurance implications. *Agric. Financ. Rev.* **2009**, *69*, 346–358. [CrossRef]

75. Benni, E.; Finfer, R. Where is the risk? Price, yield and cost risk in swiss crop production. In Proceedings of the Selected Paper, IAAE, Triennial Conference, Foz do Iguacu, Brazil, 18–24 August 2012.

76. Galton, F. The geometric mean, in vital and social statistics. *Proc. R. Soc.* **1879**, *29*, 365–367. [CrossRef]

77. McAlister, D. The law of the geometric mean. *Proc. R. Soc.* **1879**, *29*, 367–376. [CrossRef]

78. Gaddum, J.H. Log normal distributions. *Nature* **1945**, *156*, 746–747. [CrossRef]

79. Eckhard, L.; Werner, A.S.; Markus, A. Log-normal Distributions across the Sciences: Keys and Clues. *BioScience* **2001**, *51*, 341–352.

80. Arata, Y. *Income Distribution among Individuals: The Effects of Economic Interactions*; RIETI Discussion Paper Series 13-E-042; RIETI: Tokyo, Japan, 2013.

81. World Bank. *Bangladesh: Growing the Economy through Advances in Agriculture*; World Bank: Washington, DC, USA, 2016.

82. World Bank. *Bangladesh-Household Income and Expenditure Survey: Key Findings and Results 2010*; World Bank: Washington, DC, USA, 2011; pp. 153–154.

83. OECD. *Trend and Agriculture Directorate, Agriculture and Climate Change*; OECD Publishing: Paris, France, 2015.

84. UNDP. *Policy Study on the Probable Impacts of Climate Change on Poverty and Economic Growth and the Options of Coping with Adverse Effect of Climate Change in Bangladesh*; General Economics Division, Planning Commission, Government of the People's Republic of Bangladesh & UNDP Bangladesh: Dhaka, Bangladesh, 2009.

85. Dawe, D.; Moya, P.; Valencia, S. Institutional, policy and farmer responses to drought: El Niño events and rice in the Philippines. *Disasters* **2008**, *33*, 291–307. [CrossRef] [PubMed]

86. Douglas, I. Climate change, flooding and food security in south Asia. *Food Secur.* **2009**, *1*, 127–136. [CrossRef]

87. Kelkar, U.; Narula, K.K.; Sharma, V.P.; Chandna, U. Vulnerability and adaptation to climate variability and water stress in Uttarakhand State. *India Glob. Environ. Chang.* **2008**, *18*, 564–574. [CrossRef]

88. Skoufias, E.; Rabassa, M.; Olivieri, S. *The Poverty Impacts of Climate Change: A Review of the Evidence. Policy Research Working Paper 5622, Poverty Reduction and Equity Unit, Poverty Reduction and Economic Management Network*; The World Bank: Washington, DC, USA, 2010; 35p.

89. Hertel, T.W.; Burke, M.B.; Lobell, D.B. The poverty implications of climate induced crop yield changes by 2030. *Glob. Environ. Chang.* **2010**, *20*, 577–585. [CrossRef]

90. Rosegrant, M.W. Impacts of climate change on food security and livelihoods. In *Food Security and Climate Change in Dry Areas: Proceedings of an International Conference, Amman, Jordan, 1–4 February 2010*; Solh, M.,

Saxena, M.C., Eds.; International Center for Agricultural Research in the Dry Areas (ICARDA): Aleppo, Syria, 2011; pp. 24–26.

91. IPCC. *Climate Change 2013: The Physical Science Basis. Contribution of Working Group I to the Fifth Assessment Report of the Intergovernmental Panel on Climate Change*; Stocker, T.F., Qin, D., Plattner, G.-K., Tignor, M., Allen, S.K., Boschung, J., Nauels, A., Xia, Y., Bex, V., Midgley, P.M., Eds.; Cambridge University Press: Cambridge, UK; New York, NY, USA, 2013; 1535p. [CrossRef]

92. Basak, J.K.; Ali, M.A.; Islam, M.N.; Rashid, M.A. Assessment of the effect of climate change on boro rice production in Bangladesh using DSSAT model. *J. Civ. Eng.* **2010**, *38*, 95–108.

93. IPCC. *Climate Change 2007: Projections of Future Changes in Climate: Contribution of Working Group I to the Fourth Assessment Report of the Intergovernmental Panel on Climate Change*; Cambridge University Press: Cambridge, UK, 2007.

94. IPCC. *Climate Change 2014: Impacts, Adaptation, and Vulnerability. Contribution of Working Group II to the Fifth Assessment Report of the Intergovernmental Panel on Climate Change*; Field, C.B., Barros, V.R., Dokken, D.J., Mach, K.J., Mastrandrea, M.D., Bilir, T.E., Chatterjee, M., Ebi, K.L., Estrada, Y.O., Genova, R.C., Eds.; Cambridge University Press: Cambridge, UK; New York, NY, USA, 2014.

95. IPCC. *Climate Change 2014: Mitigation of Climate Change. Contribution of Working Group III to the Fifth Assessment Report of the Intergovernmental Panel on Climate Change*; Edenhofer, O., Pichs-Madruga, R., Sokona, Y., Farahani, E., Kadner, S., Seyboth, K., Adler, A., Baum, I., Brunner, S., Eickemeier, P., Eds.; Cambridge University Press: Cambridge, UK; New York, NY, USA, 2014.

96. Liu, S.; Mo, X.; Lin, Z.; Xu, Y.; Ji, J.; Wen, G.; Richey, J. Crop yield responses to climate change in the Huang-Huai-Hai Plain of China. *Agric. Water Manag.* **2010**, *97*, 1195–1209. [CrossRef]

97. Reid, S.; Smit, B.; Caldwell, W.; Belliveau, S. Vulnerability and adaptation to climate risks in Ontario agriculture. *Mitig. Adapt. Strateg. Glob. Chang.* **2007**, *12*, 609–637. [CrossRef]

98. Rogelj, J.; Perette, M.; Menon, A.; Schleussner, C.F.; Bondeau, A.; Svirejeva-Hopkins, A.; Schewe, J.; Frieler, K.; Warszawski, L.; Rocha, M. *Turn Down the Heat: Climate Extremes, Regional Impacts, and the Case for Resilience*; World Bank: Washington, DC, USA, 2013.

99. IRRI. Climate Change-Ready Rice. Available online: http://irri.org/our-impact/tackling-climate-change/developing-drought-tolerant-rice (accessed on 11 January 2018).

100. Khatun, M. Climate Change and Migration in Bangladesh: Golden Bengal to Land of Disasters. *Bangladesh e-J. Soc.* **2013**, *10*, 64–79.

101. Rahman, A.; Alam, M.; Alam, S.S.; Uzzaman, M.R.; Rashid, M.; Rabbani, G. *Risks, Vulnerability, and Adaptation in Bangladesh*; Human Development Report 2007/08, Human Development Report Office Occasional Paper, 2007/13; UNDP: New York, NY, USA, 2008.

102. Paul, B.K. Coping mechanism practiced by drought victims (1994/5) in North Bengal, Bangladesh. *Appl. Geogr.* **1998**, *18*, 355–373. [CrossRef]

103. Shahid, S.; Behrawan, H. Drought risk assessment in the western part of Bangladesh. *Nat. Hazards* **2008**, *46*, 391–413. [CrossRef]

104. Habiba, U.; Shaw, R.; Takeuchi, Y. Chapter 2 socioeconomic impact of droughts in Bangladesh, Droughts in Asian Monsoon Region Community. *Environ. Disaster Risk Manag.* **2011**, *8*, 25–48.

105. Rahman, A.; Roytman, L.; Krakauer, N.Y.; Nizamuddin, M.; Goldberg, M. Use of Vegetation Health Data for Estimation of *Aus* Rice Yield in Bangladesh. *Sensors* **2009**, *9*, 2968–2975. [CrossRef] [PubMed]

106. Alam, M.S.; Quayum, M.A.; Islam, M.A. Crop Production in the Haor Areas of Bangladesh: Insights from Farm Level Survey. *Agriculturists* **2010**, *8*, 88–97. [CrossRef]

107. Huda, M.K. Experience with modern and hybrid rice varieties in haor ecosystem: Emerging Technologies for Sustainable Rice Production. In *Twentieth National Workshop on Rice Research and Extension in Bangladesh*; Bangladesh Rice Research Institute: Gazipur, Bangladesh, 2004; pp. 19–21.

108. Needs Assessment Report. *Flooding in the North-East Bangladesh*; Rapid Repose Team, Shifting the Power, Christian Aid and NAHAB: Dhaka, Bangladesh, 2017; pp. 1–8.

109. Litchfield William Alex. Climate Change Induced Extreme Weather Events & Sea Level Rise in Bangladesh Leading to Migration and Conflict, ICE Case Studies, Number 229; 2010. Available online: http://mandalaprojects.com/ice/ice-cases/bangladesh.htm (accessed on 17 January 2018).

110. BBS. *Bangladesh Population and Housing Census*; Bangladesh Bureau of Statistics: Dhaka, Bangladesh, 2011.

111. Haque, S.A. Salinity Problems and Crop Production in Coastal Regions of Bangladesh. *Pakistan J. Bot.* **2006**,

38, 1359–1365.

112. Akter, T. Climate Change and Flow of Environmental Displacement in Bangladesh. Unnayan Onneshan-The Innovators. 2009. Available online: www.unnayan.org (accessed on 30 April 2018).

113. Nicholls, R.J.; Leatherman, S.P. The implications of accelerated sea-level rise and developing countries: A discussion. *J. Coast. Res.* **1995**, 303–323.

114. Hossain, M.A. Global Warming Induced Sea Level Rise on Soil, Land and Crop Production Loss in Bangladesh. In Proceedings of the 19th World Congress of Soil Science, Soil Solutions for a Changing World, Brisbane, Australia, 1–6 August 2010.

115. Minar, M.H.; Hossain, M.B.; Shamsuddin, M.D. Climate Change and Coastal Zone of Bangladesh: Vulnerability, Resilience and Adaptability. *Middle-East J. Sci. Res.* **2013**, *13*, 114–120. [CrossRef]

116. Xu, H.; Twine, T.E.; Girvetz, E. Climate change and maize yield in Iowa. *PLoS ONE* **2016**, *11*, e0156083. [CrossRef] [PubMed]

117. Kabir, H.M.; Ahmed, Z.; Khan, R. Impact of climate change on food security in Bangladesh. *J. Pet. Environ. Biotechnol.* **2016**, *7*, 306. [CrossRef]

14

Conservation Farming and Changing Climate: More Beneficial than Conventional Methods for Degraded Ugandan Soils

Drake N. Mubiru [1],*, Jalia Namakula [1], James Lwasa [1], Godfrey A. Otim [1], Joselyn Kashagama [1], Milly Nakafeero [1], William Nanyeenya [1] and Mark S. Coyne [2]

[1] National Agricultural Research Organization (NARO), P.O. Box 7065, Kampala, Uganda; jalianamakula@kari.go.ug (J.N.); jlwasa@kari.go.ug (J.L.); ogodfrey@kari.go.ug (G.A.O.); jkashagama@kari.go.ug (J.K.); mnakafeero@kari.go.ug (M.N.); nwilliam@kari.go.ug (W.N.)

[2] Department of Plant and Soil Sciences, University of Kentucky, 1100 S. Limestone St., Lexington, KY 40546-0091, USA; mark.coyne@uky.edu

* Correspondence: dnmubiru@kari.go.ug

Abstract: The extent of land affected by degradation in Uganda ranges from 20% in relatively flat and vegetation-covered areas to 90% in the eastern and southwestern highlands. Land degradation has adversely affected smallholder agro-ecosystems including direct damage and loss of critical ecosystem services such as agricultural land/soil and biodiversity. This study evaluated the extent of bare grounds in Nakasongola, one of the districts in the Cattle Corridor of Uganda and the yield responses of maize (*Zea mays*) and common bean (*Phaseolus vulgaris* L.) to different tillage methods in the district. Bare ground was determined by a supervised multi-band satellite image classification using the Maximum Likelihood Classifier (MLC). Field trials on maize and bean grain yield responses to tillage practices used a randomized complete block design with three replications, evaluating conventional farmer practice (CFP); permanent planting basins (PPB); and rip lines, with or without fertilizer in maize and bean rotations. Bare ground coverage in the Nakasongola District was 187 km^2 (11%) of the 1741 km^2 of arable land due to extreme cases of soil compaction. All practices, whether conventional or the newly introduced conservation farming practices in combination with fertilizer increased bean and maize grain yields, albeit with minimal statistical significance in some cases. The newly introduced conservation farming tillage practices increased the bean grain yield relative to conventional practices by 41% in PPBs and 43% in rip lines. In maize, the newly introduced conservation farming tillage practices increased the grain yield by 78% on average, relative to conventional practices. Apparently, conservation farming tillage methods proved beneficial relative to conventional methods on degraded soils, with the short-term benefit of increasing land productivity leading to better harvests and food security.

Keywords: land degradation; land management; conservation farming

1. Introduction

Land degradation arising from inefficient and unsustainable land management is reducing crop productivity across sub-Saharan Africa (SSA). Land degradation reportedly affects 67% of SSA, and in some countries, more than 30% of the land area is severely or very severely degraded [1]. This is the case despite most households overwhelmingly relying on land resources [1]. The impacts of land degradation, which are becoming increasingly severe and are accelerating, include low crop productivity leading to food insecurity and disruption of ecosystem functions, which reduces ecosystem performance, resilience, and stability. The combined effects of the land degradation impacts are poor human livelihoods and wellbeing.

The extent of land affected by degradation in Uganda ranges from 20% in relatively flat and vegetation-covered areas, to 90% in the eastern and southwestern highlands [2,3]. Earlier observations indicated that land/soil degradation and soil fertility are major impediments in all cropping systems in Uganda, especially where there has been agricultural intensification [4]. However, as elsewhere in SSA, much of the population depends on land for their livelihoods [5–7] and therefore suffers the repercussions of land degradation. Additionally, climate change is a major influence on the food security and livelihoods of households in Uganda, which mostly depend on rain-fed agriculture, and are increasingly at risk from perpetually low yields of major staples such as maize (*Zea mays*) [8–11] and common beans (*Phaseolus vulgaris* L.) [12]. Many households must deal with degraded, nutrient-starved soils, and the inability to access or purchase inputs such as improved seeds and fertilizer [13].

To its comparative advantage over the rest of SSA, Uganda has a diverse agricultural production system within 10 agricultural production zones (APZs) [14]. The zones are characterized by different farming systems determined by soil types, climate, topography, and socio-economic and cultural factors. Due to the different zonal characteristics, the APZs experience varying levels of land degradation and vulnerability to climate-related hazards, which include drought, floods, storms, pests, and disease [5].

Soil/land degradation stemming from deforestation, burning of grasslands and organic residues, and continuous cultivation with minimum soil fertility enhancement leads to soil erosion and organic matter and nutrient depletion [13,15,16]. Other unsustainable land-use practices, such as overgrazing, have produced compacted soil layers and often bare grounds in extreme cases [13]. Another underlying factor in the development of compacted soil layers is that hand-hoeing, which only disturbs the first 15 to 20 cm—or sometimes as little as 5 cm—of the topsoil, when done consistently and regularly, can potentially produce restrictive layers below 0–20 cm of the topsoil. Under these soil conditions, nutrient- and water-use efficiency is reportedly very low [17,18]. These soil layers act as barriers to root and water movement and soil water-holding capacity (WHC), making land susceptible to the frequency and intensity of rainfall. Soil compaction in these layers affects agricultural land in several ways, including inhibiting root and water movement, limiting water infiltration and retention (hence facilitating runoff), and making plowing difficult. As a consequence, this directly affects agricultural productivity and contributes to the yield gap between potential output vis-à-vis farmer outputs. In that regard, land degradation and a total dependence on rain-fed agriculture has increased the vulnerability of farming systems and predisposed rural households to food insecurity and poverty [13]. Furthermore, it has led to significant adverse impacts on smallholder agro-ecosystems, including direct damage and loss of critical ecosystem services such as agricultural land/soil and biodiversity.

Due to climate change, the frequency and severity of climate-related hazards have increased, severely affecting agricultural production and in many cases leading to instability in agricultural production systems [19,20]. For example, poor rains severely affect pastures and livestock in most pastoral areas of the country, resulting in the migration of thousands of people and animals in search of water and food [5]. Jennings and Magrath [21] noted that excessive rains, both in intensity and duration, lead to water logging and negatively affect crops and pastures.

Past climatic scenarios make the outlook for the future unsettling; empirical evidence shows that seven droughts were experienced between 1991 and 2000. This caused severe water shortages, which seriously affected the animal industry [5]. Other impacts included low crop yields and increased food prices, culminating in food insecurity and negative effects on the economy. An increase in the intensity and frequency of heavy rains, floods, and landslides in the highland areas in the eastern, western, and southwestern parts of the country has been documented [22]. The recent severe drought in 2016 affected thousands of people, mainly in the Karamoja and Teso sub-regions and Isingiro District of southwestern Uganda. This was followed by the outbreak of the fall armyworm (*Spodoptera frugiperda*), affecting thousands of hectares of maize planted in the early 2017 season. The effects of climate change and variability in Uganda are compounded by existing developmental challenges of high population growth rates, high and increasing poverty levels, and declining GDP growth rates. Thus, climate change can undermine and even undo significant gains in social and economic developments in the country.

Unsustainable land-use practices lead to land degradation, and reduce the ecological and social resilience of landscapes. The overall impact of degradation has been the disruption of ecosystem services, particularly provisioning services, due to habitat fragmentation that reduces complexity and diversity, and soil erosion with consequent declining fertility and productivity. The situation is further aggravated by high population growth rates, which have led to extensive land fragmentation—a problem for sustainable land management [23]. Average landholdings in Uganda range from 0.4 to 3 ha for each typical household of seven persons [24]. High population areas are also often associated with poverty, thus requiring improved management systems to increase food security. Without a doubt, Uganda needs meaningful mitigation measures for the protection, recovery, and rehabilitation of the ecosystem services. The viability, functionality, and quality of ecosystem services are essential in enhancing and supporting community health and wellbeing, prosperity, and sustainability [25].

Ecosystem-based land management practices, such as conservation farming, bestow adaptive benefits that reduce the negative impacts of extreme weather events by buffering temperature extremes, harvesting and conserving rainwater, reducing soil loss within the agro-ecosystem, improving soil physicochemical and biological conditions, and regulating pest and disease cycles. Conservation farming practices can potentially address the soil and water management constraints faced by smallholder farmers [26]. The conservation farming package entails dry-season land preparation using minimum tillage systems, crop residue retention, seeding and input applications in permanent planting stations, and nitrogen-fixing crop rotations [27].

Permanent planting basins (PPBs) and rip lines are two major components of the recently introduced conservation farming package for renovating degraded landscapes that are being extensively promoted for smallholder farming [26,28–31]. PPBs and rip lines, as used in conservation farming, are crop management methods that enhance the capture and storage of rainwater, and allow sustainable precision management of limited nutrient resources. Both methods reduce the risk of crop failure due to erratic rainfall and extended droughts. The use of these methods in combination with improved seed and crop residues to create a mulch cover that reduces evaporation losses has consistently increased average yields by 50–200%, depending on the amount of rainfall, soil type, and fertility [32]. PPBs are being targeted for households with limited or no access to oxen, while ripping is meant for smallholder farmers with oxen [26].

Maize and beans are major staple foods for much of the population, and are a major source of food security in Uganda. The annual per capita maize consumption is estimated to be 28 kg, and bean consumption, 58 kg [33]. Both crops are cash crops for some smallholder farmers. Maize is also an important animal feed. At the household level, the importance of maize and beans is centered on their dietary roles of supplying proteins, carbohydrates, minerals, and vitamins to resource-constrained rural and urban households with rampant shortages of these dietary elements. Reportedly, the dietary intake for the most resource-constrained households in Uganda comprises 70% carbohydrates, and these are mainly from maize, supplying 451 kcal/person/day and 11 g protein/person/day. Beans provide about 25% of the total calories and 45% of the protein intake in the diets of many Ugandans [34].

Unfortunately, due to the biophysical and socio-economic factors previously noted, the average maize and bean grain yields on smallholder farms, which on average are less than 1 ha, are less than 30% of their potentials [8–12]. The potential maize yield in Uganda is estimated to range from 3.8 to 8.0 t ha^{-1} [9], while that of beans is 2.0 t ha^{-1} [12]. Poor soil conditions (low soil fertility, compacted soils, and moisture stress) coupled with a low nutrient- and water-use efficiency are major contributing factors to this yield gap.

We postulate that employing ecosystem-based land management practices such as conservation farming will increase water- and nutrient-use efficiency, provide greater rooting depth, and improve WHC that would increase land productivity, leading to better grain harvests and food security. The long-term benefits would be an increased soil organic matter content, increased return on fertilizer use, and greater resilience of dry-land smallholder plots to erratic rainfall patterns from climate change.

This study: (i) assessed the extent of bare grounds in Nakasongola, one of the districts in the Cattle Corridor of Uganda; and (ii) evaluated yield responses of maize and beans to different tillage methods in the district.

2. Materials and Methods

2.1. Site Description

The Nakasongola District is in central Uganda, between 00°57'44.89" and 10°40'42.76" North latitude and between 310°58'03.77" and 320°48'00.29" East longitude. The district is in the Pastoral Rangelands agro-ecological zone (AEZ), which is one of the AEZs that comprise the Cattle Corridor of Uganda (Figure 1).

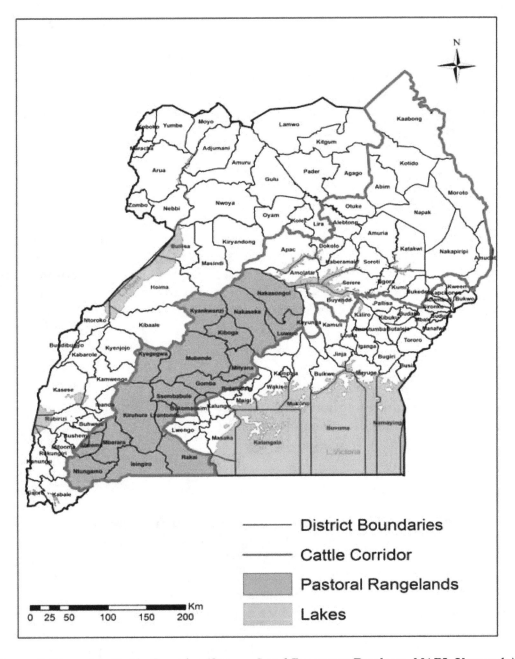

Figure 1. Uganda's Cattle Corridor (Source: Land Resources Database, NARL-Kawanda).

Constituting the country's rangelands, the Cattle Corridor has a total area coverage of 84,000 km^2, which is approximately one-third of the total 241,000 km^2 of the land area in Uganda, and is home to a population of 6.6 million people. The corridor is host to a mixed production system comprising nomadic pastoralists, agro-pastoralists, and subsistence farmers. On average, it receives 500–1000 mm of rainfall annually, which is spatially variable, from about 400 mm in some parts of the northeastern corridor, to about 1200 mm in parts of the southwestern and central corridor. The rainfall pattern is bimodal in the southwestern and central parts of the corridor, and transitions into one rainy season of about $5\frac{1}{2}$ months in the northern and northeastern areas [14]. Dry spells are frequent, and droughts of significant magnitudes occur, causing hardship to peoples' livelihoods and economy in the districts that comprise the corridor.

Specifically, the Pastoral Rangelands AEZ receives moderate rainfall, spatially varying from 915 to 1021 mm/year with a bimodal pattern [14]. The main rainy season is from March to May with a peak in April, and the secondary season is from September to December with a modest peak in November. Dry periods are from June to August and January to February. The daily average temperature ranges from 12.5 to 30 °C. Evaporation exceeds rainfall by a factor of about 6 during the dry months from June to August, while during the main rainy months (April and May), rainfall equals evaporation. Altitude in the zone spatially ranges from 129 to 1524 m ASL (above sea level), with the land characterized by rolling hills with some flat areas and moderate-to-poor soils. The farming system and socio-economic characteristics are characterized by smallholders with many communal grazing and agro-pastoral practices; low literacy levels; absentee landlords with a squatter population; and infrastructure and marketing systems that are poor to moderate [13,14].

2.2. Assessing the State of the Soil in the Nakasongola District

2.2.1. Quantification of Bare Ground Coverage

Based on the assumption that bare grounds are one of the visible indicators of extreme land degradation, the approach was to physically survey and capture, using GPS, the spatial extent of some bare grounds, and use the data to locate the same features on a satellite image captured during a fairly dry month. These points were used to develop digital signatures for searching similar features in the rest of the image, and generating coverage statistics using Geographic Information System (GIS) tools.

2.2.2. Data Sources/Analysis

A supervised multi-band satellite image classification using the Maximum Likelihood Classifier (MLC) was used [35]. A high resolution (<5 M) image from 2013, covering a greater part of the district, was used for the analysis. Bands 1, 4, and 7 of the Landsat Thematic Mapper image (p171r059_7t20011127_z36_nn10) were used.

2.2.3. Soil Physicochemical Analysis

Soil samples from depths of 0 to 20 cm were collected from geo-referenced sites in eight sub-counties comprising the Nakasongola District. The samples were dried in open air, ground to pass a 2 mm sieve, and analyzed according to Okalebo et al. [36] and Foster [37]. Texture analysis was performed by the hydrometer method [38]. Soil pH was measured with a soil/water ratio of 1:2.5. Extractable P, K, and Ca were measured in a single ammonium lactate/acetic acid extract buffered at pH 3.8 [36]. Total nitrogen (N) was determined by a micro-Kjeldahl block digestion apparatus, and soil organic matter was determined by acid-dichromate digestion. Soil samples were also collected using a double-cylinder, hammer-driven core sampler to determine the bulk density according to methods by Blake and Hartge [39].

2.2.4. Statistical Analysis

The soils' physicochemical data was subjected to Pearson's correlation to establish relationships among the parameters, using Statistix V. 2.0. Furthermore, Principal Component Analysis (PCA) was used to determine similarities among soils from different farms and sub-counties as manifested in the status of their physicochemical properties.

2.3. Sustainable Agricultural Production

Sustainable agriculture has been defined as a means of production that seeks to sustain farmers, resources, and communities by promoting farming practices and methods that are profitable, environmentally sound, and good for communities [40–44]. Sustainability rests on the principle that the present generation must meet its own needs without compromising the ability of future generations to meet their own needs [44]. In this study, we assessed how conservation farming practices could contribute to sustainable agriculture production.

Trials were conducted on 16 randomly-selected farmer fields in the first and second seasons of 2015 in two sub-counties in the Nakasongola District. The first season (A) ran from March to May with a rainfall peak in April, while the second season (B) ran from October to December with a peak in November.

2.3.1. Field Design

The experiment design was a randomized complete block with three replications. The different tillage methods under assessment were: Conventional Farmer Practices (CFPs), PPBs, and rip lines, all with or without fertilizer. CFP entailed the preparation of a seedbed followed by at least two hand-hoe weeding passes, with crop residues incorporated into the soil.

Prior to the trial's establishment, in conservation farming treatments, the fields were slashed and sprayed with glyphosate (500 mg L^{-1}) at a rate of 7.5 L ha^{-1}, two weeks after slashing. In the preceding cropping season, most fields had been used to grow maize, beans, or sweet potatoes (*Ipomoea batatas*). Due diligence was made to ensure that there was no continuous cropping of a particular crop in the same plot. The traditional crop rotations in this area are: sweet potato, bean or groundnut (*Arachis hypogea*), maize, then cassava (*Manihot esculenta*). Sweet potato is important as a first crop in the rotation because it helps to loosen, as well as increase, the soil volume, while cassava, which is tolerant to low soil nutrient levels, comes last in the rotation (Sarah Nakamya per. Comm., [45]). Due to multiple uses of crop residues, little material was laid down on the plots. In the conservation farming treatments, weeds were controlled by light weeding with a hand-hoe or by hand. A high-yielding and drought-tolerant hybrid maize variety (PH5052) and bean variety (NABE 15) were used in all treatments. The average plot size was 513 m^2.

2.3.2. Seeding Rates

Conventional Farmer Practice: Planting holes for maize were designated by planting lines and digging with a hand-hoe at a spacing of 75 cm between rows and 60 cm within rows. Each hole was seeded with two seeds, giving a total of 44,444 plants/ha. In the case of beans, spacing was 50 cm × 10 cm and each hole was seeded with one seed to give a total of 200,000 plants/ha.

Permanent Planting Basins: Basins were designated using planting lines and digging planting basins 35 cm (long) × 15 cm (wide) × 15 cm (deep), with a spacing of 75 cm between rows and 70 cm within rows from center-to-center of the PPB, before the onset of rains. Available crop residues were laid between rows to create a mulch cover. The basins were seeded with three maize seeds per basin (57,143 plants/ha) and six bean seeds per basin (114,286 plants/ha).

Rip lines: Rip lines were ripped before the onset of rains by an ox ripper set at a depth of 15 cm. Available crop residues were laid between rows to create a mulch cover. Maize was seeded at a spacing of 75 cm × 30 cm with one seed per hill (44,444 plants/ha). Beans were seeded at a spacing of 75 cm × 10 cm with two seeds per hill (266,667 plants/ha).

In the maize and bean trials, micro-doses of basal fertilizer (DAP) at a rate of 92.5 kg ha^{-1} were applied and covered with topsoil before planting the seeds. For maize, nitrogen as urea at a rate of 150 kg ha^{-1} was evenly side-dressed when the maize was at knee height, approximately at vegetative stage 9 (V9).

2.4. Statistical Analysis

Data was examined by ANOVA to determine significant ($p \leq 0.05$) treatment effects. Comparisons of means were made by LSD all-pair-wise comparisons. All analyses were done using Statistix V. 2.0.

3. Results and Discussion

3.1. Assessment of the State of the Soil in the Nakasongola District

Quantification of Bare Ground Coverage

Bare ground coverage in the Nakasongola District due to extreme cases of soil compaction was 187 km^2 (11%) of the 1741 km^2 of arable land (Figure 2 and Table 1). At present, Uganda has 7.2 million hectares of arable land under crop agriculture, which is less than 50% of the arable land estimated at 16.8 million hectares [6]. Pessimistic forecasts indicate that the available arable land for agriculture will run out in most parts of the country by around 2022. With such grim statistics, the country cannot afford to lose any arable land. It is therefore imperative that Uganda embraces sustainable land management to reverse this trend of land degradation.

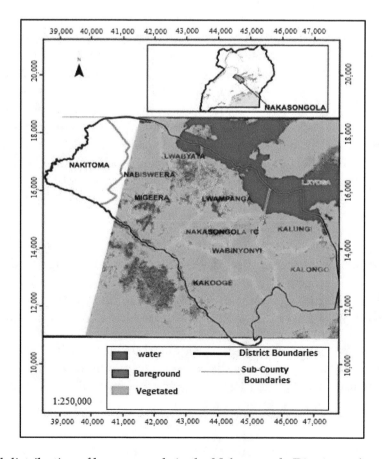

Figure 2. Spatial distribution of bare-grounds in the Nakasongola District and surrounding areas.

Table 1. Spatial distribution of different land cover classes in the Nakasongola District.

	Class	Area (km^2)	% Cover
1	Open water	233	7.9
2	Vegetated	1527	51.7
3	Bare ground	187	6.3
4	Seasonal wetland	915	31.0
5	Cloud cover	48	1.6
6	Permanent wetland	46	1.6
Total		2956	100

Pearson's correlation (Table 2) of soil physicochemical data from all sub-counties revealed that the bulk density, which was used as an indicator of soil compaction, was significantly correlated only with clay ($r = -0.54$, $p < 0.0003$) and sand ($r = 0.48$, $p < 0.002$). Therefore, clay and sand were the most important determinant parameters for bulk density or soil compaction. Observations from our study are well corroborated by several workers [46–50], who observed, from different areas and soil types, that the higher the amount of sand in the soil, the greater the bulk density, while the higher the amount of clay, the lower the bulk density.

Table 2. Pearson's correlation of soil physicochemical data from all sub-counties in the Nakasongola District.

	pH	OM $^¥$	N	P	K	Ca	Mg	Sand	Silt	Clay
OM	0.27 *									
N	0.28 *	0.97 ***								
P	0.57 ***	0.19	0.20							
K	0.42 ***	−0.05	-0.03	0.30 *						
Ca	0.82 ***	0.33 **	0.31 **	0.40 ***	0.26					
Mg	0.79 ***	0.31 **	0.30 *	0.36 **	0.38 **	0.97 ***				
Sand	−0.14	−0.48 ***	−0.49 ***	−0.07	−0.13	−0.26	−0.29 *			
Silt	0.50 ***	0.53 ***	0.52 ***	0.14	0.03	0.60 ***	0.55 ***	−0.45 ***		
Clay	−0.07	0.30 *	0.32 **	0.01	0.13	0.03	0.08	−0.92 ***	0.05	
BD †	0.23	−0.18	−0.16	0.19	0.03	0.16	0.13	0.48 ***	0.01	−0.54 ***

BD = Bulk Density;   OM = Organic Matter; * significant at 0.1 level; ** significant at the 0.05 level; *** significant at the 0.01 level.

PCA was used to determine if there were similarity clusters of soils from different farms and sub-counties with respect to soil properties. All soils from the different farms and sub-counties formed one cluster, indicating that there were no exceptional differences in the soil properties among the sub-counties. Means of all soil properties (Table 3) in the topsoils and subsoils were below normal for the soils of Uganda [36]. For example, the critical value of soil pH in Ugandan soils is 5.6, while that of organic carbon is 3.0% [51,52]; this is an indication that all soils in this study were, to some extent, chemically and physically degraded. On analyzing the properties of the topsoil and subsoil, the average pH of the subsoil was slightly higher than that of the topsoil, which was unusual. However, the concentration of calcium in the subsoil was also higher than in the topsoil, which might explain this phenomenon.

Although there was no distinct differentiation for soils from the different farms and sub-counties, separately, soils from the Kalungi sub-county had the highest average bulk density (Table 4), which was significantly different ($p \leq 0.05$) from the other sub-counties, except Kalongo and Lwampanga. Soils from the Wabinyonyi and Kakoge sub-counties had the lowest ($p \leq 0.05$) average bulk densities compared to the other sub-counties. Correspondingly, the Wabinyonyi and Kakoge sub-counties also had a higher ($p \leq 0.05$) percentage of clay and a significantly lower percentage of sand than all the other sub-counties, with a few exceptions.

Table 3. Cluster means of soil properties for soil samples from sub-counties of the Nakasongola District.

Soil Layer	BD [†]	pH	OM [*]	N	P	K	Ca	Mg	Sand	Clay	Silt
	(g/cc)		(%)			(ppm)				(%)	
Topsoil [¥]	1.4	4.4	2.2	0.2	6.3	98.8	459	283	51	41	8
Subsoil [¥]	-	4.6	2.1	0.1	3.1	45.4	571	217	50	42	8
Critical levels		5.6	3.0	0.2	35.5	72.5	1640	87			

[†] BD = Bulk Density; [*] OM = Organic Matter; [¥] Topsoil = Top layer of soil collected at 0–20 cm depths; Subsoil = Soil samples collected at 20–40 cm depths.

Table 4. Soil properties well-correlated with bulk density from the different sub-counties.

Sub-County	Soil Property [1]		
	Bulk Density (g/cc)	Clay (%)	Sand (%)
Kalungi	1.58a	42bc	51ab
Kalongo	1.57ab	38c	57a
Lwampanga	1.56ab	40c	50ab
Rwabyata	1.49bc	38c	53a
Nakitoma	1.47c	37c	56a
Nabisweera	1.44c	37c	54a
Wabinyonyi	1.34d	47ab	44b
Kakooge	1.33d	50a	44b
SE	0.04	3	4

[1] Different letters within each column indicate significant differences between treatments at the $p \leq 0.05$ level, using the LSD method.

3.2. Sustainable Agricultural Production

3.2.1. Bean Grain Yield Response to Tillage Practices and Fertilizer

There were no significant seasonal differences in the bean grain yield (Table 5). There were also no significant season × tillage interactions, indicating that treatment effects on the grain yield were independent of seasonal characteristics. Since the season × tillage interactions were not significant, the yield means were averaged across the seasons (Table 5). All tillage practices, whether conventional or the newly introduced conservation farming practices in combination with a fertilizer increased bean grain yield. However, the increases were only significantly different between rip lines with and without fertilizer. On average, fertilizer use in combination with the tillage practices increased the bean grain yield from 436 kg ha^{-1} to 743 kg ha^{-1}, a 70% increase. Separately, the highest average percentage yield increase (102%) was between rip lines with and without fertilizer; this was followed by conventional practices without and with fertilizer (56%), and lastly between PPBs without and with fertilizer (53%). The average bean grain yield from conventional practices was 460 kg ha^{-1}; from PPBs, 648 kg ha^{-1}; and from rip lines, 661 kg ha^{-1}. Apparently, the newly introduced conservation farming tillage practices increased the bean grain yield relative to conventional practices by 41% in PPBs, and 43% in rip lines.

Table 5. Average bean and maize grain yields as a response to different tillage practices [†].

Tillage Practice	Bean Yield		Maize Yield	
	(kg ha^{-1})	SE	(kg ha^{-1})	SE
Conventional	359c	±138	1536b	±879
Conventional + fertilizer	560abc	±138	2481ab	±879
PPB	512abc	±138	3328ab	±918
PPB + fertilizer	784ab	±138	4963a	±918
Rip line	438bc	±148	2086b	±963
Rip line + fertilizer	884a	±148	3921ab	±963

[†] Yield means for a particular crop followed by the same letter are not significantly different according to LSD at $p = 0.05$.

3.2.2. Potential versus Actual Bean Grain Yield

The potential bean grain yield in Uganda is about 2.0 t ha^{-1} [12]. In our study, the response of bean grain yields to fertilizer and the newly introduced conservation farming tillage practices was below the yield potential, notwithstanding the remarkable increase. Other workers [53–56] have observed that yields from on-farm trials were enhanced by using improved seeds and fertilizers, but yields still remained below the genetic potential. This has been attributed to management factors that contributed to poor early-season vigor, in-season plant loss, and environmental stresses.

The tillage effects increased the bean grain yield in the newly introduced conservation farming practices relative to conventional practices. However, the yield differences between rip lines and PPBs could partly be attributed to differences in plant population; that is, 266,667 plants/ha in rip line tillage vis-à-vis 114,286 plants ha^{-1} in PPBs. In an earlier study (not published) conducted to determine the optimum seeding rates in PBBs, it was established that six bean seeds per basin, as were used in the current study, was optimal. It is plausible that increasing the seeding rate in PPBs creates competition among the plants, thus affecting productivity.

Ghaffarzadeh et al. [57] observed that the potential for stress could be increased when crops compete among themselves. Ghaffarzadeh et al. [58] further intimated that competition for resources might develop because of root growth patterns and/or different resource demands, although they acknowledged that there is limited information available about light, water, and nutrient competition in regard to plant position. Some studies suggest that spatial and temporal arrangement of crops may influence competition for water and light [59,60]. Under water-limiting conditions, production advantages could diminish [61–63].

3.2.3. Maize Grain Yield Response to Tillage Practices and Fertilizer

Unlike for beans, there were significant seasonal differences in the maize grain yield (Table 5). In the first season (2015A), the maize grain yield was 2113 kg ha^{-1} (106%) greater than in the second season (2015B). It is plausible that the yield difference was a result of water stress experienced in the 2015B season. This effect was more pronounced in maize than in beans because beans are short-term compared to maize, and it is likely that by the time drought manifested, the bean crop was already in advanced stages of development.

Although there were significant seasonal yield differences, the season × tillage interactions were not significant. As was the case with beans, this indicated that the tillage effects on the maize grain yield were independent of the seasonal characteristics. Correspondingly, the yield means were averaged across seasons (Table 5). As would be expected, there were yield responses to fertilizer applications in all tillage practices, however, the differences between particular tillage practices without and with fertilizer were minimally significant. Suffice to note also that the newly introduced conservation farming practices, on average, increased the grain yield more than the conventional practice, by 78%. In their study spanning three seasons, Mazvimavi et al. [64] realized that maize in conservation farming tillage practices out-yielded that in conventional tillage practices by 59%.

When each season was critically examined, this demonstrated the performance differences between the two conservation farming tillage practices. In season 2015A, which was deemed to have normal rainfall, rip line tillage had a higher maize grain yield compared to the PPBs. Conversely, in 2015B, which is believed to have had below-average rainfall, the PPBs had a higher maize grain yield compared to rip lines. Although it cannot be conclusively concluded from our study results, it is plausible that in years with below-average rainfall, the PPBs are better at harvesting and conserving rainwater than rip lines, and are thus the superior performer. In their study on conservation tillage for soil water management, Mupangwa et al. [26] concluded that planting basin tillage is better at controlling water losses than ripper, double, and single conventional ploughing techniques.

3.2.4. Potential versus Actual Maize Grain Yield

The potential maize yield in Uganda is estimated to range from 3.8 to 8.0 t ha^{-1} [8–11], with the open pollinated varieties (OPV) being on the lower end compared to hybrid varieties. However, according to the Food and Agriculture Organization Statistical Database (FAOSTAT), the actual maize productivity is stagnant, at a low level of between 1.5 and 2.5 t ha^{-1} [11]. The yield gap is attributed to the limited use of inputs such as improved seed and fertilizer, now coupled with soil moisture stress due to climate variability. In the current study, the newly introduced conservation farming practices apparently brought the maize grain yield within the potential yield range, although there was still room for improvement.

4. Conclusions

This study showed that 11% of the arable land in the Nakasongola District is bare ground, an extreme case of soil compaction and land degradation. Because this is not an isolated case, it is imperative that the country embraces sustainable land management and agricultural production to meet the food needs of its people and to spur economic development, while at the same time conserving the environment.

The newly introduced conservation farming tillage practices increased the bean grain yield relative to conventional practices by 41% in PPBs and 43% in rip lines. For maize, the newly introduced conservation farming tillage practices on average increased the grain yield by 78%, relative to the conventional practices. Conservation farming tillage methods, that is, PPBs and rip lines, proved to be more beneficial than conventional methods for degraded soils, with a short-term benefit of increasing land productivity, leading to better harvests and food security. The long-term benefits are expected to be an increased soil organic matter content, an increased return on fertilizer use, and a greater resilience of dryland smallholder plots to erratic rainfall patterns, occasioned by climate change. Conservation farming practices, as empirically tested in this study, facilitated the rehabilitation and recovery of degraded farmer fields, as evidenced by increased grain yields, thus fitting well within the league of sustainable agricultural production practices.

Long-term studies are needed to establish the effects of variable rainfall on the performance of planting basins vis-à-vis rip lines. Furthermore, considering the variable costs of inputs and the variability of outputs among the different tillage practices, there is a need to conduct a cost-benefit analysis to determine the cost effectiveness of each tillage practice.

Acknowledgments: The authors are grateful to the participating farmers and local government of the Nakasongola District. The research was supported by the National Agricultural Research Organization, the National Agricultural Research Laboratories–Kawanda, the International Maize and Wheat Improvement Centre (CIMMYT), and the Australian Centre for International Agricultural Research (ACIAR) through the Sustainable Intensification of Maize-Legume cropping systems for Food Security in Eastern and Southern Africa (SIMLESA) program. M.S. Coyne was supported by the USDA National Institute of Food and Agriculture Hatch project KY007090, with additional support from a Natural Resources Conservation Service Conservation Innovation Grant, and a USDA-ARS specific cooperative agreement.

Author Contributions: Drake N. Mubiru was the principal investigator and lead person in the experimentation process and preparation of the manuscript. Jalia Namakula, Godfrey Otim, Joselyn Kashagama, and Milly Nakafeero established and managed the trials, and collected the yield data. James Lwasa led the team in assessing the state of the soils in Nakasongola District using, Global Positioning System (GPS) and Geographic Information System (GIS) tools. William Nanyeenya collated all the socio-economic and biophysical information of Nakasongola District. Mark S. Coyne made significant intellectual contributions during the designing of the trials and for improving the technical content of the manuscript.

References

1.   FAO. World Soils Report. No. 90. 2000. Available online: www.fao.org/soils-portal/resources/world-soil-resources-reports/en (accessed on 13 December 2016).

2.   Magunda, M.K.; Tenwya, M.M. Soil and water conservation. In *Agriculture in Uganda*; Mukiibi, J.K., Ed.; Uganda National Agricultural Research Organization (NARO): Entebbe, Uganda, 2001; Volume I, pp. 145–168.

3.   Nakileza, B.; Nsubuga, E.N.B. *Rethinking Natural Resource Degradation in Semiarid Sub-Saharan Africa: A Review of Soil and Water Conservation Research and Practice in Uganda, with Particular Emphasis on the Semiarid Areas*; Overseas Development Institute: Kampala, Uganda, 1999.

4.   Nkonya, E.; Pender, J.; Jagger, P.; Sserunkuma, D.; Kaizzi, C.K.; Ssali, H. *Strategies for Sustainable Land Management and Poverty Reduction in Uganda*; Research Report 133; IFPRI: Washington, DC, USA, 2004.

5.   Government of Uganda (GOU). *Climate Change: Uganda National Adaptation Programmes of Action*; Environmental Alert, GEF, UNEP, Ministry of Water and Environment: Kampala, Uganda, 2007.

6.   National Environment Management Authority (NEMA). *State of the Environment Report for Uganda 2006/2007*; NEMA: Kampala, Uganda, 2007.

7.   Uganda Bureau of Statistics (UBOS). Statistical Abstract. 2015. Available online: http://www.ubos.org (accessed on 11 February 2016).

8.   Otunge, D.; Muchiri, N.; Wachoro, G.; Anguzu, R.; Wamboga-Mugirya, P. *Enhancing Maize Productivity in Uganda Through the WEMA Project*; A Policy Brief; National Agricultural Research Organisation of Uganda (NARO)/African Agricultural Technology Foundation (AATF): Entebbe, Uganda, 2010.

9.   Semaana, H.R. *Crop Production Handbook for Good Quality Cereals, Pulses and Tuber Crops*; Ministry of Agriculture Animal Industry and Fisheries/Sasakawa Africa Association (SAA): Entebbe, Uganda, 2010.

10.   Regional Agricultural Expansion Support (RATES). *Maize Market Assessment and Baseline Study for Uganda*; Regional Agricultural Expansion Support Program: Nairobi, Kenya, 2003.

11.   Okoboi, G.; Muwanga, J.; Mwebaze, T. Use of improved inputs and its effects on maize yield and profit in Uganda. *Afr. J. Food Agric. Nutr. Dev.* **2012**, *12*, 6932–6944.

12.   Sebuwufu, G.; Mazur, R.; Westgate, M.; Ugen, M. Improving the Yield and Quality of Common Beans in Uganda. Available online: www.soc.iastate.edu/staff/.../CRSP (accessed on 9 September 2014).

13.   World Bank. *Uganda Sustainable Land Management: Public Expenditure Review*; Report No. 45781-UG AFTAR; Sustainable Development Department Country Department 1, Uganda Africa Region: Kampala, Uganda, 2008.

14.   Government of Uganda (GOU). *Increasing Incomes through Exports: A Plan for Zonal Agricultural Production, Agro-Processing and Marketing for Uganda*; MAAIF: Entebbe, Uganda, 2004.

15.   Magunda, M.; Majaliwa, M. *Soil Erosion in Uganda: A Review*; Nile Basin Initiative Issue Paper; NBI: Kampala, Uganda, 2006.

16.   Zake, J.; Magunda, M.; Nkwiine, C. Integrated soil management for sustainable agriculture and food security: The Uganda case. Presented at the FAO Workshop on Integrated Soil Management for Sustainable Agriculture and Food Security in Southern and Eastern Africa, Harare, Zimbabwe, 8–12 December 1997.

17.   Ewel, J.J. Nutrient Use Efficiency and the Management of Degraded Lands. In *Ecology Today: An Anthology of Contemporary Ecological Research*; Gopal, B., Pathak, P.S., Saxena, K.G., Eds.; International Scientific Publications: New Delhi, India, 1988; pp. 199–215.

18.   Fatondji, D.; Martius, C.; Bielders, C.L.; Vlek, P.L.G.; Bationo, A.; Gerard, B. Effect of planting technique and amendment type on pearl millet yield, nutrient uptake, and water use on degraded land in Niger. *Nutr. Cycl. Agroecosyst.* **2006**, *76*, 203. [CrossRef]

19.   Mubiru, D.N.; Komutunga, E.; Agona, A.; Apok, A.; Ngara, T. Characterising agrometeorological climate risks and uncertainties: Crop production in Uganda. *S. Afr. J. Sci.* **2012**, *108*. [CrossRef]

20.   Ogallo, L.A.; Boulahya, M.S.; Keane, T. Applications of seasonal to interannual climate prediction in agricultural planning and operations. *Int. J. Agric. For. Met.* **2002**, *103*, 159–166. [CrossRef]

21.   Jennings, S.; Magrath, J. What Happened to the Seasons? 2009. Available online: http://www.oxfam.org.uk/resources/policy/climate_change/ (accessed on 11 August 2014).

22.   National Environment Management Authority (NEMA). *National State of the Environment Report for Uganda 2014*; NEMA: Kampala, Uganda, 2016.

23.   United Nations Development Programme (UNDP). *Uganda Strategic Investment Framework for Sustainable Land Management*; UNDP: Kampala, Uganda, 2014.

24.   Okwi, P.O.; Hoogeveen, J.G.; Emwanu, T.; Linderhof, V.; Begumana, J. Welfare and environment in rural Uganda: Results from a small-area estimation approach. *Afr. Stat. J.* **2016**, *3*, 135–188. [CrossRef]

25.   Serrao-Neumann, S.; Turetta, A.P.; Prado, R.; Choy, D.L. Improving the management of climate change impacts to support resilient regional landscapes. In Proceedings of the Conference of the Ecosystem Services Partnership Local Action for the Common Good, San Jose, Costa Rica, 7–12 September 2014.

26. Mupangwa, W.; Twomlow, S.; Walker, S. Conservation Tillage for Soil Water Management in the Semiarid Southern Zimbabwe. Available online: http://www.cgspace.cgiar.org/bitstream/handle/ (accessed on 10 December 2014).

27. Haggblade, S.; Tembo, G. Early Evidence on Conservation Farming in Zambia. Presented at the International Workshop on Reconciling Rural Poverty and Resource Conservation: Identify Relationships and Remedies, Ithaca, NY, USA, 2–3 May 2003.

28. Hove, L.; Twomlow, S. Is conservation agriculture an option for vulnerable households in Southern Africa? Presented at the Conservation Agriculture for Sustainable Land Management to Improve the Livelihood of People in Dry Areas Workshop, Damascus, Syria, 7–9 May 2007.

29. Twomlow, S.; Urolov, J.; Oldrieve, J.C.; Jenrich, B. Lessons from the Field: Zimbabwe's Conservation Agriculture Task Force. *J. SAT Agric. Res.* **2008**, *6*, 1–11.

30. Mazvimavi, K.; Twomlow, S. Socioeconomic and institutional factors influencing adoption of conservation farming by vulnerable households in Zimbabwe. *Agric. Syst.* **2009**, *101*, 20–29. [CrossRef]

31. Pedzisa, I.; Minde, I.; Twomlow, S. An evaluation of the use of participatory processes in wide-scale dissemination of research in micro dosing and conservation agriculture in Zimbabwe. *Res. Eval.* **2010**, *19*, 145–155. [CrossRef]

32. Twomlow, S. *Integrated Soil Fertility Management Case Study*; SLM Technology, Precision Conservation Agriculture: Bulawayo, Zimbabwe, 2012.

33. Soniia, D.; Sperling, L. Improving technology delivery mechanisms: Lessons from bean seed systems research in eastern and central Africa. *Agric. Hum. Values* **1999**, *16*, 381–388.

34. National Agricultural Research Organization (NARO). *Annual Report*; NARO: Entebbe, Uganda, 2000.

35. Lillesand, T.M.; Kiefer, R.W. *Remote Sensing and Image Interpretation*; John Wiley & Sons: New York, NY, USA, 1987; p. 669.

36. Okalebo, J.R.; Gathau, K.W.; Woomer, P.L. *Laboratory Methods of Soil and Plant Analysis: A Working Manual*; TSBF-CIAT and SACRED Africa: Nairobi, Kenya, 2002.

37. Foster, H.L. Rapid routine soil and plant analysis without automatic equipment: I. Routine soil analysis. *E .Afr. Agric. For. J.* **1971**, *37*, 160–170.

38. Gee, G.W.; Bauder, J.W. Particle-size analysis. In *Methods of Soil Analysis, Part 1*; SSSA Book Series 5; Klute, A., Ed.; Soil Science Society of America, American Society of Agronomy: Madison, WI, USA, 1986; pp. 383–411.

39. Blake, G.R.; Hartge, K.H. Bulk density. In *Methods of Soil Analysis, Part 1*; SSSA Book Series 5; Klute, A., Ed.; Soil Science Society of America, American Society of Agronomy: Madison, WI, USA, 1986; pp. 363–375.

40. National Sustainable Agriculture Coalition. Available online: http://sustainableagriculture.net/about-us/what-is-sustainable-ag/ (accessed on 31 May 2017).

41. Union of Concerned Scientists. Science for a Healthy Planet and Safer World. Available online: http://www.ucsusa.org/food-agriculture/advance-sustainable-agriculture/what-is-sustainable-agriculture#.WS6WkpKGPIU (accessed on 31 May 2017).

42. UWestern SARE. Sustainable Agricultural Research and Education. Available online: http://www.westernsare.org/About-Us/What-is-Sustainable-Agriculture (accessed on 31 May 2017).

43. Grace Communications Foundation. Available online: http://www.sustaianble.org/246/sustainable-agriculture-the-basics (accessed on 31 May 2017).

44. UCDAVIS Agricultural Sustainability Institute. Available online: http://asi.ucdavis.edu/programs/sarep/about/what-is-sustainable-agriculture (accessed on 31 May 2017).

45. Musiitwa, F.; Komutunga, E. Agricultural systems. In *Agriculture in Uganda*; Mukiibi, J.K., Ed.; Uganda National Agricultural Research Organization (NARO): Entebbe, Uganda, 2001; Volume I, pp. 220–230.

46. Nelson, L.B.; Muckenhirn, R.J. Field percolation rates of four Wisconsin soils having different drainage characteristics. *J. Am. Soc. Agron.* **1941**, *33*, 1028–1036. [CrossRef]

47. Fritton, D.D.; Olson, G.W. Bulk density of a fragipan soil in natural and disturbed profiles. *Soil Sci. Soc. Am. Proc.* **1972**, *36*, 686–689. [CrossRef]

48. Dawud, A.Y.; Gray, F. Establishment of the lower boundary of the sola of weakly developed soils that occur in Oklahoma. *Soil Sci. Soc. Am. J.* **1979**, *43*, 1201–1207. [CrossRef]

49. Larson, W.E.; Gupta, S.C.; Useche, R.A. Compression of agricultural soils from eight soil orders. *Soil Sci. Soc. Am. J.* **1980**, *44*, 450–457. [CrossRef]

50. Yule, D.F.; Ritchie, J.T. Soil shrinkage relationships of Texas Vertisols: I. Small cores. *Soil Sci. Soc. Am. J.* **1980**, *44*, 1285–1291. [CrossRef]

51. Ssali, H. Soil fertility. In *Agriculture in Uganda*; Mukiibi, J.K., Ed.; Uganda National Agricultural Research Organization (NARO): Entebbe, Uganda, 2001; Volume I, pp. 104–135.

52. Hazelton, P.; Murphy, B. *Interpreting Soil Test Reults: What do the Numbers Mean?* CSIRO: Collingwood, Australia, 2010.

53. Namugwanya, M.; Tenywa, J.S.; Etabbong, E.; Mubiru, D.N.; Basamba, T.A. Development of common bean (*Phaseolous Vulgaris* L.) production under low soil phosphorus and drought in sub Saharan Africa: A review. *J. Sustain. Dev.* **2014**, *7*, 128–139.

54. Goettsch, L.H.; Lenssen, A.W.; Yost, R.S.; Luvaga, E.S.; Semalulu, O.; Tenywa, M.; Mazur, R.E. Improved production systems for common bean on Phaeozem soil in south-central Uganda. *Afr. J. Agric. Sci.* **2016**, *11*, 4797–4809.

55. Kalyebara, R. *The Impact of Improved Bush Bean Varieties in Uganda*; Network on Bean Research in Africa, Occasional Publication Series 43; Chartered Institute of Architectural Technologists (CIAT): Kampala, Uganda, 2008.

56. Sibiko, K.W.; Ayuya, O.I.; Gido, E.O.; Mwangi, J.K. An analysis of economic efficiency in bean production: Evidence from eastern Uganda. *J. Econ. Sustain. Dev.* **2013**, *4*. Available online: www.iiste.org (accessed on 2 May 2017).

57. Ghaffarzadeh, M.; Garcia, F.; Cruse, R.M. Grain yield response of corn, soybean, and oat grown in strip intercropping system. *Am. J. Altern. Agric.* **1994**, *9*, 171–177. [CrossRef]

58. Ghaffarzadeh, M.; Garcia, F.; Cruse, R.M. Tillage effect on soil water content and corn yield in a strip intercropping system. *Agron. J.* **1997**, *89*, 893–899. [CrossRef]

59. Shaw, R.H.; Felch, R.E.; Duncan, E.R. *Soil Moisture Available for Crop Growth*; Special Report; Iowa State University: Ames, IA, USA, 1972.

60. Hulugalle, N.R.; Willatt, S.T. Seasonal variation in the water uptake and leaf water potential of intercropped and monocropped chillies. *Exp. Agric.* **1987**, *23*, 273–282. [CrossRef]

61. Francis, C.A.; Jones, A.; Crookston, K.; Wittler, K.; Goodman, S. Strip cropping corn and grain legumes: A review. *Am. J. Altern. Agric.* **1986**, *1*, 59–164.

62. Hulugalle, N.R.; Lal, R. Soil water balance in intercropped maize and cowpea in a tropical hydromorphic soil in western Nigeria. *Agron. J.* **1986**, *77*, 86–90. [CrossRef]

63. Fortin, M.C.; Culley, J.; Edwards, M. Soil water, plant growth, and yield of strip-intercropped corn. *J. Prod. Agric.* **1994**, *7*, 63–69. [CrossRef]

64. Mazvimavi, K.; Ndlovu, P.V.; Nyathi, P.; Minde, I.J. Conservation agriculture practices and adoption by smallholder farmers in Zimbabwe. Presented at the 3rd African Association of Agricultural Economists (AAAE) and 48th Agricultural Economists Association of South Africa (AEASA) Conference, Cape Town, South Africa, 19–23 September 2010.

Permissions

The contributors of this book come from diverse backgrounds, making this book a truly international effort. This book will bring forth new frontiers with its revolutionizing research information and detailed analysis of the nascent developments around the world.

We would like to thank all the contributing authors for lending their expertise to make the book truly unique. They have played a crucial role in the development of this book. Without their invaluable contributions this book wouldn't have been possible. They have made vital efforts to compile up to date information on the varied aspects of this subject to make this book a valuable addition to the collection of many professionals and students.

This book was conceptualized with the vision of imparting up-to-date information and advanced data in this field. To ensure the same, a matchless editorial board was set up. Every individual on the board went through rigorous rounds of assessment to prove their worth. After which they invested a large part of their time researching and compiling the most relevant data for our readers.

The editorial board has been involved in producing this book since its inception. They have spent rigorous hours researching and exploring the diverse topics which have resulted in the successful publishing of this book. They have passed on their knowledge of decades through this book. To expedite this challenging task, the publisher supported the team at every step. A small team of assistant editors was also appointed to further simplify the editing procedure and attain best results for the readers.

Apart from the editorial board, the designing team has also invested a significant amount of their time in understanding the subject and creating the most relevant covers. They scrutinized every image to scout for the most suitable representation of the subject and create an appropriate cover for the book.

The publishing team has been an ardent support to the editorial, designing and production team. Their endless efforts to recruit the best for this project, has resulted in the accomplishment of this book. They are a veteran in the field of academics and their pool of knowledge is as vast as their experience in printing. Their expertise and guidance has proved useful at every step. Their uncompromising quality standards have made this book an exceptional effort. Their encouragement from time to time has been an inspiration for everyone.

The publisher and the editorial board hope that this book will prove to be a valuable piece of knowledge for researchers, students, practitioners and scholars across the globe.

List of Contributors

Carla Asquer, Emanuela Melis and Efisio Antonio Scano
Biofuels and Biomass Laboratory, Renewable Energies Facility, Sardegna Ricerche – VI strada ovest Z.I. Macchiareddu, 09010 Uta, Italy

Gianluca Carboni
Agris Sardegna, Viale Trieste 111, 09123 Cagliari, Italy

Xueqiong Wei, Xiuqi Fang and Yikai Li
School of Geography, Faculty of Geographical Science, Beijing Normal University, Beijing 100875, China

Yu Ye
School of Geography, Faculty of Geographical Science, Beijing Normal University, Beijing 100875, China
Key Laboratory of Environment Change and Natural Disaster, Ministry of Education, Beijing Normal University, Beijing 100875, China

Temitope S. Egbebiyi, Chris Lennard, Olivier Crespo, Phillip Mukwenha, Shakirudeen Lawal and Kwesi Quagraine
Climate System Analysis Group (CSAG), Department of Environmental and Geographical Science, University of Cape Town, Private Bag X3, Rondebosch, 7701 Cape Town, South Africa

Lisandro Roco
Department of Economics and Institute of Applied Regional Economics (IDEAR), Universidad Católica del Norte, Antofagasta 1240000, Chile

Boris Bravo-Ureta
Department of Agricultural and Resource Economics, University of Connecticut, Storrs 06269, CT, USA
Department of Agricultural Economics, Universidad de Talca, Talca 3460000, Chile

Alejandra Engler and Roberto Jara-Rojas
Department of Agricultural Economics, Universidad de Talca, Talca 3460000, Chile
Center for Socioeconomic Impact of Environmental Policies (CESIEP), Talca 3460000, Chile

Jon Hellin
Sustainable Impact Platform at the International Rice Research Institute (IRRI), Metro Manila 1301, Philippines

Eleanor Fisher
School of Agriculture, Policy and Development at the University of Reading, Reading RG6 6AH, UK

Honest Machekano and Casper Nyamukondiwa
Department of Biological and Biotechnological Sciences, Botswana International University of Science and Technology, P. Bag 16, Palapye, Gaborone 0267, Botswana

Brighton M. Mvumi
Department of Soil Science and Agricultural Engineering, Faculty of Agriculture, University of Zimbabwe, Mt. Pleasant, 00263 Harare, Zimbabwe

Aymar Yaovi Bossa and Jean Hounkpè
West African Science Service Centre on Climate Change and Adapted Land Use (WASCAL), Ouagadougou, Burkina Faso
National Water Institute, University of Abomey Calavi, Cotonou, Benin

Luc Olivier Sintondji
National Water Institute, University of Abomey Calavi, Cotonou, Benin

Yacouba Yira
West African Science Service Centre on Climate Change and Adapted Land Use (WASCAL), Ouagadougou, Burkina Faso
Applied Science and Technology Research Institute–IRSAT/CNRST, Ouagadougou, Burkina Faso

Georges Serpantié
Institute for Research and Development — IRD-UMR GRED-UPV, 34090 Montpellier, France

Bruno Lidon and Jean Louis Fusillier
Centre for International Cooperation in Agronomic Research for Development — CIRAD-UMR G-eau, 34090 Montpellier, France

Jérôme Ebagnerin Tondoh
UFR des Sciences de la Nature, Université Nangui Abrogoua, 02 BP 801 Abidjan 02, Cote D'Ivoire

Bernd Diekkrüger
Department of Geography, University of Bonn, Meckenheimer Allee 166, 53115 Bonn, Germany

Anita Lazurko
Department of Environmental Sciences and Policy, Central European University, Budapest 1051, Hungary

Henry David Venema
Prairie Climate Centre, International Institute for Sustainable Development, Winnipeg, MB R3B 0T4, Canada

Lauren E. Parker
USDA California Climate Hub, Davis, CA 95616, USA
John Muir Institute of the Environment, University of California, Davis, CA 95616, USA

John T. Abatzoglou
Department of Geography, University of Idaho, Moscow, ID 83844, USA

Tianyi Zhang
State Key Laboratory of Atmospheric Boundary Layer Physics and Atmospheric Chemistry, Institute of Atmospheric Physics, Chinese Academy of Sciences, Beijing 100029, China

Jinxia Wang
School of Advanced Agricultural Sciences, Peking University, Beijing 1000871, China

Yishu Teng
BICIC, Beijing Normal University, Beijing 1000875, China

Behnam Mirgol
Department of Water Engineering, Faculty of Engineering and Technology, Imam Khomeini International University, 3414896818 Qazvin, Iran

Meisam Nazari
Department of Crop Sciences, Faculty of Agricultural Sciences, Georg-August University of Göttingen, Büsgenweg 5, 37077 Göttingen, Germany

Department of Soil Science, University of Kassel, Nordbahnhofstr. 1a, 37213 Witzenhausen, Germany

Elaine Wheaton
Department of Geography and Planning, University of Saskatchewan, Saskatoon, SK S7N 5A8, Canada

Suren Kulshreshtha
Department of Agricultural and Resource Economics, University of Saskatchewan, Saskatoon, SK S7N 5A8, Canada

Mostafiz Rubaiya Binte
University of Tsukuba, Tsukuba, Ibaraki 305-8577, Japan

Md. Shah Alamgir
University of Tsukuba, Tsukuba, Ibaraki 305-8577, Japan
Sylhet Agricultural University, Sylhet 3100, Bangladesh

Jun Furuya and Shintaro Kobayashi
Japan International Research Center for Agricultural Sciences; Tsukuba, Ibaraki 305-8686, Japan

Md. Abdus Salam
Bangladesh Rice Research Institute, Gazipur 1701, Bangladesh

Drake N. Mubiru, Jalia Namakul, James Lwasa, Godfrey A. Otim, Joselyn Kashagama, Milly Nakafeero and William Nanyeenya
National Agricultural Research Organization (NARO), Kampala, Uganda

Mark S. Coyne
Department of Plant and Soil Sciences, University of Kentucky, 1100 S. Limestone St., Lexington, KY 40546-0091, USA

Index

Printed in the USA
CPSIA information can be obtained
at www.ICGtesting.com
JSHW051414091023
49903JS00006B/413